Cell Hybrids

Nils R. Ringertz

Institute for Medical Cell Research and Genetics
Karolinska Institute
Stockholm, Sweden

Robert E. Savage

Department of Biology
Swarthmore College
Swarthmore, Pennsylvania

ACADEMIC PRESS   New York   San Francisco   London   1976

A Subsidiary of Harcourt Brace Jovanovich, Publishers

ACADEMIC PRESS, INC.
111 Fifth Avenue, New York, New York 10003

United Kingdom Edition published by
ACADEMIC PRESS, INC. (LONDON) LTD.
24/28 Oval Road, London NW1

Library of Congress Cataloging in Publication Data

Ringertz, Nils,   Date
      Cell hybrids

      Bibliography: p.
      Includes index.
      1.   Cell hybridization.   2.   Somatic hybrids.
I.   Savage, Robert E., joint author.   II.   Title.
QH451.R56   574.8'76   76-2947
ISBN 0–12–589150–4

PRINTED IN THE UNITED STATES OF AMERICA

Contents

Preface

The technique of inducing fusion of human, animal, and plant cells of different origins to produce cell hybrids has rapidly become a widely used one, with important applications to somatic cell genetics and to fields of practical concern such as medicine and agriculture. In this volume the methodology of cell fusion is summarized and the main applications and current findings of the hybridization technique are surveyed. Because the volume of literature involving somatic cell hybrids has increased at an explosive rate in recent years, we have had to focus our attention on major developments, sacrificing some niceties of detail in the process. Although we have cited close to a thousand articles in the Bibliography, many papers on cell hybrids are noted only briefly in the text or listed in tables summarizing the properties of different types of hybrid cells. For example, chromosome mapping is now one of the most important applications of cell hybridization. While this book describes the techniques used and summarizes gene assignments based on cell hybrids, detailed references to individual papers have not always been included. These omissions seem justified both because the status of human gene mapping is now being summarized annually at international meetings and because cell hybridization is only one of several methods for gene assignment. Thus, a detailed discussion of the human genome per se is not included and the number of references is accordingly curtailed.

One aim of this book is to present the cell hybrid field in such a way that students with basic training in biology can read it. To help in this respect, we have included a glossary of terms that are not used by every biologist or physician. To each reader, some terms will seem too elementary to have been included. Presumably, though, different groups of readers will disagree as to which are the inappropriate terms. When in the text we have first had occasion to use any such term, it usually is defined explicitly or by implication. However, it has been our intention that the chapters need not be read in sequence, but that the book could be opened to any chapter and read with reasonable comprehension. Thus, a reader who browses through may very well encounter an unfamiliar term whose context gives no clue to its meaning. When this occurs, the reader should turn to the glossary for help. We hope we have anticipated the needs of a diverse readership.

The same encouragement should be given to those who encounter an unfamiliar abbreviation or acronym. In general, these are defined only the first time they are used in the text. There is, however, a List of Abbreviations, following this preface, in which these symbols are defined; it should prove helpful to the browsers and the uninitiated.

Many chapters contain introductory sections which define biological problems and terms which may be unfamiliar. Experts can easily skip these sections and proceed to the following ones which discuss the results of cell hybrid studies. We hope, however, that these introductory sections will make it possible for readers whose experience is limited in the fields presented to better understand the problems discussed.

Many of our friends have read one or more chapters, given us advice, and thereby saved us from making many serious errors. We are particularly grateful to Drs. Eric Sidebottom and Richard Davidson for reading the entire manuscript and giving us detailed comments and criticisms. Other very useful comments were made by Drs. Ernest Chu, Thorfinn Ege, Waheb Heneen, Bob Johnson, Agnes Kane, Dietrich Kessler, George Klein, Peter Reichard, Dan Röhme, Barbara Stewart, and Francis Wiener, and by our colleagues of the Cell Research and Biochemistry Departments at Karolinska Institutet.

For providing us many excellent figures and tables, we would like to thank Drs. M. Bramwell, F. Constabel, H. Coon, R. Davidson, J. Ericsson, L. C. Fowke, R. D. Goldman, S. Gordon, M. Harris, I. Hosaka, J. Jami, R. Johnson, D. Kessler, H. Koprowski, G. Melchers, O. Miller, B. Mintz, D. Patterson, D. Prescott, D. Röhme, F. Ruddle, A. A. Sandberg, E. A. de Weerd-Kastelein, M. Weiss, L. Zech, and Y. Zegeye.

We are also very much indebted to Mrs. Vivi Jacobson who patiently and carefully typed and retyped the chapters and to Mrs. Ulla Krondahl who rendered much of the artwork.

Nils R. Ringertz
Robert E. Savage

List of Abbreviations*

1 s	diploid chromosome number or DNA content
2 s	tetraploid chromosome number or DNA content
3T3	mouse cell line, contact inhibited
3T6	mouse cell line, spontaneously transformed
Å	angstrom unit
A9	cell line derived from L cells (HGPRT⁻)
A9HT	malignant subline of A9 cell line
AAT	alanine aminotransferase
AChE	acetylcholinesterase
ACO	aconitase
ACP_1	acid phosphatase-1 (red cell)
ACP_2	acid phosphatase-2
ADA	adenosine deaminase
Ade⁻	adenine-requiring mutant cell line
Ade⁻A	adenine-requiring auxotrophic mutant belonging to complementation group A
Ade⁺B	gene complementing adenine B auxotrophs of Chinese hamster cells
ADH	alcohol dehydrogenase
AG	azaguanine
AGᴿ	azaguanine resistent
AGMK	cell line from African Green Monkey Kidney
AICAR	5-aminoimidazole-4-carboxamide ribonucleotide
AIR	5-aminoimidazole ribonucleotide
AMP	adenosine 5'-phosphate
APRT	adenine phosphoribosyltransferase
Ara-C	cytosine arabinoside
ATP	adenosine 5'-triphosphate

* See also Table XIII-4 for abbreviations of gene markers.

ATPase	adenosinetriphosphatase
AU	arbitrary unit
AUC	human cell line from a case of orotic aciduria
B82	BUdR^R subline of mouse L cells
B82HT	malignant subline of B82 mouse cells
BCdR	5-bromodeoxycytidine
BHK	cell line from baby hamster kidney (Syrian hamster)
BUdR	bromodeoxyuridine
BUdR^R	bromodeoxyuridine resistant
C4	complement factor 4
C5	complement factor 5
Ca^{2+}	calcium ion
CAP	chloramphenicol
CAP^R	chloramphenicol resistant
cDNA	complementary DNA
cRNA	complementary RNA
CHO	Chinese hamster ovary cell line
Cl	clone
Con A	concanavalin A
D98	aneuploid human cell line × 10
D98/AH-2	human cell line HGPRT$^-$
DMSO	dimethyl sulfoxide
DNA	deoxyribonucleic acid
DON	Chinese hamster cell line
EBV	Epstein Barr Virus
EDTA	ethylenediaminetetraacetic acid
F$_1$	first generation progeny
FI	fusion index
FGAM	α-N-formylglycine-amidine ribonucleotide
Fu5	mouse hepatoma cell line
Fu5-5	mouse hepatoma cell line, subclone 5 of Fu5
G$_1$	gap 1, prereplication phase of cell cycle
G$_2$	gap 2, postreplication phase of cell cycle
G6PD	glucose-6-phosphate dehydrogenase
GAR	β-glycineamide ribonucleotide
β-Glu	β-glucuronidase
Glu$^-$	glucose-requiring mutant cell line
Gly$^-$	glycine-requiring mutant cell line
Gly$^+$A	gene complementing Gly$^-$ mutants of complementation group A (serine hydroxymethylase deficient)
Gly A$^-$	glycine-requiring mutant belonging to complementation group A
h	hour
H-2	main histocompatibility antigen complex in mice
^{3}H	tritium-labeled
HAM	selective medium containing hypoxanthine, aminopterin, and 5-methyl deoxycytidine

HAT	selective medium containing hypoxanthine, aminopterin, and thymidine
HAU	hemagglutinating unit
Hb	hemoglobin
HeLa	cell line of human origin
HGPRT	hypoxanthine guanine phosphoribosyltransferase
HL-A	main histocompatibility antigen complex in man
HnRNA	high molecular nuclear RNA
HVJ	hemagglutinating virus of Japan; synonym: Sendai virus
Ig	immunoglobulin
Ino$^-$	inositol-requiring mutant cell line
IPO-A	indophenol oxidase-A (dimer)
IPO-B	indophenol oxidase-B (tetramer)
KB	human cell line
L929	mouse cell line; synonym: L cell
L1210	mouse leukemia
LDH	lactate dehydrogenase (tetrameric)
LDH-1	lactate dehydrogenase-1 (consisting of 4 A-subunits)
LDH-5	lactate dehydrogenase-5 (consisting of 4 B-subunits)
LDH-A	lactate dehydrogenase-A
LDH-B	lactate dehydrogenase-B
LM (TK$^-$)Cl ID	mouse L cell mutant
LN	Lesch Nyhan's syndrome or cells from such cases
LN SV40	human Lesch Nyhan cell line transformed by SV40 virus
MDH	malate dehydrogenase
MDH-1	malate dehydrogenase (cytoplasmic)
MDH-2	malate dehydrogenase (mitochondrial)
MDR	measles–distemper–rinderpest virus
ME-1	malic enzyme (cytoplasmic)
ME-2	malic enzyme (mitochondrial)
Mg^{2+}	magnesium ion
min	minute
mM	millimolar
MM	minimal medium
mRNA	messenger RNA
MSUD	maple syrup urine disease
μm	micrometer
NAD	nicotinamide adenine dinucleotide
NADP	nicotinamide adenine dinucleotide phosphate
NDV	Newcastle disease virus
nm	nanometer
p	short arm of a chromosome
PCC	premature chromosome condensation
PEG	polyethylene glycol
PGK	phosphoglycerate kinase
Pha	phytohemagglutinin
Pro$^-$	proline-requiring mutant cell line

PRPP	phosphoribosyl pyrophosphate
r	roentgen units (rads)
RAG	cell line from mouse renal adenocarcinoma
Rh	Rhesus blood group in man
RNA	ribonucleic acid
rRNA	ribosomal RNA
RNase	ribonuclease
RSV	Rous sarcoma virus
S	Svedberg units
S	cell cycle phase during which nuclear DNA is synthesized
SAICAR	5-aminoimidazole-4-(N-succinocarboximide) ribonucleotide
SDS	sodium dodecyl sulfate
SEM	standard error of the mean
Ser$^-$	serine-requiring mutant cell
SOD-1	superoxide dismutase-1 (cytoplasmic)
SOD-2	superoxide dismutase-2 (mitochondrial)
SV40	Simian virus 40
T	tumor-specific antigen in cells transformed by tumor viruses
TAT	tyrosine aminotransferase
TG	thioguanine
TG^R	thioguanine resistent
TK	thymidine kinase
ts	temperature sensitive
UK	uridine kinase
UMPK	uridine:monophosphate kinase
WI38	human diploid cell strain
XP	xeroderma pigmentosum
μ	micron
μCi	microcurie, unit of radioactivity

I

Introduction

Somatic cell hybridization is a technique which has been introduced relatively recently but which has already proved to be an extremely powerful experimental procedure with applications in cell biology, genetics, developmental biology, tumor biology, and virology. Basically the technique involves the spontaneous or induced fusion of different cells to give a *cell hybrid*. A wide variety of animal, human, and even insect and plant cell types have been used as parental cells in these fusions. When cells of different organisms are fused (e.g., mouse + man, hen + rat, mosquito + man) interspecific hybrid cells are produced. In these cases, the parental cells differ at least with respect to genotype if not phenotype as well. Intraspecific hybrids are obtained by fusing two different cell types from the one species (e.g., mouse fibroblasts + mouse lymphoblasts). In these instances the parental cells differ in phenotype, that is in morphological, biochemical, immunological, or functional properties.

Spontaneous fusion of cells occurs infrequently under laboratory culture conditions (*in vitro*) and also occasionally in living organisms (*in vivo*). The present standard laboratory procedure for inducing cell hybrids *in vitro* involves addition of inactive Sendai virus to a culture containing two different cell types. Using this procedure, two groups of multinucleate cells (polykaryons) are generated. The first group contains nuclei from only one parental type (homokaryons), while the second group contains nuclei from both parental types (heterokaryons). Only the latter are hybrid cells.

1

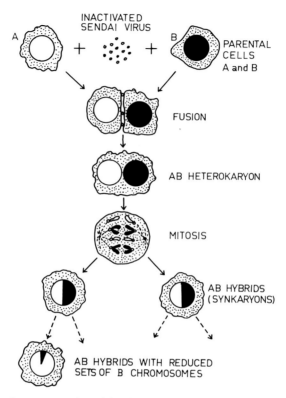

Fig. I-1. Schematic representation of Sendai virus induced fusion of two mononucleate cells (A and B) from different species into a binucleate heterokaryon which then divides and gives rise to two mononucleate hybrid cells (synkaryons). These AB hybrids then continue to divide while gradually eliminating most of the chromosomes originating from parental cell B.

A heterokaryon, once formed, has two alternatives open to it: after an interim in culture as a polykaryon it may die, or it may survive and give rise to two mononucleate hybrid cells called synkaryons (Fig. I-1). The latter term is not used very extensively in current literature. Instead the expression "hybrid cell" has become more or less synonymous with the synkaryotic condition. It should be remembered, though, that heterokaryons are in fact also hybrid cells. Synkaryons probably arise most frequently as a result of synchronous division of nuclei in heterokaryons, but direct nuclear fusion within polykaryons has also been observed. Many synkaryons have been found to be capable of cell multiplication over prolonged periods of time.

Individual heterokaryons and synkaryons may be subjected to analysis by cytochemical and immunocytochemical methods applicable to single cells. For some analyses of biochemical or other properties, however, it is necessary to prepare large numbers of hybrid cells. In a few instances, pure populations of hybrid cells are obtained because the hybrid cells proliferate more rapidly than either of the parental cell types (probably a type of "hybrid vigor") and therefore overgrow them. In many cases, however, the hybrid cell populations grow slowly and tend instead to be overwhelmed by the parental cells. Under these circumstances, homogeneous populations of synkaryons can be isolated by resorting to one of two general approaches: selective or nonselective. Nonselective isolation involves mechanically separating single hybrid cells from the remainder of the cells in the culture. These isolated cells are then allowed to proliferate, each producing a cell clone. A difficulty here is that in most experiments the hybrid cells represent such a small minority of the cells present that they are not always easily isolated. And even after these single cells are isolated, they may fail to proliferate for various reasons. This in turn may require isolating a very large number of hybrid cells in order to obtain but a few clones. On the other hand, selective isolation involves the culturing of mixed cell cultures on a medium which only allows hybrid cells to multiply. Both parental cell types are overgrown or die. When only synkaryons remain, single cells may be cloned in order to ensure that the population consists of only one type of hybrid cell.

Armed with such techniques, biologists have already found that the ability to make and isolate hybrid cells has opened new avenues to the analysis of gene regulation in eukaryotic cells. The use of parental cells with unique marker properties and the subsequent examination of hybrid cells for the expression of these properties allow questions of gene regulation to be analyzed. These studies will no doubt help in the understanding of mechanisms which govern cell growth and differentiation during normal embryonic development, wound healing, and tumor formation. It is still not known, for example, whether the cell specialization that occurs during the maturation of some tissues, such as those responsible for production of immunoglobins, is due to changes in patterns of gene blocking and gene expression or to changes in DNA. Cell fusion with embryonic cells exhibiting or lacking properties characteristic of differentiated tissue (histotypic markers) promises to help clarify problems like this.

The question of the compartmentalization of metabolic regulators within cells can also be approached through virus-induced fusion: anucleate cytoplasms can be fused with intact cells, and nuclei virtually free of cytoplasm—even subsets of chromosomes—may be fused with anucleate cytoplasms or whole cells. It may prove possible to introduce specific soluble proteins by fusing recipient cells with erythrocyte ghosts or lipid vesicles that have been preloaded with the protein. By such means, the developmental roles of the nucleus, the cytoplasm, and perhaps even the separate cytoplasmic or nuclear components may be investigated.

Among long-lived, proliferating synkaryons there often is a reduction in chromosome number. While intraspecific hybrids usually retain all or most of the parental chromosomes, interspecific synkaryons tend to lose chromosomes as they divide. Furthermore, there seem to be rules governing chromosomal loss in different interspecific hybrid cells. For example, in a mouse + human synkaryon, the human chromosomes, not those of the mouse, are gradually eliminated with time. The loss of one species' chromosomes from a hybrid in this way results in chromosomal segregation. This phenomenon has made it much easier to associate genetic function with specific chromosomes in eukaryotic cells. The mouse + human synkaryon which has retained all its mouse chromosomes and lost all but one of its human chromosomes is the ideal synkaryon for assigning genes. By analyzing such a cell for human proteins and identifying the retained human chromosome, it is possible to locate the structural genes for the proteins on that specific chromosome. The medical significance of this technique becomes apparent when one realizes that marker properties of various inherited diseases are now quickly being assigned to specific chromosomes. A practical consequence of this may be improved genetic counseling and better antenatal diagnosis of inherited errors of metabolism.

Fusion of normal cells with tumor cells followed by analysis of the *in vitro* and *in vivo* properties of the resulting hybrids has begun to provide insight into the mechanisms involved in neoplastic processes. Synkaryons formed from fused normal and neoplastic cells frequently give rise to tumors when injected into appropriate experimental animals. There are, however, exceptions, and the question of whether or not a hybrid cell is malignant appears to be intimately linked to which chromosomes it has retained and which it has eliminated.

Another promising application for cancer research is the use of cell hybridization as an instrument for detection of latent tumor viruses. It has been shown that some virus-induced mouse tumors consist of cells which multiply for many cell generations, show malignant properties, and yet do not contain any detectable quantities of tumor virus. However, when some of these tumor cells are fused with normal cells that permit virus multiplication the resulting hybrid cells produce virus. This means that the confrontation of the tumor cell nucleus with the cytoplasm of the sensitive cell somehow awakens ("rescues") dormant virus in the tumor nucleus. In this way, cell fusion can be used in some cases as an analytic tool for screening unknown tumor cells for the presence of oncogenic virus.

From examples such as these, it is clear that cell hybridization techniques have a number of useful and interesting applications and that new ones are likely to develop. In this book we attempt to present the most important of these applications and, occasionally, to give some indication of likely developments.

Fortunately our task in writing has been made lighter by the fact that there have been several excellent earlier reviews of the somatic cell hybridization field. These include two monographs: Professor H. Harris' "Cell Fusion" (1970b), and Professor B. Ephrussi's "Hybridization of Somatic Cells" (1972). Other reviews have been penned by Harris (1968, 1970a, 1974), Okada (1969), Barski (1970), Grzeschik (1973a), and Davidson (1974). There are in addition many very helpful reviews dealing with subdivisions of the cell hybridization field, and these are referred to in the appropriate chapters of this volume.

II

History

A. EARLY OBSERVATIONS OF MULTINUCLEATED CELLS

Suggestions that fusion could occur between somatic cells probably began with the discovery of multinucleated cells in the early decades of the nineteenth century. Credit for the first observation is given to Johannes Müller, the German biologist who could count among his students some of the most distinguished scientists of the time (e.g., Schwann, Virchow, Henle, Remak, Kölliker, and Helmholtz). Most of the early reports of multinucleate cells, including Müller's, were made in connection with histological studies of pathological conditions. Hence, by the turn of the century the medical literature contained several reports of "polykaryocytosis" as symptomatic of a variety of diseases, including tuberculosis, variola, varicella, and rubeola (Fig. II-1). The existence of these polykaryons raised the question of whether they originated from successive mitoses (nuclear divisions) without concomitant cytoplasmic divisions or from fusion of preexisting mononucleated cells (Langhans, 1868). W. H. Lewis, one of the pioneers in cell culture techniques, reported in 1927 that both mechanisms were observable *in vitro* (Lewis, 1927). He tended to doubt, though, that the nuclear divisions, when they occurred, were strictly mitotic—despite his wife's inclination to think they were. Whichever the case there, he clearly established fusion as a mode in a nicely documented case of giant cell formation in a culture of Warren sarcoma cells. Subsequently, it became clear that both mechanisms are found in na-

6

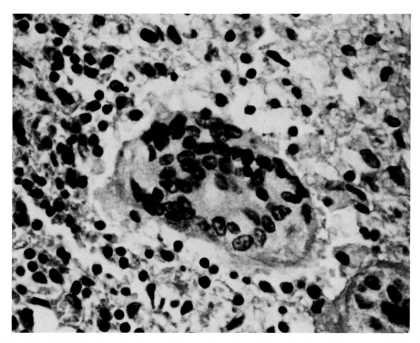

Fig. II-1. Multinucleated giant cell in tuberculous-tissue lesion. (Courtesy of Dr. J. Ericsson).

ture as well as in tissue culture. The mitotic route is seen, for example, in the slime mold, *Physarum*, where it results in a polykaryon, termed plasmodium (Fig. II-2). Among higher organisms, the fusion route is not often found, except in association with disease (see Chapter III). To distinguish fused polykaryons from plasmodia, the former have been dubbed syncytia.

It is an interesting historical parenthesis to this book that the discovery of polykaryons provided data apparently supporting Schleiden's erroneous proposal that new cells originate as vesicles within the plasma membranes of parent cells. On this point, Virchow in 1851 presented a drawing of a multinucleate tumor cell in the belief that the nuclei represented the endogenous formation of new cells. In addition, polykaryons supplied fuel for the anti-cell-theory fires which burned into the twentieth century: they seemed to provide evidence for the hypothesis that the common organismal construction is a single tissue mass in which cytoplasm is continuous. In contrast, the cell theory maintains that an organism consists of many united tissues each built of separated bits of cytoplasm, each bit with a nucleus. Historically, of

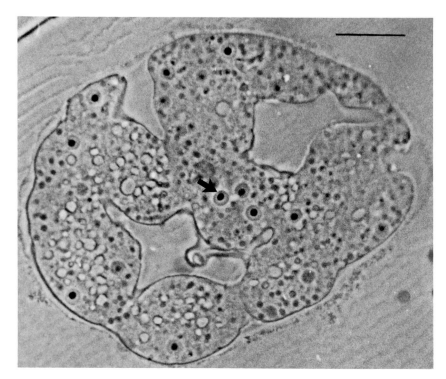

Fig. II-2. Microplasmodium of *Physarum polycephalum*. When grown in a shake-flask culture, the plasmodium breaks up into small pieces, or "microplasmodia." There are 12 nuclei in this section, each with a large darkly stained central nucleolus (arrow). Section photographed under phase-contrast microscopy. Bar = 10 μm. (Courtesy of Y. Zegeye and D. Kessler.)

course, the cell idea prevailed, and polykaryons became interesting exceptions.

B. DISCOVERY OF HYBRID CELLS

Despite the nineteenth century beginnings, hybrid somatic cells were not discovered until the 1960's (Table II-1). In the first years of that decade, Barski, Sorieul, and Cornefert (1960) reported success in isolating a hybrid strain of cells. From mixed cultures of two sarcoma-producing cell lines derived several years earlier from a single mouse cell (Sanford *et al.*, 1954), they isolated hybrid cell lines after about 3 months of culture (Barski *et al.*, 1960, 1961; Barski and Cornefert, 1962). The original cell lines could be distinguished from one another by two character-

TABLE II-1

Chronology of Highlights in the History of Somatic Cell Hybridization

1960	Barski, Sorieul, and Cornefert (1960, 1961) discover cell hybrids formed *in vitro* by spontaneous cell fusion.
1957–1962	Okada (1958, 1962a,b) demonstrates that UV-inactivated Sendai virus can be used to fuse cells *in vitro* to generate homokaryons. Sorieul and Ephrussi (1961) confirm the results of Barski *et al.*
1963	Gershon and Sachs (1963) confirm the results of Barski *et al.*
1964	Littlefield (1964a) introduces the use of mutant cells and selective media for the isolation of hybrid cells.
1965	Harris and Watkins (1965) and, independently, Okada and Murayama (1965) produce the first interspecific heterokaryons using inactivated Sendai virus. Ephrussi and Weiss (1965) produce the first interspecific proliferating synkaryons by spontaneous fusion.
1966	Yerganian and Nell (1966) demonstrate that Sendai virus can be used to produce proliferating hybrids. Davidson, Ephrussi, and Yamamoto (1966), using melanoma and unpigmented cells, demonstrate that a phenotypic marker of histiotypic differentiation (i.e., pigmentation) can be selectively extinguished in hybrids.
1967	Weiss and Green (1967) discover preferential elimination of human chromosomes in man–mouse hybrids and demonstrate that this phenomenon can be used for gene assignment in man. Watkins and Dulbecco (1967) and, independently, Koprowski *et al.* (1967) demonstrate virus rescue when active SV40 particles emerge from hybrids derived from transformed cells and permissive ones.
1969	Harris, Klein, and their colleagues (Harris *et al.*, 1969a) begin experiments showing that malignancy need not dominate in synkaryons derived from oncogenic and normal cells.
1970	Lucy and his colleagues (Lucy, 1970) begin experiments with lysolecithin-induced fusion. Cocking and colleagues successfully fuse plant protoplasts (Power *et al.*, 1970).
1971	Ruddle, Bodmer, Miller, Siniscalco, Bootsma, and their co-workers begin the systematic use of cell hybridization as experimental system for mapping human chromosomes (for references see Chapter XIII).
1972	Carlson, Smith, and Dearing (1972) produce an interspecific hybrid plant from fused protoplasts.

istics: (1) the ability to produce tumors when injected into histocompatible mice and (2) the total number and morphology of chromosomes. The karyological character of the resulting hybrid was very close indeed to what could have been expected of a synkaryon derived from the two

TABLE II-2

Properties of Parental Cells (N1 and N2) and the First Hybrid Cells (M)[a]

	N1	N2	Hybrid (M)
Modal chromosome number	55	62	115–116
Number of metacentric chromosomes	0	9–19	9–15

[a] Discovered by Barski *et al.* (1960).

lines (Table II-2). The possibility that the new cells were in fact derived
from polyploidization of one original line was made unlikely by the observation that when polyploids were found in unmixed cultures, they
were karyologically distinct from the new line which itself occurred
only in mixed cultures. The first reports from Barski's group were received with a great deal of skepticism mainly because of the rarity of the
appearance of hybrid cells and the technical problems involved in the
karyological analysis. Barski's results were, however, repeated and confirmed in 1961 by Sorieul and Ephrussi and again in 1963 by Gershon
and Sachs using two different cell lines. In the latter study, histocompatibility antigens of the parental and hybrid lines were studied by
transplanting the cultured cells into mice. The parental lines produced
tumors only in the strain of mice from which the cells had originally
been derived, but regressed when injected into the other strain. The
hybrid line did not survive in either strain, but it did in the offspring of
genetic crosses between the two strains (Table II-3). These results
suggested that the hybrid cells expressed surface antigens characteristic
of both parental cells, thus accounting for their rejection by each

TABLE II-3

Oncogenicity of Hybrid Cells[a]

		Oncogenic in		
Cells	Derived from	C3H mice	SWR mice	F$_1$(C3H × SWR) mice
L	C3H mice	Yes	No	Slightly
MT 1	SWR mice	No	Yes	Yes
Hybrid	Cultures of C3H + SWR cells	No	No	Yes

[a] Produced by Gershon and Sachs (1963).

parental mouse strain. The hybrid cells, however, would not be antigenic in interstrain F_1 mice. These suggestions were confirmed by Spencer *et al.* (1964) when they reported that antigens characteristic of both parental cells were expressed in somatic cell hybrids.

C. HYBRID SELECTION

While the hybrid cells produced in these early experiments were discovered because they happened to grow more vigorously than the parental lines, it became evident to Littlefield (1964a) that careful selection of cell lines and media could enable far more frequent isolation of hybrids. He accomplished this by selecting two cell lines, one of which (A9) lacked the enzyme hypoxanthine guanine phosphoribosyltransferase (HGPRT) and the other (B82) lacked the enzyme thymidine kinase (TK). He then treated these cells with a drug, aminopterin, whose effect is to block the main synthetic pathways to purines and pyrimidines. In normal cells in which TK and HGPRT are found, the toxic effect of aminopterin can be bypassed if the culture medium provides thymidine and hypoxanthine which are metabolized directly by the enzymes to nucleotides (the salvage pathways). However, TK^- and $HGPRT^-$ cells lack the salvage pathways. Yet, Littlefield showed that when mixed cultures were grown in a medium supplemented with hypoxanthine, aminopterin, and thymidine (the so-called HAT medium; Szybalski *et al.*, 1962), clones of surviving cells appeared. Their chromosomal number was approximately the sum of the A9 and B82 chromosomal modes. He had, thus, selected for hybrid cells which were capable of survival because the enzymes missing in the parental strains were present in the hybrid by complementation. A more detailed review of this and other selective methods is given in Chapter IX. Davidson and Ephrussi (1965) subsequently found that Littlefield's concept could be used in a simple "half-selective" system. When large numbers of A9 cells were mixed with far fewer normal mouse cells and cultured in HAT medium, the sensitive A9 cells degenerated, leaving after about 2 weeks a thin monolayer of normal cells overlaid with several clones of fast-growing hybrids. As with the A9 cells, the hybrids proved to be a permanent cell line.

D. INTERSPECIES HYBRIDS AND THE USE OF VIRUS

In 1965, three groups (Harris and Watkins, 1965; Okada and Murayama, 1965; Ephrussi and Weiss, 1965) demonstrated that it is possible to form hybrid cells by fusing cells from different animal species. Ephrussi and Weiss mixed rat embryo cells with mouse L cells and using the half selective system isolated mononucleated (synkaryotic) interspecific hybrid cells capable of proliferation. On the other hand, with their papers, Harris and Watkins and Okada and Murayama introduced into common usage a new technique. It was based upon the observation of Okada and co-workers (Okada, 1958, 1962a,b) that a myxovirus called HVJ or Sendai, dead or alive, could be used to form polykaryons from cells in culture (see Chapter IV). Using this method Harris and Watkins used the UV-inactivated virus to generate heterokaryons from HeLa cells of human origin and Ehrlich ascites tumor cells from mice. Okada and Murayama (1965) produced heterokaryons by fusing human with mouse, and human with pig cells. When, a year later, Yerganian and Nell (1966) showed that such heterokaryons could also generate long-lived lines of synkaryons, the use of Sendai virus became the standard method of fusing cells.

Two very important findings came out of the discovery that interspecific proliferating hybrid cells could be obtained: first, both genomes may be expressed in the hybrid (Weiss and Ephrussi, 1966a,b); and second, in long-lived interspecies hybrid lines, chromosomes of one species tend to be preferentially eliminated (Weiss and Green, 1967). The first has facilitated the study of fundamental developmental phenomena (see Chapter XI). The chromosome segregation phenomenon has enabled mapping of human chromosomes and this in turn has established a subfield in cell hybrid study of enormous significance to biology and medicine (see Chapter XIII and Kucherlapati *et al.*, 1974; Ruddle and Creagan, 1975).

In the interest of historical completeness, it should be noted that interspecies heterokaryons had actually already been experimentally generated nearly 3 decades earlier. In 1937, Michel reported using the mechanical technique developed by af Klercker (1892) for isolating living plant protoplasts to make very short-lived heterokaryons from as distant species as cabbage and onion or cabbage and the alga *Chara*. While this historical precedent can be seen to have influenced the current interest in and techniques for plant cell fusion, it played no role in the background of those engaged in animal cell hybridization. Yet, it is clear from Michel's writing that both he and his mentor Küster were well aware of the kind of scientific questions that experiments with heterokaryons could provide answers for, answers which were not obtained until the development of animal cell fusion techniques in the 1960's and 1970's.

E. LATER DEVELOPMENTS

Since 1970, a number of attempts have been made to discover agents other than Sendai virus which can fuse cells. Experiments with high calcium-ion concentrations and various lipid derivatives and polyethylene glycol (see Chapter V) have proved sufficiently successful that there is some hope that, in the future, fusion may be effected at will among all sorts of cells without risking infection from incompletely inactivated virus.

Nineteen seventy also saw the rekindling of interest in plant cell fusion (Power *et al.*, 1970) that culminated in the generation of an interspecific hybrid plant 2 years later (Carlson *et al.*, 1972) (see Chapter XVI). Fusion in the case of plant cells is not effected by virus but, instead, by treatment with inorganic salts or other agents after removal of the cell wall.

Experiments performed in the 1960's led to the conclusion that synkaryons derived from the fusion of malignant and normal cells were always malignant. Beginning about 1970, however, a series of experiments performed in Klein's and Harris' laboratories (Harris *et al.*, 1969a; see Chapters XIV and XV) showed that when such synkaryons retained most of the parental chromosomes, malignancy was not dominant. The frequency of oncogenic hybrids increased as chromosomes were lost. Lately this interpretation has been challenged by Croce and Koprowski (1975) who, on the basis of their work with hybrids between normal and virus-induced tumor cells, maintain that malignancy is dominant. Although there is no simple explanation at hand the results of both groups illustrate the potential of cell hybridization in the analysis of malignancy. An interesting finding of the Harris–Klein group is that fusion of tumor cells with normal host cells *in vivo* occurs with a remarkably high frequency (Wiener *et al.*, 1972a) suggesting that this sort of event might even be part of the normal development of malignancy (see Chapter XIV).

The early years of the 1970's also saw the development of technology enabling routine fusion of cells with enucleated cytoplasms (Ladda and Estensen, 1970; Poste and Reeve, 1972; Ege *et al.*, 1973) (see Chapter IX). Enucleated cells are prepared by removing cell nuclei in the presence of the drug, cytochalasin B, and their fusion with other cells is accomplished by Sendai virus. Because the nuclei removed by these techniques are surrounded by a narrow rim of cytoplasm and a plasma membrane (Ege *et al.*, 1973), they also are amenable to virus fusion (Ege *et al.*, 1974b; Veomett *et al.*, 1974; Ege and Ringertz, 1975). This raises the possibility of reconstituting cells from nuclei (or, more properly, *"minicells"*) of one species or state of differentiation with cytoplasms of another. The promise that reconstituted cells hold for analyzing developmental questions is very large indeed.

Thus, within a decade of its origin, somatic cell hybridization has become an important tool in cell biology, genetics, virology, developmental biology, and tumor biology. It is noteworthy that in each of these disciplines hybridization has been used in combination with other techniques to good effect. For example, it only became possible to use the chromosomal segregation seen among interspecific cell hybrids to map the human genome after Caspersson *et al.* (1968, 1970) developed the first method which unequivocally identified each of the human chromosomes. And the techniques for enucleation of cells on a large scale developed by Prescott *et al.* (1972), based on the original observation of Carter that the drug cytochalasin causes nuclear extrusion (Carter, 1967), has enabled the development of fusion with enucleated cytoplasms. The likelihood that imaginative persons will continue to conceive of "fusions" of techniques such as these surely means that other disciplines will join the list of those served by somatic cell hybridization and that a "historical" summary such as that in Table II-1 will soon be too short.

III

Spontaneous Cell Fusion

"Spontaneous" events in biology may be defined as those for which a cause is not known. This chapter briefly reviews instances of cell fusion which have been detected *in vivo* and *in vitro*, and even selected for, but which occurred without deliberate addition of some fusing agent. While all these fusions may thus be considered spontaneous, some were not unexpected. They were fusions which are evolutionarily programmed for, involving cells which invariably fuse with one another during the ontogeny of the organisms in which they are found. Other cases, though, were unexpected. They involved cells which apparently do not normally fuse in nature.

A. FUSION OF INTRACELLULAR MEMBRANES

One great biological puzzle is why membranes frequently fuse intracellularly, but rarely intercellularly. Cinematographic and autoradiographic studies show clearly that cells which engage in endocytosis or in secretion provide many examples of fusing membranes: primary lysosomes join with phagocytic vacuoles; endoplasmic vesicles coalesce with elements of the Golgi apparatus while transferring newly synthesized proteins to them; vesicles derived from Golgi apparatuses coalesce to form the cell plate during cytokinesis in plants; secretory granules are apparently built up from fusion of smaller vesicles; and the granules themselves join with the plasma membrane dur-

15

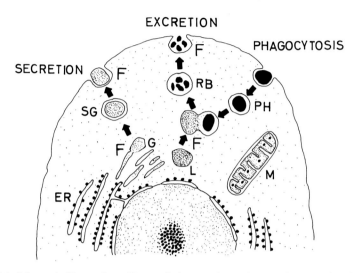

Fig. III-1. Schematic illustration of intracellular membrane fusions (F). Vesicles of the endoplasmic reticulum (ER) fuse with the Golgi apparatus (G). Secretory granules (SG) and residual bodies (RB) fuse with the cell membrane during secretion and excretion. Phagocytic vesicles (phagosomes, PH) fuse with lysosomes (L) during the process of phagocytosis. Mitochondrion (M).

ing release of their contents (Fig. III-1). But with the notable exception of fertilization, healthy cells rarely fuse with other cells in nature.

B. FERTILIZATION

The phenomenon of fertilization, the fusion of two cells, sperm with egg, compounds the puzzle. Obviously it is not a phenomenon confined to a few unique species, but is ubiquitous, suggesting that organisms retain a genetic capability for effecting cell fusion, yet with this exception forego the ability. The membranes which fuse during fertilization may be closely related structurally to those which coalesce within cells. Several of the intracellular fusions cited above involve membranes which at some time have been part of the Golgi apparatus. This is true also of at least some of the membranes involved in fertilization. In many species the head of a sperm is capped with a complex organelle, the acrosome, which is derived during spermiogenesis from the Golgi apparatus (Burgos and Fawcett, 1955). During fertilization the acrosomal membrane first interdigitates and then coalesces with the egg membranes (Colwin and Colwin, 1960), probably releasing in that instant digestive enzymes which aid the penetration of the egg cytoplasm by the sperm nucleus.

While normal fertilization produces "hybrids," although rarely inter-specific ones, these are beyond the scope of this book. Nevertheless, it should be noted that virus-induced cell fusion has now been used to generate hybrids in which sperm cells are one of the parental types, thus making possible interspecies "fertilizations" and anomalies like triploid cells. These are within the scope of the book and are discussed in Chapter VII.

C. PLASMOGAMY IN FUNGI

But even if fertilization is set aside, there still exist a few special cases in which fusion of cells occurs. In higher fungi, for example, there is the phenomenon of plasmogamy, in which mononucleated haploid cells join to produce binucleate cells (dikaryons). These cells proliferate mi-totically, always remaining dikaryotic, resulting *inter alia* in the familiar sporophores of culinary delight. In terms of the fungal life cycle, plas-mogamy is ultimately followed by nuclear fusion, meiosis, and sporula-tion. Among fungi exhibiting plasmogamy, there clearly are instances of production of cells hybrids—still not interspecific—and their genetics has been extensively studied and reviewed (Pontecorvo, 1956; Raper, 1966; Ling and Ling, 1974).

D. FUSION OF MYOBLASTS

Moreover, there are examples of natural cell fusion in mammals as well: in the formation of myotubes, of osteoclasts, and of foreign body giant cells and in the events accompanying embryo implantation in the uterus. In regard to the first of these, it had been observed in the nine-teenth century that striated myofibrils developed in large elongate cells, myotubes, which were polykaryons. Early suggestions that mitotic pro-liferation accounted for the multinuclearity lost support by failure to observe mitosis in myotubes. Other hypotheses involving amitotic di-visions were finally overcome in about 1960 by demonstrations that mononucleated myoblasts fuse to form myotubes, at least *in vitro* (Holtzer *et al.*, 1958; Capers, 1960; Konigsberg *et al.*, 1960).

Like many embryonic cells, myoblasts from prenatal rats or chicks are easy to culture. A primary culture of myoblasts may be prepared by ex-cising skeletal muscle tissue from 10- to 13-day-old rat embryos, digesting briefly with weak trypsin so as to break down intercellular adhesions and disperse the cells and then transferring them to tissue culture media. After 3 to 4 days of growth, large numbers of mobile

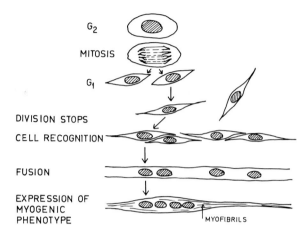

G_2

MITOSIS

G_1

DIVISION STOPS

CELL RECOGNITION

FUSION

EXPRESSION OF
MYOGENIC
PHENOTYPE

MYOFIBRILS

Fig. III-2. Schematic summary of myogenesis *in vitro*. Mononucleate myoblasts after a period of proliferation enter a G_1 state, aggregate with each other, and fuse into giant multinucleate myotubes. In these structures large quantities of contractile proteins and enzymes characteristic of skeletal muscle are synthesized.

myoblasts align with their long axes parallel to each other, and within a period of a few hours they fuse into a network of multinucleated myotubes. Single myoblasts or even other newly formed myotubes may subsequently fuse with the myotubes, adding even more nuclei and cytoplasm. The myotubes develop myofibrils of actin, myosin, and related proteins (for a review see Yaffe, 1969) and subsequently the fibrils engage in muscular twitching so vigorous that they not infrequently pull the myotubes free from the substrate. The formation of myotubes is illustrated in Figs. III-2 and III-3.

1. Myoblast Fusion in Chimeric Mice

Normally a myotube is not a hybrid cell since the myoblasts which fuse to make it are genetically identical. But in interesting experimental situations hybrid myotubes have been generated both *in vivo* and *in vitro*. The experiments performed *in vivo* at the same time have demonstrated that myoblast fusion occurs in organisms as well as in culture. These experiments involve the production of mice containing in their tissues cells of more than one genotype (chimeric or allophenic animals). The techniques for producing chimeric mice were developed independently by Tarkowski (1961) in Warsaw and by Mintz (1962) in Philadelphia. Fertilized eggs which have cleaved to about eight blastomeres are surgically removed from pregnant donor females (Fig. III-4). Blastomeres from two dissimilar genotypes, as, for example, from white coated and black coated mice, are "denuded" (i.e., their zona pellucida

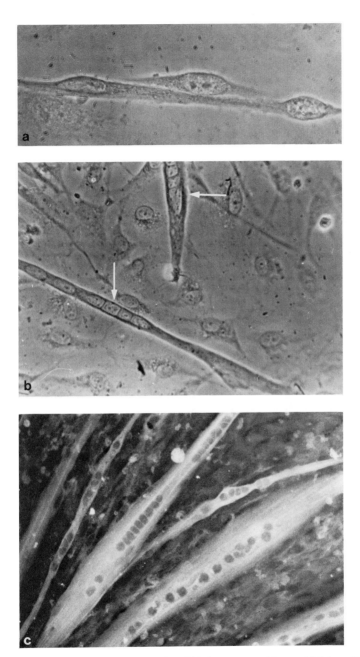

Fig. III-3. Phase contrast photomicrographs of (a) three rat myoblasts which have lined up and are ready to fuse, (b) two rat myotubes (arrows), unfused myoblasts, and non-myogenic mononucleate cells, and (c) fluorescence photomicrograph of rat myotubes stained with fluorescein-conjugated antibodies in order to demonstrate rat myosin. The cytoplasm of the myotubes shows a strong reaction while the unfused mononucleate cells in the background are negative.

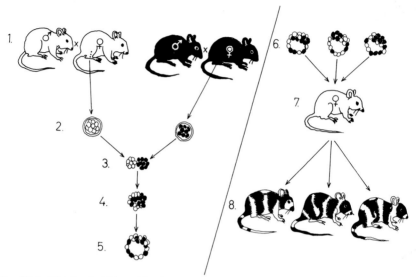

Fig. III-4. Allophenic (chimeric) mice are produced by removing two fertilized eggs at the 8-cell stage and allowing the cells to aggregate into a 16-cell stage. After these composite embryos have developed to blastocysts they may be implanted in the uterus of a hormone-treated foster mother who then gives birth to allophenic mice. If the parents of one of the embryos have white fur while the parents of the other have black fur the allophenic mice will get a striped or patched black and white fur.

are removed, usually by pronase digestion), placed in direct contact with each other at 37°C, and allowed to adhere into a double-sized early embryo. This is cultured to a blastocyst of 128–260 cells in which "black" and "white" cells are intermingled. The blastocyst is then surgically introduced into the uterus of a foster mother. This mouse has in advance been made pseudopregnant by treatment with hormones to allow implantation of the chimeric embryo. Apparently normal *in utero* embryogenesis then proceeds, and the foster mother eventually gives birth to a healthy individual. When the newborn mouse develops fur, it is not homogeneous white or black like that of its real parents, but is patched or striped black and white (Fig. III-5). This indicates that the tissues of the chimeric individual are mosaics of "white" and "black" cells.

In a similar but not so apparent way the internal tissues are also mosaics consisting of cells with "black" or "white" genotypes. There may be, for example, a difference between the two parental types in the proteins comprising an enzyme function—the reaction catalyzed is the same in both mice, even to the cofactors required, but the proteins, while often similar, are not identical. Such enzymes are called iso-

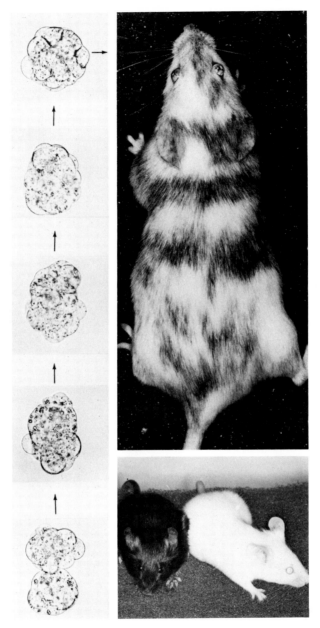

Fig. III-5. Allophenic mouse (upper right) with a patched or striped black and white fur. The fur of the two mothers from whose embryos the allophenic mouse was made is illustrated in the bottom right picture. Sequence of photographs to the left illustrate the aggregation of two embryos (bottom left) into one cell aggregate (top left). (Courtesy of Dr. B. Mintz. From *Methods in Mammalian Embryology* edited by Joseph C. Daniel, Jr. W. H. Freeman and Company. Copyright © 1971.)

21

zymes. They are easily separable from one another by electrophoresis. Mintz and collaborators (Mintz and Baker, 1967; Mintz, 1971) showed that chimeric mice produced from parents (DBA/2 and C3H) which had different isozymes for isocitrate dehydrogenase (IDH-1) or malate dehydrogenase (MDH) exhibited both forms in most of their tissues. This was true also for liver tissue in which there are many binucleate cells. In skeletal muscle, however, Mintz also found a third type of enzyme, a "hybrid" or heteropolymeric enzyme in addition to the enzymes found in the parents.

Both of the enzymes Mintz studied consist of several polypeptide chains. IDH is a dimer consisting of two protein subunits while MDH is a tetramer with four. Each of the isozyme subunits is normally identical, so that in the case of IDH, the enzyme in one parental animal could be represented as aa, while in the other as bb. Parental forms of MDH could analogously be represented as cccc and dddd. Interestingly, these subunits are compatible enough that given the opportunity a can combine with b and c's with d's to generate new isozymes which are heteropolymeric but equally effective catalysts. Such opportunity occurs when the parental strains are mated, producing an F_1 generation, but more relevant here is the fact that it also arose in skeletal muscle of the chimeric mice when myoblasts of both parental strains fused to form myotubes. In the case of IDH, Mintz identified the ab isozyme as well as parental forms (aa and bb). This indicates that when residing in a common cytoplasm the two dissimilar nuclei coded for proteins that combined into a new heteropolymeric enzyme that was not possible while the nuclei were isolated in separate cells. Mintz also found that all the five possible isozymic forms of MDH occurred, but in widely varying proportions in different samples. Probably this reflects the varying ratios of parental nuclei in myotubes which could be expected to follow from random fusions of myoblasts (Mintz and Baker, 1967; Carlsson *et al.*, 1974a). Not only does this provide evidence for *in vivo* fusions of myoblasts in the generation of the multinucleate myotube, but it provides a case of *in vivo* generation of heterokaryons, which, as Mintz and Baker (1967) have pointed out, may allow investigation of causes of genetic diseases in muscle by such classical genetic techniques as complementation analysis (see Chapter XIII).

2. Myotube Heterokaryons *in vitro*

In primary cultures of myoblasts there are found a great many cell types other than myoblasts. All of these cells exhibit motility, and as a result, mononucleated myoblasts collide with fibroblasts and epithelial

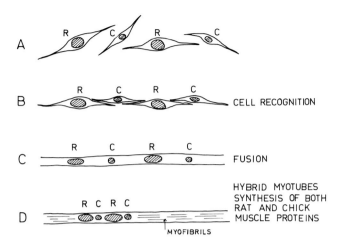

Fig. III-6. Spontaneous fusion of chick myoblasts (C) and rat myoblasts (R) resulting in the formation of chick + rat hybrid myotubes.

cells as well as with each other. In spite of this, only the myoblasts fuse. Furthermore, even among myoblasts there are restrictions as to which cells fuse. Only those which have reached a particular time point in G_1 phase (postmitotic interphase) participate in fusion (Okazaki and Holtzer, 1966; Bischoff and Holtzer, 1969; Buckley and Konigsberg, 1974). Myoblasts in S phase (the period of DNA synthesis), G_2 phase (premitotic interphase), or mitosis never fuse with the G_1 myoblasts (Fig. III-2). These cells thus have a highly developed capacity to recognize other cells on contact (cell recognition), determining not only whether the touched cell is a myoblast but also the stage of the cell cycle it occupies. It is therefore rather remarkable that myoblasts from entirely different animal species can fuse with each other if they belong to the right stage of development (Yaffe and Feldman, 1965). If, for instance, chick myoblasts are mixed with rat myoblasts, myotubes containing both rat and chick nuclei are formed (Figs. III-6 and XII-1). The proportion of chick nuclei in the myotubes will in general depend on the ratio of chick to rat myoblasts in the mixed culture. In these hybrid myotubes, myofibrils containing both chick and rat muscle proteins are formed, and similarly the proportion of chick proteins is a function of the ratio of chick to rat nuclei in the particular cell (Fig. III-7) (Carlsson et al., 1974a.)

Two factors which affect the ability of myoblasts to fuse should be mentioned. The first is the finding of Shainberg et al. (1969), that unless the culture medium contains sufficient Ca^{2+}, fusion does not occur. This is reconsidered in a theoretical way in Chapter V, but here it

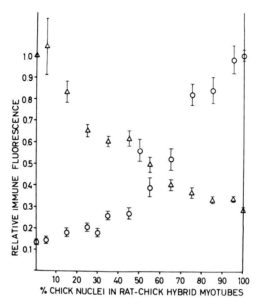

Fig. III-7. Accumulation of rat (△) and chick (○) myosin in hybrid myotubes as measured by microfluorometry after immune staining. The amount of chick and rat myosin accumulating reflects the proportion of chick and rat nuclei. (Carlsson *et al.,* 1974a.)

may be noted that myoblasts so prohibited from their "social" activity seem not to be drastically thwarted otherwise. Although there was decreased leucine uptake, less creatine phosphokinase activity, and a somewhat slower cell cycle among those grown in low calcium, the effects were immediately reversible upon addition of calcium. Furthermore thymidine and uridine incorporation was not adversely affected (Shainberg *et al.,* 1969).

The second factor is the possibility that the age of the myogenic culture affects the ability to fuse. Bischoff and Holtzer (1969) presented evidence which suggests that both myoblasts and myotubes from old primary cultures (e.g., 8-day-old) have greatly reduced capability for further fusion. Perhaps this suggestion is consistent with the observed decrease in morphogenetic capabilities of cultured cells taken from older and older embryos (for a review see M. Harris, 1964) and may indicate a change in surface properties accompanying development. Or possibly the evidence must be seen only in the context of the culture conditions. It is clear that 8-day primary cultures often exhibit degenerative symptoms and that the supply of mononucleated myogenic cells is rapidly depleted in cell cultures after the commencement of fusion. Konigsberg (1963) succeeded in growing up a clone of myoblasts which continued to proliferate and fuse into an extensive myo-tube network over a period of at least 13 days in culture. This success, he found, was very much a function of the condition of the culture medium.

E. FUSION OF OTHER CELLS IN VIVO

1. Heterokaryon Formation during Implantation?

Another case of cell fusion which yields heterokaryons *in vivo* under nonpathologic circumstances may occur during implantation of the blastocyst in the uterine wall during mammalian reproduction. In the process there seem to be fusions between at least two different groups of cells, and since one of these involves young embryo cells with ma-ternal cells, somatic cell hybrids could therefore be generated. Shortly after implantation the blastocyst wall, the trophoblast, consists of two layers of cells. The outer, which is comprised of fused cells and is called the syncytiotrophoblast, is in direct contact with the uterine epithe-lium.

In the rabbit, at least, extensions from the syncytium make their way through the epithelium, apparently not by mechanical intrusion between cells, but by fusion with the apical ends of epithelial cells so that their cytoplasms become part of the cytoplasm of the syncytiotro-phoblast (Enders and Schlafke, 1971). Some of the epithelial cells are already multinucleate at this stage, allowing for the possibility that prior fusion has occurred within the epithelium. If so, these coalescings and those which form the syncytiotrophoblast do not form hetero-karyons, but the fusion of the syncytiotrophoblast with epithelial cells clearly may. The membrane of the embryo which then is in contact with maternal tissue and blood supply presumably contains fetal histocom-patibility antigens as well as maternal ones. As pointed out by Enders and Schlafke (1971), this fact raises immunological questions. For ex-ample, does the presence of its own antigens somehow mitigate what could be expected to be maternal reactivity against the foreign (pa-ternal) antigens of the embryo? [For a discussion of this problem, see Billington (1969).]

2. Giant Cells and Osteoclasts

The term "giant cell" is applied to polykaryons found in association with a variety of inflammations. While cell fusion is generally accepted as the most important mechanism for their generation, instances of mitosis within giant cells have also been reported. Moreover, among those cases where fusion is clearly recognized, there are some which are suspected not to be "spontaneously" induced but virus mediated. These include giant cells associated with variola (smallpox), varicella (chicken pox), vaccinia (cow pox), measles, mumps, herpes simplex, and parainfluenza. Under other pathological conditions, though, giant cells are the specific result of macrophage fusion, and, as in the case of myoblast fusion, the coalescence here seems to be genetically programmed and thus "spontaneous." Typical of these are the giant cells seen in association with tuberculosis. These polykaryons contain upwards of 200 nuclei and are called "foreign body giant cells" or "Langhan's giant cells" depending on their morphology. The latter are more organized and characteristically contain peripherally arranged nuclei (Fig. II-1). The parental macrophages which form these polykaryons migrate from the circulatory system to sites where foreign particulate matter has accumulated. There they fuse with macrophages already *in situ* to form foreign body giant cells. It seems quite likely that the Langhan's type are derived from these first polykaryons by maturational processes (Mariano and Spector, 1974). Their function at sites of alien invasion apparently is to ingest the foreign material by phagocytosis.

Another polykaryon engaged in consuming external material is the osteoclast. Its preferential diet is bone, and it is the cell system accounting for bone resorption under both normal developmental and pathologic circumstances. The cells contain from one to several dozen nuclei and have brush borders which, in electron micrographs, are seen to be in intimate contact with regions of bone which are undergoing decalcification and resorption. These polykaryons are also the result of cell fusion; stem cells which otherwise line the surface of resting bone seem to be the precursor line for osteoclasts (Ham, 1974).

3. Possible Fusion of Erythroid Cells

The examples of fusion presented in the preceding pages constitute instances of cell fusion which seem to be genetically programmed for in mammals. Although they generate polykaryons, they are not normally events which lead to hybrid cells. There exist, though, a few reports which suggest that from time to time there are chance intercellular fusions which do generate hybrid cells. One case involved the spontaneous change in blood type of a bull which at birth had been a chimera in regard to its erythrocytes. Reports of this kind of mosaicism among twins, of which this bull was one, are not rare. It may come about as a result of *in utero* exchanges between nonidentical twins of primordial erythropoietic tissues which then generate two types of red cells. While this bull exhibited at birth both paternal and maternal erythrocytes, after 8 years it had but one type, a hybrid, which presumably could only have been produced by some sort of genetic exchange between the two parental strains. Somatic cell fusion was suggested as the mechanism allowing the exchange (Stone *et al.*, 1964).

Such hybrid cell production must necessarily be rare, since it depends in the first place upon chance production of a chimera. It is worth mentioning, though, that chimerism in blood types is often effected when human fetuses are transfused *in utero* (Nuzzo *et al.*, 1971). At least one case of hybrid leukocyte formation following such transfusion has been reported (Turner *et al.*, 1973).

Second, there may well have to be an agent other than mere juxtaposition to cause the fusion of these dissimilar cells. In the case of the bull, it was reported that he suffered severe symptoms of a gastrointestinal disease ("scours") in his first month. If this disease were of viral etiology, it may have provided the exogenous means for cell fusion (Stone *et al.*, 1964). Consistent with this possibility is the subsequent observation of Karakoz *et al.* (1969) that hybrid erythrocytes were inducible in chimeric chickens by infecting them with western equine encephalomyelitis virus. This virus was chosen both because it is lipid containing (see Chapter IV) and because its initial replication was known to be in erythropoietic tissues.

4. Spontaneous Fusion in Tumors

As mentioned at the beginning of Chapter II, the earliest cases of multinucleate cells were discovered during histological examination of tumors. Interest in tumor-related polykaryons has been rekindled by the discovery that tumor cells sometimes fuse with normal cells *in vivo*. Such fusions are extremely rare, but important because the chromosomal instability of the resulting hybrid cells may increase the genetic variability in tumor cell populations. This matter is discussed in more detail in Chapter XIV.

F. FUSION OF OTHER CELLS IN CULTURE

Because tumor cells are now known to fuse spontaneously *in vivo*, it is hardly surprising that they do so in culture as well. It is noteworthy that Lewis' early observations of cell fusion *in vitro* were made in cultures of rat sarcoma cells (Lewis, 1927). In the same paper, Lewis also reported that "the binucleate condition is fairly common in cultures of embryonic mesenchyme, liver, endothelium, ectoderm, endoderm, smooth muscle, heart muscle and kidney epithelium, and of macrophages, epithelioid and ectodermal cells." None of these was a neoplastic line nor a cell whose genetic programming would lead to expectations of its participation in fusion. Indeed, Lewis believed that the binucleate condition seen *in vitro* results from cells in which mitosis is not followed by cytokinesis.

Nonetheless, it may well be that some of the dikaryons recorded by Lewis and other cell culturists were instead the consequences of cell fusions. Doubts that fusion could occur among such cells in culture ended in 1960 with the discovery of cell hybrids (Barski *et al.*, 1960) (see Chapter II). The event, though, is a rare one. Estimations of the frequency of spontaneous fusion *in vitro* have varied between 10^{-2} and 10^{-6}. In regard to experimentation with cell hybrids, if the frequency of their occurrence had not been raised, progress would have been slow. But as already noted in the preceding chapter, higher frequencies were achieved by virus- and chemical-induction of fusion. These two processes are discussed in some detail in the next two chapters.

G. CONCLUDING REMARKS

The really puzzling thing about spontaneous cell fusion is its rarity. It is clear that membrane architecture allows for fusion, for intracellular membrane coalescence is commonplace. Moreover, the few specialized instances of intercellular fusion recounted here make obvious that when appropriate, the cell can be programmed for it. Then, too, there seems to be lack of uniformity among the barriers to fusion in those instances where it occurs. While fertilization is a highly species- and tissue-specific cell fusion, fusions with myoblasts, stem cells, and white cells seem to recognize only a tissue barrier; and it is not clear that the fusions accompanying implantation or occurring *in vitro* recognize even that. Perhaps part of the answer lies in the fact that most intracellular membrane fusions involve what may be described topologically as a meeting of inner membrane surfaces, while intercellular fusions involve outer surfaces. Fusion of membranes from inner surfaces may be as easy as fusion from outer surfaces is difficult. Species or tissue barriers observed in intercellular fusion are likely to be a function of outer surface constituents, probably glycoproteins. In Chapter V, membrane architecture is considered in relation to fusion, and in the course of that, these problems are discussed again, but in more detail.

IV

Virus-Induced Cell Fusion

Even though many of the historically important somatic cell hybrid studies were based upon spontaneous cell fusion, the enormity of the past decade's literature on cell hybrids really stemmed from the discovery that cell fusion can be induced more or less at will. That the techniques involved in Sendai virus-induced fusion are in fact simple, accounts, at least partly, for their widespread use. Sometimes, however, the simplicity proves to be deceptive, for the commonly used methods occasionally fail to fuse some cell combinations for no apparent reason. At such times it may be useful to consider what is known about the factors which influence virus-induced fusion and cell fusion in general. To that end, this chapter summarizes factors which have been reported to affect cell fusion by viruses; the next chapter discusses cell fusion from a more theoretical view point.

A. STRUCTURE OF PARAMYXOVIRUSES

Several different viruses have been reported to induce formation of polykaryons among cells in culture (see Table IV-1), but of these the most useful in making cell hybrids are the *paramyxoviruses*. Until recently these were classified with influenza viruses as myxoviruses. This group was first characterized by Andrewes *et al.* (1955) as those exhibiting affinity for cell membrane glycoproteins, which they attack enzymatically, and sensitivity to ether, an indication that these viruses are surrounded by a lipid-containing envelope. However, the larger size of

TABLE IV-1

Some Viruses Reported to Induce Cell Fusion

DNA-containing viruses
  Herpesvirus
    Varicella (Weller *et al.*, 1958)
    Herpes simplex (Hoggan and Roizman, 1959)
  Poxvirus
    Rabbitpox (Appleyard *et al.*, 1962)
RNA-containing viruses
  Paramyxovirus
    Mumps (Henle *et al.*, 1954)
    Newcastle disease (Kohn, 1965)
    Parainfluenza types, e.g., Sendai (HVJ), SV5 (Okada,
      1958; Compans *et al.*, 1964)
    Measles [a] (Enders and Peebles, 1954; Cascardo and Karzon, 1965)
    Respiratory Syncytia [a] (Morris *et al.*, 1956)
  Oncornavirus
    Rous sarcoma (Moses and Kohn, 1963)
    Visna (Harter and Choppin, 1967)
  Coronavirus
    Avian infectuous bronchitis (Akers and Cunningham, 1968)

---

[a] Measles and its relatives, rinderpest and distemper, and the respiratory syncytial virus, all of which have been reported to induce polykaryocytosis, are in some ways intermediate between paramyxoviruses and orthomyxoviruses.

the viruses now classified as paramyxoviruses, the higher molecular weight of their ribonucleoprotein (RNP) core, and differences in their envelopes have led to their separation from influenza viruses, which now are classified as *orthomyxoviruses* (Melnick, 1973).

Because of the exceptionally large variety of cell types susceptible to its attack, the paramyxovirus that has come to be used as a standard fusing agent for generating cell hybrids is the Sendai virus. It is known also as parainfluenza virus 1 and as the hemagglutinating virus of Japan (HVJ). The first name derives from the fact that it was first isolated from mice by N. I. Ishida and his colleagues at Tohoku University School of Medicine in Sendai, Japan.

Like other paramyxoviruses, the morphology of Sendai is characterized by (1) pleomorphic particles of up to about 5000 Å in diameter; (2) a lipid-containing envelope which stains like eukaryotic membranes, but which is decorated with "spikes"; and (3) a helical RNP core of about 180 Å in diameter (Horne *et al.*, 1960) (Fig. IV-1).

The membranous character of the envelope stems from the fact that as Sendai virus particles leave a parasitized host cell, they are surrounded by a piece of the plasma membrane. Electron micrographs show filamentous or, more commonly, spheroid protrusions of host cell membrane during the release of paramyxoviruses from infected cells (e.g., Blough, 1964; Howe *et al.*, 1967). Moreover, Rott *et al.* (1966) showed that when glycolipid and glycoprotein antigens (blood group substances A, B, F, H, Forssman, and mononucleosis antigens) were found on host cells, myxovirus liberated from these cells exhibited the same serologically defined membrane structures. Neurath and his colleagues (Neurath and Sokol, 1963; Neurath, 1965) demonstrated the presence of an ATPase on the surface of Sendai virus which, because of its occurrence on even incomplete virus particles, is probably host-cell derived.

The presence of host membrane may partly account for the fact that the host can exert a modifying effect upon virus phenotype. For example, the abilities of Sendai virus to hemolyze and fuse have been shown to be host dependent (Homma, 1972; Homma and Tamagawa, 1973). Sendai virus populations grown in fertile chick eggs are capable of lysing red cells, while those grown on L cells or chick embryo fibroblasts are not. Homma (1972) restored hemolytic and fusing capacity of the Sendai virus grown on L cells by very short tryptic digestion, suggesting that some additional protein component had been masking its ability to hemolyze. Subsequently, Scheid and Choppin (1974) showed that batches of Sendai virus grown in bovine kidney cells, which also lacked fusing and hemolytic functions, possessed a glycoprotein of 65,000 daltons which was diminished to 53,000 daltons by the tryptic digestion that restored the two functions. Sendai virus grown in eggs already possessed the 53,000 dalton form, as well as the fusing and hemolytic capacities. The host cell's influence, thus, seems to be to determine whether a larger ineffective or a smaller potent glycoprotein is synthesized in the course of virus assembly.

The plasma membrane is modified, however, before it pinches free from the host cell to become the viral envelope. The most obvious change is the addition of "spikes." These are 80- to 100-Å-long protrusions having a center-to-center spacing of 70 to 80 Å on the envelope. Each has been shown to be a univalent hemagglutinin, which taken together make the virus a multivalent agglutinating agent. Their addition to the envelope accounts for the virus' effectiveness in aggregating not only red cells, but the many other agglutinable cells as well. In addition, the spikes exhibit neuraminidase activity. This is an enzyme which hydrolyzes terminal sialic acid residues from the oligosaccharide moiety of glycoproteins. Since there is a good deal of exposed glycoprotein on cell plasma membranes, the effect of this viral enzyme is to attack cell surfaces. Curiously the receptor sites to which viruses attach are among the surface loci digested. Other less visible changes include insertion of some other virus-specific proteins, including the one

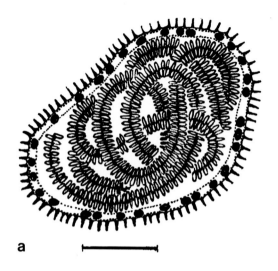

a

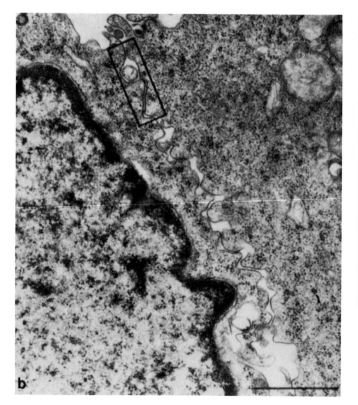

b

c

already noted which determines the virus' ability to fuse or hemolyze cells (Scheid and Choppin, 1974). (For a review of virus envelope properties see Lenard and Compans, 1974).

B. FACTORS AFFECTING VIRAL FUSION OF CELLS

While phages attack bacteria like syringes with hypodermic needles, animal viruses generally are phagocytosed and gain entry to host cells from vacuoles. Paramyxoviruses enter by coalescing with host plasma membrane, either while it surrounds the cell or when it is part of a phagocytic vacuole (e.g., Hosaka and Koshi, 1968; Morgan and Howe, 1968; Dales, 1973). The coalescence appears, under the electron microscope, to begin with a disorganization of adjacent regions in both viral and host membranes (Fig. IV-1c). Not infrequently the formation of continuous membrane between virus and host entraps small vesicles whose membranes must be part viral and part host in derivation (e.g., Morgan and Howe, 1968). Left unanswered at the moment are the questions of the fate of these hybrid vesicles and of any remnants of viral membrane in the host plasma membrane and the role these viral pieces may play in host properties or in viral reproduction.

The paramyxoviruses also share with other viruses the capability of agglutinating cells: one side of a virus particle affixes itself to the surface of one cell and the other side sticks to another, thus "stapling" the two cells into very close contact (Fig. IV-1b and c). But unlike other viruses, after this aggregation, paramyxoviruses effect fusion of the agglutinated cells. Because both capabilities—to coalesce with cell membranes and to agglutinate cells—are found in many viruses, the ability to fuse cells cannot be linked solely to these properties. Many of the factors which are involved in the virus' ability to fuse cells have been studied, and among these, important ones appear to be some particular properties of both the host cell and virus, virus multiplicity,

Fig. IV-1. Sendai virus induced cell fusion. (a) Schematic representation of the structure of the virus showing the lipid-containing envelope, spikes, and a helical RNP core. Scale: 0.1 μm. 狐狐狐狐, "chromosome," 18 nm × ca. 5000 nm long. ::::::, viral envelope derived from lost cell. ●●●, 53,000 dalton glycoprotein enabling virus to affect cell fusion and hemolysis. ⋀⋀⋀⋀, "spikes" showing neuraminidase and hemagglutining activity, 10 nm long. (b) Close association of adjacent membranes near an adsorbed virion, observed after 1 min. at 37°C. The closely opposed cell membranes form a contact zone of two membranes. Scale: 1 μm (c) An enlargement of the enclosed area in (b). Two unit membranes of the contact zone are about 150 Å apart, along a distance of about 400 nm. A discontinuity in the two cell membranes is seen nearest an adsorbed virion. Scale: 0.1 μm [from Hosaka and Koshi (1968)].

presence of ions, pH, and a source of energy. These factors are elaborated in the following paragraphs.

1. Host Cell Characteristics

The host specificity of most viruses accounts for the variation among cell lines with regard to their susceptibility to virus-induced fusion. The specificity stems from the chemical nature of receptor sites on the surfaces of cells and means that only some viruses react with a given cell. As noted already, the composition of the receptor site is glycoprotein, and in consequence, cells may be made insensitive to viral attack by neuraminidase digestion. Parenthetically it may be noted that the neuraminidase which is part of the influenza viral envelope is so effective in destroying influenza receptor sites that the virus quickly elutes from its attachment, allowing agglutinated red cells to disaggregate (Hirst, 1941). The rapidity of this neuraminidase activity perhaps partly accounts for the inability of influenza virus to effect cell fusion.

The widespread use of Sendai virus for inducing cell fusion is very much related to the large number of cell types, both of established and primary cultures, which have receptors for it. Although the ability of cells to fuse is only an indirect indication of the presence of receptors, the paucity of reports of total inability of Sendai virus to effect fusion among cultured cells suggests that receptor sites for the virus must be widespread. Early reports of failure to generate homokaryons from leukocytes with Sendai virus (Okada and Tadokoro, 1963) have been qualified by subsequent reports that Sendai virus has induced formation of homokaryons and heterokaryons from various white cells (e.g., Harris, 1965; Gordon and Cohn, 1970). It has also been shown that conditions employed for the fusing process often determine whether or not there is success (e.g., Velazquez *et al.*, 1971). Perhaps then, the general rule of rigorous host specificity must be modified to take into account the apparent fact that cell receptors and viral "antireceptors" can be so altered by environmental conditions as to extend the range of susceptibility. But within a set of standard conditions there seems to be greater susceptibility of malignant cells to nonmalignant, established culture lines to primary or secondary cultures, and "younger" cells to "older" cells in any given passage level of culture [see Poste (1970) for a review of these generalizations and their exceptions].

The behavior of host cells in mixed cultures also plays a role in the frequency of homokaryon versus heterokaryon generations. It is now well established that, in culture, cells of the same histiotype tend to find each other during their movements and preferentially to aggregate (e.g., Moscona, 1966). Such histiotypic aggregation occurs even across species boundaries. For example, chick embryo myoblasts aggregate readily with rat myoblasts even though many other chick cell types may offer their favors in a culture. There is evidence suggesting that histiotypic aggregation of this sort tends to favor Sendai virus induction of homokaryons, or of heterokaryons of the same histiotype, in preference to more divergent combinations (Mukherjee *et al.*, 1971). Moreover, Koprowski (1971) has shown that, given a choice of fibroblasts from many species, hamster fibroblasts prefer intra- to interspecies fusions (Table IV-3).

There are also data consistent with the hypothesis that the capacity of Sendai virus to induce fusion varies with stages in the cell cycle. Although one study showed no preference of G_1, S, or G_2 cells participating in fusion (Westerveld and Freeke, 1971), another, using a culture-adapted line of mouse leukemic cells showed that among cell populations synchronized with colcemid or thymidine block, the ability to fuse is low during G_1 and S periods and high in late G_2 and mitosis (Stadler and Adelberg, 1972). These data nicely parallel observations of periodic glycoprotein and glycolipid secretion by cells which is also maximal at mitosis. In some cell lines there is a similar periodicity in changes of cell surface antigenicity or response to wheat germ agglutinin.

2. Fusion Factor

As already noted, paramyxoviruses are able to agglutinate erythrocytes and other cells and to hemolyze and fuse cells. Because the influenza viruses can hemagglutinate but neither lyse erythrocytes nor fuse other cells, it is natural to suppose that the ability of paramyxoviruses to fuse cells is related to their ability to hemolyze. For this and other reasons, it has been proposed that fusion is nothing more than a controlled lysis of agglutinated cells (Yanovsky and Loyter, 1972). A difficulty in subscribing to this hypothesis is that a variety of treatments (heating, sonicating, freezing and thawing, and trypsin digestion) affect hemolysis and fusion in different ways (Okada and Tadokoro, 1962; Kohn, 1965; Koprowski, 1971) (Fig. IV-2). In order to make these data compatible with the proposal, then, one may suppose that a change in some part of the virus causes it to react with red cells (where hemolysis is observed) in one way and with other cells (where controlled lysis leads to fusion) in another, or else that the process of virus-induced lysis in red cells is unrelated to lysis in other cells.

While either of these ideas is plausible, a number of workers have instead opted for the possibility that paramyxoviruses have a specific "fusion factor" (e.g., Kohn, 1965; Kohn and Klibansky, 1967; Guggenheim

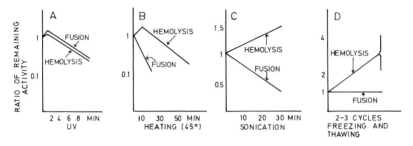

Fig. IV-2. The effects of various treatments upon the hemolytic and fusing capabilities of paramyxoviruses [modified from Okada and Tadokoro (1962) and Kohn (1965)].

et al., 1968; Okada, 1969; Koprowski, 1971). Their studies have shown that the fusing capacity of these viruses lies either in the phospholipid moiety of the viral envelope or at least in something which is altered by attacks on the phospholipid.

What is clear is that the fusion factor is not related to the activity of the viral nucleic acid. Okada and Tadokoro (1962) showed that infectivity of Sendai virus dropped to almost zero within a one-minute exposure to UV radiation, but that its capacity to fuse Ehrlich's ascites cells remained at 100% through 2 min of UV and was still about 20% after 6 min. Much the same results were reported by Kohn (1965) for NDV. Indeed, it is customary to profit from these observations by inactivating the infectivity of Sendai virus preparations with UV (Harris and Watkins, 1965) or by a β-propiolactone treatment (Neff and Enders, 1968) before use. In this way the virus serves only as an agent for fusing cells without bringing along the complications of infection.

The studies of Scheid and Choppin (1974), however, make it unlikely that a fusion factor exists separate from the virus' ability to hemolyze. As discussed above (Section A), Sendai virus particles which are active in fusing cells and in hemolysis possess a 53,000 dalton glycoprotein. When this component is absent, so also are both capabilities. As discussed in the next chapter, there are a number of membrane-related components which participate in cell fusion and which might account for the distinctions between hemolysis and fusion. While some of these components are found as constituents of viral envelopes, others are associated only with the host cell plasma membrane. Collectively these components account for the series of events which effect either cell fusion or cell lysis.

3. Virus Multiplicity

It is also clear from studies on virus-induced cell fusion that virus multiplicity is an important factor in successful fusion. Kohn (1965) has estimated for NDV that the ratio of 150 viral particles per cell is minimal for fusing monolayers of cultured cells into polykarons. The minimal number for suspended cells may be somewhat higher: Okada and Murayama (1968) extrapolated their data to show that Ehrlich ascites cells in suspension were not fused until the ratio of virus to cell had risen above 250. These results indicate that relatively large numbers of virus particles are necessary.

By way of comparison, under ideal conditions a single virion causes an infection in one cell. But since not all potentially infectious viruses in any sample result in recognizable infection, virologists empirically determine what is called "the number of infectious units." In a given sample the number of infectious units is somewhat smaller than the absolute number of virus particles. In the case of influenza virus, for example, the ratio of virus particles to infectious units is between 7 and 10 to 1. Thus, if paramyxoviruses and influenza are comparable, the minimal number of paramyxovirus particles necessary for fusion is 15 to 300 times that necessary for infection.

These figures for virus multiplicity are pertinent to cell fusion in which replication of virus plays no role—in which irradiated, inactive viruses suffice. This kind of virus-induced fusion has been described by virologists as "fusion from without," or FFWO (Bratt and Gallaher, 1972). Some paramyxoviruses, though, are capable of effecting changes in host cell membranes during processes accompanying their reproduction inside the cells that cause host cells to fuse with one another. This is known as "fusion from within" (FFWI) (Kohn, 1965). Unlike FFWO which can be detected within a short time after addition of virus, FFWI is first seen only after 6 h of infection with live virus. Because FFWI is a property of viral reproduction, the required number of virus particles per cell is no more than that necessary for infection. But of course it is FFWO which is routinely employed when paramyxoviruses are used for generating hybrid cells, and it requires high multiplicity of virus.

In practice it is not feasible to calculate the ratio of virus particles per cell in each fusion experiment. Instead it is common to express viral quantitation as the hemagglutinating capacity per volume of virus suspension. The relative simplicity of estimating virus concentration by this means accounts for the fact that in so many studies using Sendai virus to produce cell hybrids, viral multiplicity is reported as some number of hemagglutinating units (HAU) per ml. HAU are empirically derived measurements of the ability of a given virus suspension to provide sufficient number of virions to agglutinate a standard suspension (ca. 10^7 cells/ml) of erythrocytes. The magnitude of HAU/ml of any given suspension is the number of times it may be diluted and still effect hemagglutination.

As may be obvious, the optimum HAU/ml in any experiment depends upon any or all of the fusion-influencing factors discussed here. In consequence, the appropriate HAU is determined experimentally with each new cell combination and condition. Furthermore, the ability of Sendai virus to fuse cells is not linear with increasing HAU/ml (Okada, 1962a). Indeed, the two properties are separable on several counts. Okada and Tadokoro (1962) showed, for example, that the two abilities were differently sensitive to UV irradiation and to heating; the virus' capacity to hemagglutinate was retained longer than its capacity to fuse. Because of these observations and because the expression of Sendai multiplicity in HAU/ml is useful for cell fusion only relative to specific cell pairings under specific conditions, many workers have decided that it is misspent time to calculate and report hemagglutinating titers for cell hybrid studies. Instead, they work out the volume of Sendai virus suspension which, for each batch of virus, for each cell type, and for each set of fusing conditions, proves to be appropriate to their experimentation.

4. Calcium Ion Concentration

Like many other phenomena related to the interaction of cell surfaces, virus-induced fusion is influenced by the presence of calcium ions. Okada and Murayama (Murayama and Okada, 1965; Okada and Murayama, 1966; Okada, 1969) showed that while agglutination is not dependent on calcium, fusion is. When insufficient calcium or a competing chelating agent like EDTA was present in the medium, Ehrlich ascites cells tended to lyse instead of fuse in the presence of Sendai virus. But with concentrations between 0.1 and 1 mM $CaCl_2$, the frequency of cell fusion increased proportionately. Moreover, in order to achieve maximum frequency of fusion, the presence of 0.42 mM calcium was required for more than 10 min after the start of fusion (Okada, 1969).

The usual recipes for Sendai virus-induced cell fusion call for the use of Hanks' or Earle's basal salt solution (BSS) during the fusion process. They meet the requirement for calcium, for their concentrations, as $CaCl_2$, are, respectively, about 1.27 and 1.8 mM.

5. pH of the Medium

Okada (1969) also showed that fusion of ascites cells using Sendai was pH-dependent. While the earlier studies showed the optimum pH was between 7.3 and 7.8, subsequent investigations (Croce *et al.*, 1972; Yanovsky and Loyter, 1972) have shown that it lies higher in the pH range (7.8 to 8.0). The disparity may be explained, according to Yanovsky and Loyter (1972), by the fact that at high pH's the buffering anions, phosphate and carbonate, precipitate calcium ions. Thus, in the earlier studies, the apparent decrease in fusion was not a direct function of low hydrogen ion concentration, but of unavailability of calcium ions. The pH of the media in the later investigations was determined by "Good" buffers (Good *et al.*, 1966), which have nontoxic, univolved, nonprecipitating anions. With these, noticeably higher frequencies of fusion were obtained near pH 8 than in the mid-7 range. Croce and coworkers (1972) also showed that when long-lived hybrid lines are being selected, maintaining the cultures at near pH 8 for some days after Sendai- or lysolecithin-induced fusion very markedly increased the number of colonies obtained. As noted in the next chapter, high pH and high calcium ion concentrations, without addition of virus, have several times been reported to suffice in generating polykaryons.

6. Energy Requirement

Unless cells which are undergoing fusion are provided with metabolizable substrate and allowed to respire, they are likely to lyse rather than to form stable polykaryons. Although earlier studies suggested that respiration had to be aerobic (Okada *et al.*, 1966), glycolysis was later shown to be equally effective (Yanovsky and Loyter, 1972), at least with some cell lines if not all (Okada, 1969). Moreover, Okada (1969) showed that aeration of Ehrlich ascites tumor cells can improve the subsequent susceptibility to virus-induced fusion and that for optimum results aerobic respiration must continue during the first 5 to 10 minutes of the process. While these observations may not be applicable to all cell lines, they nonetheless emphasize the importance of energy input for cell fusion. How the energy is used is discussed in the following chapter.

Agglutination of cells by virus, on the other hand, seems not to require enzyme activity. When using Sendai virus to fuse cells, it is usual to allow agglutination to proceed by exposing the cell and inactivated virus mixture to a period at 4°C, during which time respiration and other metabolic activities are slowed. Little if any fusion occurs during the cold treatment but subsequent warming of the preparation to 37°C initiates it. Warming also speeds up the viral envelope neuraminidase activity, which is held in check by the cold. To some degree raising the temperature to 37°C starts a race between successful fusion of aggregated cells and disaggregation by neuraminidase destruction of receptor sites.

C. TECHNIQUES FOR VIRUS-INDUCED FUSION

Detailed accounts of the propagation of Sendai virus and its use in fusing cells are given by Watkins (1971), Rao and Johnson (1972a), and Giles and Ruddle (1973a,b). The use of this virus to induce fusion of Ehrlich ascites cells in suspension was introduced by Okada (1958) and used by Harris and Watkins (1965) and Okada and Murayama (1965) to generate interspecific heterokaryons. For cells adapted to growth on solid substrate, the technique in principle involves four steps: getting the cells into suspension, agglutinating them with Sendai virus, allowing them to fuse, and restoring the cultures to optimal conditions for attachment and growth. Kohn (1965) and Davidson (1969) described methods which used cultured cell lines while still growing in monolayers for experiments with virus-induced fusion. Since then, numbers of workers have used Sendai virus to fuse heterogeneous cells growing in monolayer or with one parental line attached to substrate, another in suspension. In general, fusion with at least one parental cell line in monolayer is more successful than with all cells in suspension; higher yields of both polykaryons and synkaryons are obtained (Klebe *et al.*, 1970; Hitchcock, 1971). Where parental lines are cocultured and grown attached to substrate, virus is simply added to the culture and the surplus eventually removed. When one parental cell is attached and the other is in suspension, the virus and the suspended line are added to the culture either simultaneously or sequentially. If the virus alone is first added to the culture the surplus unattached virus may be removed either before adding the cell suspension or after fusion.

The techniques outlined above are generally appropriate, but special demands of experimental design or use of untried cells, as well as the many factors influencing fusion, may necessitate deviation from these methods. These methods are standard only because they have been used frequently during the 10 years or so of experimentally induced cell fusion. Each new system requires preliminary experimentation to find the optimum method for fusion.

D. THE QUANTITATION OF FUSION

The variety of factors influencing the frequency of virus-induced fusion suggests that a standard means of quantifying fusion would be useful. Unfortunately, such a means is not available. While a number of different ways have been used by authors, there is no agreement on a standard method. In 1962, Okada and Tadokoro introduced the term "fusion index" (FI) to express quantitation, and it has subsequently been adopted and modified by other authors, sometimes in combination with percent (FI%). Table IV-2 borrows these terms and for the sake of uniformity applies them to the several formulas which have been used for quantitation. The implications of these formulas, some of which may not be immediately obvious, are discussed in what follows. The term polykaryon, as used in the formulas and discussion, refers to all cells with more than one nucleus even though it means "cells with many nuclei" in the strict sense. Most experimentally induced fusion produces more dikaryotic cells than any other kind of polykaryon, as is illustrated in the *nucleogram* from the work of Koprowski (1971) (Table IV-3). But these, along with all cells with more than two nuclei, are classified here as polykarons.

Probably the most obvious way to quantitate fusion is simply to express the number of polykaryons in a culture as a percentage of the total number of cells [Table IV-2 (a)]. If for example one's interest is to define the best conditions for producing the largest possible number of heterokaryons of any size, then this quantitation suffices. But it can misrepresent the absolute number of cell fusion events which have occurred because it neglects the extra fusions that trikaryons and higher order polykaryons represent. As Fig. IV-3 shows, in some experiments there are many of these.

The original formulation for FI is given in (b) of Table IV-2. In it the "control culture" may be either the experimental dish before addition of a fusing agent or a comparable culture treated as the experimental one except for the addition of fusing agent. The latter control is preferable in that it takes into consideration any distortion stemming from cell damage or loss during the procedures and from cell multiplication up to the time of assay. A theoretical disadvantage with this FI is that it does not distinguish well between cases where some cells fuse into a small number of very large polykaryons and where more cells fuse into a large number of small polykaryons. For example, if 51 of 100 original cells fused into one big polykaryon, while 49 cells remained mononucleate, the FI would equal 1. But FI would also equal 1 if all 100 original cells fused into 50 binucleate cells.

The advantages of this FI include the fact that determining it only requires recognizing and counting cells, yet the first term (number of cells in control culture/number of cells in experimental culture), in fact, is equivalent to the mean number of nuclei per cell in the experimental

TABLE IV-2

Formulations for Fusion Index

(a) $\text{FI}\% = \left(\dfrac{\text{Number of polykaryons}}{\text{Number of cells}} \right) \times 100$

(b) $\text{FI} = \dfrac{\dfrac{\text{Number of cells in control culture}}{\text{Number of cells in experimental culture}} - 1}$ (Okada and Tadokoro, 1962)

(c) $\text{FI} = \dfrac{\text{Number of nuclei in all cells}}{\text{Number of cells}}$ (Kohn, 1965)

(d) $\text{FI} = \dfrac{\text{Number of nuclei}}{\text{Number of cells}} \text{ in experimental culture} - \dfrac{\text{Number of nuclei}}{\text{Number of cells}} \text{ in control culture}$ (Hosaka, 1970)

(e) $\text{FI} = \dfrac{\text{Number of nuclei in polykaryons}}{\text{Number of polykaryons}} \text{ in experimental culture} - \dfrac{\text{Number of nuclei}}{\text{Number of cells}} \text{ in control culture}$ (Rao and Johnson, 1972a)

(f) $\text{FI}\% = \left(\dfrac{\text{Number of nuclei in polykaryons}}{\text{Number of nuclei in all cells}} \right) \times 100$ (Velazques et al., 1971; Homma and Tamagawa, 1973)

(g) $\text{FI}\% = \dfrac{\text{Number of nuclei in polykaryons}}{\text{Number of nuclei in all cells}} \text{ in experimental culture} - \dfrac{\text{Number of nuclei in polykaryons}}{\text{Number of nuclei in all cells}} \text{ in control culture} \times 100$

TABLE IV-3

"Nucleogram" of 1348 Cells Resulting from Sendai Virus-Induced Fusion of Two Cell Types[a]

B-Cell nuclei	A-Cell nuclei											
	0	1	2	3	4	5	6	7	8	9	10	11–20
0		466	56	21	8	2	—	3	—	—	—	—
1	421	96	28	20	5	5	—	2	—	—	—	—
2	54	27	13	9	4	—	1	—	—	—	—	1
3	23	5	7	3	2	2	—	—	—	—	—	—
4	10	4	2	5	1	2	1	—	—	—	—	—
5	6	5	—	2	—	—	—	—	—	—	—	—
6	5	—	1	—	1	—	—	1	—	—	—	—
7	3	1	—	1	1	1	—	1	—	—	—	—
8	—	—	—	—	1	—	1	—	—	—	—	—
9	—	2	—	—	1	—	—	—	—	—	—	—
10	—	—	1	—	—	—	—	—	—	—	—	—
11–20	—	1	—	—	1	—	—	—	—	—	—	—
21–30	—	—	—	—	—	—	—	—	—	1	—	1

[a] The table shows the number of mononucleated cells containing A- and B-nuclei and the number of multinucleated cells containing the two types of nuclei. (Modified from Koprowski, 1971.)

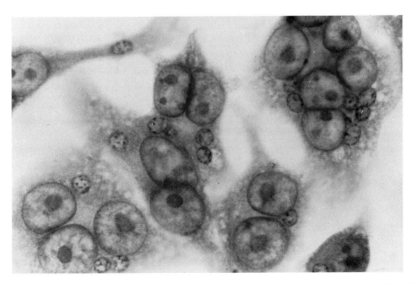

Fig. IV-3. Heterokaryons formed by fusing HeLa cells (large nuclei) with nucleated chick erythrocytes (small nuclei). Most cells are heterokaryons and contain 2–10 nuclei [from Bolund *et al.*, (1969a)].

culture and the second term (1) is equal to the mean number of nuclei per cell in the control culture. Hosaka (1970) modified the FI in Table IV-2 (b) to a form which explicitly spells out that equivalence [Table IV-2 (d)]. In addition, the second term in Table IV-2(d) (number of nuclei/number of cells) allows for the possibility that some cells may fuse spontaneously in control and experimental cultures, whereas (b) assumes no significant frequency of spontaneous fusion.

Kohn (1965), in effect, simplified formula (b), to just the first term [Table IV-2 (c)], and Rao and Johnson (1972a) further modified it to (e) in Table IV-2. In (c), it is understood that when FI equals 1, no cell fusion has taken place. FI becomes large as the number of cells fused increases. In (e), unfused cells of the experimental cultures are disregarded, and, as in (d), the second term takes into account the fact that polykaryons occur spontaneously in many cell culture lines. Both (b) and (e) require recognizing and tallying nuclei as well as cells. While (e) improves upon (b) in providing a range of values for the first term

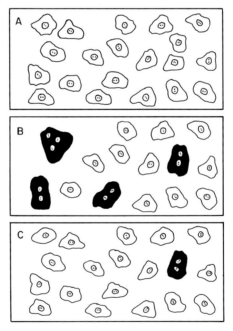

Fig. IV-4. Hypothetical cell fusion experiment in which 20 cells (A) in the experimental culture fuse into 15 cells (B) in the presence of Sendai virus. One spontaneous fusion in the control culture (no virus) (C). Fusion index (FI) has been calculated according to the 7 different formulas of Table IV-2 (a–g). Depending on which method is used the following fusion indices are obtained: (a) 27%; (b) 0.3; (c) 1.3; (d) 0.25; (e) 1.2; (f) 45%; (g) 35%.

where (b) does not, there still exists a distortion that stems from using the mean number of nuclei per polykaryon. For example, if in three separate experiments starting with 100 cells each, 6 cells fuse into 2 polykaryons in the first, 45 cells into 15 in the second, and 90 cells into 30 in the third, the FI as calculated by (e) is in every case the same.

Because of difficulties like these, the generally most satisfactory method for quantitating fusion is to express the number of nuclei in polykaryons as a percentage of the total number of nuclei [Table IV-2 (f)], preferably with a correction for control cultures [Table IV-2 (g)]. Where (d) requires recognizing and tallying both nuclei and cells (f) and (g) require recognizing both but tallying only nuclei. Moreover, the use of percentage as units may have the advantage of better matching an intuitive notion of extent fusion.

To illustrate their differences, the seven formulae for FI given in Table IV-2 are applied to a hypothetical fusion experiment in Fig. IV-4. The numerical diversity in the way fusion in this single case is quantitated emphasizes the need for standardization. Since (f) or (g) seem most accurately to represent the extent of fusion, they are undoubtedly the best candidates for a standard method.

V

Mechanism of Cell Fusion

A. NONVIRAL AGENTS THAT INDUCE FUSION

In the light of observations of occasional spontaneous fusion among cells *in vivo* and *in vitro* (Chapter III), it is interesting to note that a number of physical and chemical processes which affect cell surfaces and cell aggregation are known to result in formation of polykaryons. This is especially true among progenitors of foreign body giant cells (Haythorn, 1929; Roizman, 1962). Paper, lint, coal dust, India rubber, sutures, silica, cellophane, and many other substances when lodged in tissue in such a way as to cause a lesion all elicit the formation of giant cells. Unfortunately, the mechanism for coalescence of macrophages under these stimuli is not at all understood, as is indicated in Chapter III. But these cells are unique in that they are genetically programmed for this activity.

1. Calcium Ions

Other cells which are not programmed in this way have nonetheless been fused *in vitro* by manipulating ionic conditions. There are a number of reports of cell fusion being effected by appropriate calcium ion concentrations. Keller and Melchers (1973) reported that the incubation of plant protoplasts at 37°C with 40 to 50 mM $CaCl_2$ induced extensive fusion. Toister and Loyter (1971, 1973) found that chicken erythrocytes fuse when they are incubated first at 37°C at high pH (10.5), then cooled and exposed to 40 mM $CaCl_2$, and finally rewarmed. Using simi-

46

lar conditions, di- and tetrakaryons have also been generated from the fusion of the protozoan *Actinosphaerium* (Vollet *et al.*, 1971). In the latter two cases the effect was reported to be specific for calcium, since magnesium ions were ineffective. One is reminded by these data that for Sendai virus to fuse two cells or for myoblasts to fuse *in vitro*, an adequate supply of calcium ions is required. It has also been known for some time that the stability and plasticity of membranes depends upon adequate concentrations of calcium (Chambers, 1938). Are calcium ions then involved in fusion in a general way? It seems quite likely they are for the reasons discussed in several of the following sections.

2. Lysolecithin

Polykaryons have also been induced by applying lipids and lipid-related compounds to mononucleate cells. Among surface-active lipoidal substances, lysolecithin is probably the best studied with regard to its fusing capabilities. It is the product of enzymatic degradation of lecithin (i.e., phosphatidylcholine or more properly choline phosphoglyceride). The enzyme involved is phospholipase A, which may be prepared in quantity from various snake venoms. Its action is to hydrolyze the ester linkage between the fatty acid chain and carbon 2 of the phosphoglyceride (Fig. V-1). The resulting lysolecithin is extremely injurious to membranes and, as might be surmised, therefore toxic to living systems. Nonetheless, Poole *et al.* (1970), showed that at low concentrations (in the order of 100–1000 μg/ml), lysolecithin could be employed with moderate success to form homokaryons from hen red cells or heterokaryons from fibroblasts and red cells. Its cytotoxicity can be reduced by including defatted serum proteins or albumin during treatment (Poole *et al.*, 1970; Croce *et al.*, 1971) or by using it as an emulsion (Ahkong *et al.*, 1972), and its effect as a membrane fusing and destabilizing agent is neutralized afterwards by fetal calf serum (Croce *et al.*, 1971; Ahkong *et al.*, 1972).

This property of lysolecithin makes it interesting to note that phospholipase A is found as one of the lysosomal enzymes (Tappel, 1969). Perhaps as a consequence the lysolecithin content of lysosomal membranes is unusually high compared with other cell membranes. Since lysosomes are themselves engaged in so many processes involving membrane fusion, it is tempting to speculate that natural fusion, including fertilization, may involve phospholipase and lysolecithin activity [for example, Ahkong *et al.* (1972) suggest that exocytosis of milk fat may be related to its high lysolecithin content].

Lucy and his colleagues (Ahkong *et al.*, 1973, 1975; Bruckdorfer *et al.*, 1974; Cramp and Lucy, 1974) have also tested several other lipoidal sub-

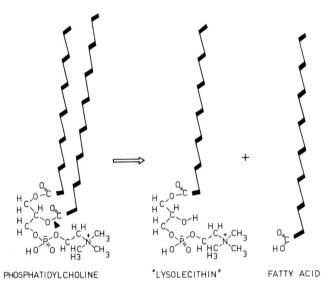

PHOSPHATIDYLCHOLINE   "LYSOLECITHIN"   FATTY ACID

Fig. V-1. The generation of "lysolecithin" from phosphatidylcholine by the action of phospholipase A. Jagged lines represent the hydrocarbon tails of fatty acid residues. The triangle indicates the site of the lytic activity of the enzyme. One theory as to the destabilizing effect of lysolecithin on membranes is that its more conical geometry fits lipid bilayers less well than do cylindrical molecules like phosphatidylcholine.

stances for their fusing capacities including fatty acids, fatty acid alcohols and amines, and glycerol derivatives. Among these they have reported that glycerol monooleate is quite promising. It induces hybrid cell formation in tissue culture at a rate four to seven times greater than the spontaneous frequency and is considerably less damaging than lysolecithin, especially when osmotic shock to the cells is reduced by using dextrans. Still other agents that affect membrane lipids which have been used for fusing cells are small lipid vesicles (Papahadjopoulos *et al.*, 1973) and phospholipase C (Sabban and Loyter, 1974).

3. Polyethylene Glycol

In 1974, polyethylene glycol (PEG) (Fig. V-2) was introduced as an agglutinating agent for plant protoplasts which leads to extensive cell fusion if subsequently diluted away slowly with a growth-supporting culture medium (Bonnett and Eriksson, 1974; Constabel and Kao, 1974; Kao and Michayluk, 1974). Pontecorvo (1975) tested whether or not this agent could be used to the same end with animal cells. His results with several cell lines and lymphocytes were sufficiently encouraging to raise the possibility that PEG might become a standard fusing agent for both plant and animal cells (Fig. V-2).

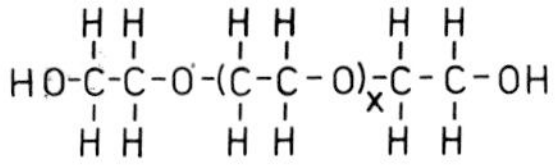

$$HO-\overset{\displaystyle H}{\underset{\displaystyle H}{\overset{|}{\underset{|}{C}}}}-\overset{\displaystyle H}{\underset{\displaystyle H}{\overset{|}{\underset{|}{C}}}}-O-(\overset{\displaystyle H}{\underset{\displaystyle H}{\overset{|}{\underset{|}{C}}}}-\overset{\displaystyle H}{\underset{\displaystyle H}{\overset{|}{\underset{|}{C}}}}-O)_x-\overset{\displaystyle H}{\underset{\displaystyle H}{\overset{|}{\underset{|}{C}}}}-\overset{\displaystyle H}{\underset{\displaystyle H}{\overset{|}{\underset{|}{C}}}}-OH$$

Fig. V-2. Polyethylene glycol structural formula. When the molecular weight of the polymer is about 7500 daltons, x ≅ 170.

B. MEMBRANE STRUCTURE IN RELATION TO FUSION

In the light of what is known about fusion effected by virus and the nonviral agents, is it not possible to derive some general principles about how cells fuse? Obviously an affirmative answer to this question would depend in the first place upon a knowledge of the molecular architecture of the plasma membrane. Unfortunately, despite enormous amounts of published work, a complete understanding is not at hand.

What is clear is that at a gross level the plasma membrane is 60–80% protein and 40–20% lipid, and models of membranes must obviously incorporate these compounds. In the 1950's, a model known as the Davson–Danielli–Robertson model was widely accepted. It had evolved over a period of 30 years and consisted of a bimolecular leaflet of lipid molecules, with their hydrophilic ends oriented outward, sandwiched between two layers of protein molecules which were in the extended β-configuration (Fig. V-3a). During the 1960's, though, the popularity fastened to this model began to wane (Kavanau, 1965; Korn, 1966). Among the criticisms leveled at it which are particularly relevant here are its inability to provide any explanation for phenomena like fusion, its relative neglect of the carbohydrate component of membranes, its rigid bilateral symmetry, and its picturing the protein moiety in the β-configuration. (The discovery that membrane proteins were for the most part globular was ironic because Danielli's and Davson's original 1935 model showed them that way.) Most of the models which have been proposed subsequently have been characterized by some sort of subunit, either of lipoprotein or of lipid. The subunit structure allows greater plasticity than does the bilayered sheet. Two of the newer models (Kavanau, 1965; Lucy, 1970) have been detailed specifically for their abilities to account for fusion. Both models suppose that the lipid moiety may exist in two alternative forms, either as a bimolecular leaflet or as micelles (Fig. V-3b). The micellar form in one model pictures lipid molecules oriented in cylinders at right angles to the plane of membrane (Kavanau, 1965). In the other, they are oriented radially into 40-Å spheres which sit in the plane of the membrane (Lucy, 1970). Both postulate that the two forms exist in dynamic equilibrium and that coalescence involves the micellar form. Kavanau argues that maintenance of

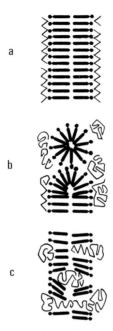

Fig. V-3. Models for cell membrane structure (viewed in cross section). (a) A bimolecular layer of lipid molecules sandwiched between layers of protein in extended configuration. (b) Lipid molecules arranged partly in micelles and partly in bilayer sandwiched between layers of globular proteins. (c) The fluid mosaic model, mostly globular proteins associated in various ways with a lipid bilayer.

membrane integrity during fusion is a function of the protein sandwich that transforms from an extended form to a globular one (as a result of removal of a "hydration crust" when the membranes approach each other). The protein then realigns into the new membrane arrangement, and finally reextends. While subsequent evidence suggested there is little extended protein in membranes to begin with, the hypothesis is still interesting in the light of protein changes which are presumed to accompany fusion (see Sections D and F). Lucy's model does not consider what may happen prior to the juxtaposition of the lipid core of two membranes but focuses on forces which would tend to increase the micellar content, at the expense of the bimolecular leaflet, thereby favoring coalescence.

Despite the conceptual advantages of subunit models, the basic element of the model now most widely accepted is again the lipid bilayer. However, in the new model, membrane proteins are not everywhere excluded from the lipid layer, but instead may be completely or partly imbedded in it. Moreover, the lipid is looked upon as existing in a very fluid state, thereby allowing extensive movement of membrane proteins under many stimuli. The model has been called "the fluid mosaic model" (Singer and Nicolson, 1972), and a great deal of evidence consistent with its postulates has accumulated in a short time (Fig. V-3c).

However, no model has yet been deemed totally satisfactory by all of the biochemists, biophysicists, and biologists concerned with membrane architecture. Perhaps there is too much variety among membranes to hope for concurrence on one model. Nonetheless, there seem to be some useful generalizations about fusion which can be set down without regard to a specific model.

Hence, there follows a discussion of a series of events which may be necessary, but separately insufficient, steps leading toward the fusion of animal cells. The steps include bringing cells into close contact (agglutination), clearing away protective coats from the lipid of the plasma membranes, the physical rearrangement of lipid molecules, effecting the coalescence of cytoplasms, and, finally, stabilizing the merged cells.

C. AGGLUTINATION

The first step is a close apposition of adjacent membranes (Fig. V-4A). They must be brought to within a few angstroms of each other so that interactions like hydrophobic bonds can come into play. Agents which cause cell agglutination apparently are good candidates for this role. Myxoviruses, like Sendai, along with a number of other viruses which do not effect cell fusion, first cause cells to agglutinate, thereby stapling them sufficiently close together to allow the next steps toward fusion to be taken. Polyethylene glycol also causes cell aggregation, although how it brings it about is not certain. The polymeric molecule has been used as a successful agglutinin when its molecular weight lies in the range of 1500 to 7500 daltons (Fig. V-2). Since PEG is slightly negatively charged in aqueous solution, perhaps molecules of this size are sufficiently large so as to form electrostatic links between cells. If so, the enhancing effect of calcium ions on PEG agglutination may be understood in the way Kao and Michayluk suggested: the divalent ions could form bridges between PEG and negatively charged carbohydrates on cell surfaces (Kao and Michayluk, 1974).

Among other agglutinating agents which have been explored somewhat for their abilities to fuse cells are plant lectins and antibodies. In experiments with lectins, Reeve *et al.* (1974) showed that phytohemagglutinin enhances NDV-induced fusion, while concanavalin A (Con A) does not. But, since 1971, there have been a few reports of suc-

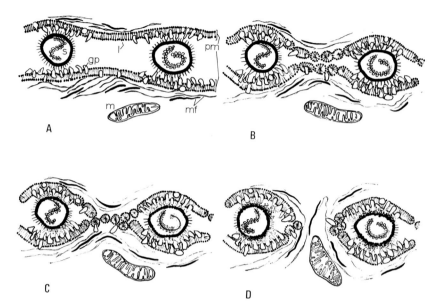

Fig. V-4. Hypothetical steps in Sendai virus-induced cell fusion. (A) Local agglutination of two cells by two virus particles. Glycoprotein surface material has begun to be cleared from portions of the plasma membranes lying between the virions by attraction to adsorbing sites of the virions. pm, plasma membranes; l, lipid molecules; gp, glycoprotein molecule; mf, microfilaments; s, Sendai virus particle; m, mitochondrion. (B) Micellization of exposed lipid. The process is enhanced by high pH and Ca^{2+} concentration. Both cells have begun to endocytose the virus particles. (C) Membrane coalescence. Ca^{2+} bridges decrease the size of micelles. Microfilaments are activated by ATP and Ca^{2+}. (D) A nascent cytoplasmic bridge, stabilized and enlarged by microfilament activity. ATP is regenerated by mitochondrial activity.

cessful lectin-induced fusion in the absence of virus. Smith and Goldman (1971) reported fusion of human macrophages after Con A or phytohemagglutinin treatment. Although macrophages are known to form giant cells *in vivo* under a variety of stimuli (see Chapter III), this polykaryon formation *in vitro* seems to be tied *specifically* to the lectin treatment because fusion was not observed in the presence of lectins pretreated with agents that block their binding sites. Rizki *et al.* (1975) credit Becker with the first fusion of *Drosophila* embryo cells with Con A. Their own experiments seem to confirm these results and to show that wheat germ agglutinin works as well. The syncytia formed from *Drosophila* cells, when viewed by scanning electron microscopy, are unique among heterokaryons in not forming a unified protoplasmic mass. The parental cells appear to remain *in situ* on the glass substrate and to fuse only peripherally (Rizki *et al.*, 1975). One hopes to see transmission electron micrographs to confirm the actual continuity of cytoplasm in these networks.

As far as antibodies are concerned, part of the cell recognition involved in the specific approach of sperm to egg is a reaction having an antibody–antigen-like component. Fertilization may therefore be thought of as an example of natural cell fusion following an antibody-like agglutination. In 1973, Hartmann *et al.* showed that antibodies directed against plant protoplasts were successful agglutinins for intra- and interspecies pairings. When serum complement was included in the incubation mixture, protoplast lysis occurred, but no proof for the existence of heterokaryons was obtained.

Both lectins and antibodies are effectively di- or multivalent. Their role as agglutinating agents is therefore undoubtedly a function of the ability of one molecule to combine with a receptor on two cell surfaces, thereby forming a link between the cells. Sufficient numbers of these molecular links can overcome the tendency of the cells to move apart, either by random (Brownian) motion, electrostatic repulsion, or active migration. It may be that calcium ions serve a similar role. As noted earlier, calcium salts alone have been reported to induce fusion in some cell systems, while in addition they are required for the naturally occurring myoblast fusion. Furthermore, many tissues respond to chelating agents by dissociating into individual cells, thus suggesting that divalent cations like calcium are essential in normal cell adhesion. It may be rare for calcium ions alone to suffice as agglutinating agents, but in combination with other links they undoubtedly play some role in holding cells together in preparation for fusion.

Other means which can be employed for bringing cells very close to one another are the polycationic molecule, polylysine (Sabban and Loyter, 1974), low speed centrifugation such as that used in the lysolecithin fusion technique outlined by Croce *et al.* (1971), and micromanipulation, that has been used to push together and fuse two or more cells (Diacumakos and Tatum, 1972; Diacumakos, 1973).

All of the agglutinating agents and even centrifugal packing or manipulation are, however, insufficient means for effecting fusion. Agglutinating agents recognize carbohydrate groups (glycoproteins and glycolipids) which are part of the outermost layers of plasma membrane (Winzler, 1970) and as a consequence they may in fact fail to bring lipids of the two membranes close enough to allow the interactions necessary for fusion. That there are agents which agglutinate but do not fuse cells and that even in studies with Sendai virus the ability to agglutinate is separable from the ability to fuse (Okada and Tadokoro, 1962) suggests that there must be more to cell fusion than merely ensuring the proximity of plasma membrane surfaces.

D. CHANGES IN MEMBRANE CARBOHYDRATES

Until recently the carbohydrate components have been the most neglected of cell surface constituents. They are found in membranes as neutral sugars (galactose, mannose, and fucose are common), acetyl amino sugars, or sialic acid. While some sugars are simply polymerized into free polysaccharides, frequently they are covalently linked to lipids or proteins, and it is in these forms that carbohydrates play extremely important roles in many biological activities (Cook, 1968; Winzler, 1970; Whaley *et al.,* 1972). That carbohydrates are found outermost on cell surfaces implies a role for them in cell fusion and this idea is reinforced by the facts that (a) they are the determinants of cell antigenic specificity (e.g., the determinant moieties for A–B and for M–N blood groups are glycoproteins); (b) they carry the characteristic negative charge born by cells at physiological pH; (c) they provide the means for cell to cell recognition and adhesion (seen in aggregation during development, in contact inhibition, and in recognition of gametes and mating strains); (d) they determine receptor sites for viruses, bacteria, agglutinating agents, and inducers of endocytosis; (e) they play a part in the control of ion permeability—all of which may well be involved in cell fusion.

Among agglutinating agents known to react with cell surface carbohydrates are the myxoviruses and plant lectins. It is likely that if calcium ions serve as agglutinating agents, as suggested above, they do so by acting as counterions between the negatively charged carbohydrates of two cells. In effect, calcium ion collaboration in PEG agglutination would be analogous, as suggested in the previous section. In the case of the natural cell fusion, fertilization, surface carbohydrates are also reported to be involved in sperm penetration of egg (Soupart and Clewe, 1965).

But aside from their involvement in agglutination, carbohydrates may normally play a negative role in fusion; their presence on cell surfaces may inhibit fusion. Quite conceivably, after agglutination the carbohydrate-containing coat must be moved aside or eliminated so that the underlying hydrophobic layers of two membranes can become contiguous. Either of these processes may suffice. The fluid mosaic model (Singer and Nicolson, 1972) (Fig. V-3c) allows for just the kind of lateral movement which may be required. Evidence that glycoprotein agglutinating sites do move was first provided by an experiment using

mouse fibroblasts and the plant lectin, Con A (Nicolson, 1971). Con A
does not agglutinate fibroblasts unless they have been transformed into
malignant cells by an oncogenic virus like SV40. Nonetheless, lectin receptor sites seem to be present on the cell before transformation because a brief digestion of normal cells with trypsin renders them as agglutinable as transformed cells. Singer and Nicolson suggested that the
receptor sites on normal cells were randomly and singly distributed,
but that virus transformation or proteolytic digestion brings about a
clustering which enhances agglutination. Nicolson's experiment using
ferritin-labeled Con A indeed showed that the transformed fibroblasts have clumped lectin-binding sites.

Singer and Nicolson also used as an argument the work of Frye and
Edidin (1970) who showed that after fusion of human and mouse cells
with Sendai, there was a rapid intermixing of the cell surface antigens.
While it is possible that the new antigen distribution was brought
about by removal and reinsertion, the authors favored diffusion as the
means for intermixing. As is discussed in the next section, Kornberg
and McConnell (1971a,b) showed that lateral displacement of lipids in
artificial membranes is rapid. Certainly that observation is compatible
with the kind of antigen intermixing envisioned by Frye and Edidin, as
also is the claim cited by Dingle (1968) that some mobility of lipid molecules is desirable for stability of bubbles.

While these observations provide evidence for movement of cell surface glycoprotein, they do not suggest what the mechanism for movement may be. However, several possibilities come to mind. Perhaps divalent agents (such as calcium ions, lectins, or antibodies) or multivalent ones (such as viruses) effect crowding of mobile receptors into
some localities, leaving others free of them. Consistent with this possibility is the observation that when Sendai virus induces cell fusion, the
formation of cytoplasmic bridges between the agglutinated cells is not
through the virus particles, but instead in regions a bit removed from
each virion (see Fig. IV-1b and c) (Schneeberger and Harris, 1966; Cassone *et al.*, 1973). The glycoproteins may have been pulled from the adjacent regions by their having been bound to viral adsorbing sites (Fig.
V-4A and B), facilitating *cell* fusion at these unobstructed places. Then,
too, several authors have described a phenomenon known as "capping" in which cell surface sites under the influence of antibodies
directed against them (e.g., Taylor *et al.*, 1971; Edidin and Weiss, 1972)
or of lectins (e.g., Edelman *et al.*, 1973; Rosenblith *et al.*, 1973) move
from a random distribution to a clustering at one side of the cell. These
studies have not only demonstrated that at least some glycoprotein can
be removed from specific regions of cell surface, they also have shown

that the distribution of glycoproteins in plasma membranes is dependent upon cell structures underlying the membrane–specifically microtubules and microfilaments. As noted in Section G of this chapter, these fibrous structures are often agents for motility. Movement of surface glycoproteins, either during capping or fusion, may therefore be a consequence of their activity. Even the negative charge of glycoprotein molecules may contribute to surface clearing; if the viscosity of membranes is as low as current estimates suggest, the multiple negative charges of sialic acid residues may tend to move unconstrained glycoproteins away as membranes are brought within a few angstroms of each other during fusion.

Conceptually just as easy, though, is the idea that the carbohydrate coat is digested away. Both myxoviruses and lysosomes are equipped with enzymes which can degrade glycoproteins, principally neuraminidase (see Chapter IV). Hirst (1941) first obtained evidence for the presence of neuraminidase in virus when he showed that influenza virus caused chick red cells to agglutinate at 4°C, but that after a short time at 37°C the cells separated and were subsequently incapable of further agglutination. However, with fresh cells, the eluted virus retained its activity. It was later shown that purified neuraminidase was equally effective in destroying the virus receptor sites on the red cells and that the viral envelopes contained that enzyme. Lysosomes from a variety of mammalian sources also exhibit neuraminidase activity together with a host of other glycosidases (e.g., galactosidase, mannosidase, fucosidase). It therefore is reasonable to suppose that lysosomal enzymes may also be involved in the removal of inhibitory carbohydrate groups in some fusions (Weiss, 1967; Dingle, 1969). Weiss (1967) suggested that lysosomal activity which does not result in cell death might properly be referred to as "sublethal autolysis." Poste (1971) subsequently succeeded in showing by ellipsometry that when lysosomes of cultured cells were induced to release their contents extracellularly, sublethal autolysis resulted in a marked decrease (as much as fivefold) in cell coat thickness. Thus, Dingle (1969) proposed that coalescence of lysosomes or Golgi vesicles into the plasma membrane causes local release of glycosidases which keeps regions of the cell surface free of carbohydrate and therefore receptive to fusion. Such lysosomal activity might be programmed in cells which normally undergo fusion (gametes, myoblasts, macrophages, etc.) or might be elicited in instances of experimentally induced fusion.

Again, however, while removal or displacement of the outer layers of the membrane may be necessary for fusion, it is not sufficient, as, for example, Hirst's early observation shows. Instead of effecting fusion of

the agglutinated cells, the glycoprotein digestion by viral neuraminidase simply freed the virus from its moorings on the erythrocytes and
allowed the system to disaggregate. Moreover, using an electron microscopic technique to localize sialic acid, Benedetti and Emmelot (1967)
showed that tight junctions of rat tissues, where plasma membranes of
adjacent cells are particularly tightly apposed and where permeability is
reported to be high, are free of sialic acid. Other regions of the contiguous membranes, including specialized loci like intermediate junctions and desmosomes, were all coated with the carbohydrate. Yet the
integrity of the plasma membrane of the two cells was not lost by fusion
at the tight junctions.

E. CHANGES IN MEMBRANE LIPIDS

As already noted, Lucy suggested that a change in membrane
lipid—from bimolecular leaflet to a planar array of micelles—would
favor membrane fusion (Fig. V-4B and C). Lysolecithin effects such micellization (Lucy, 1970); by itself it forms globular micelles in water and
it forces other lipids, even myelin, into micellar configurations. The lysolecithin molecule is wedge-shaped (Fig. V-1) (Haydon and Taylor,
1963), and its incorporation into lipoidal sheets probably renders them
less stable, more micellar, and "leaky," because of its inability to be
packed tightly in a plane. And, as already noted, lysolecithin effects fusion of membranes.

Conceivably the high calcium ion concentrations reported to cause
fusion do so by influencing micelle formation in membrane lipid. In the
first place, phospholipase A, the enzyme which generates lysolecithin,
has an absolute and specific requirement for calcium and, incidentally,
exhibits an activity optimum at pH 8.0 to 8.5 (Wells, 1972). Thus, wher-
ever the enzyme is involved in fusion, high calcium concentrations and
high pH are favorable conditions. Second, divalent cations have a
strong compacting effect on micelles brought about by bridging phos-
phoryl groups of adjacent lipid molecules (Kavanau, 1965). The resulting smaller units of juxtaposed membranes may more easily interca-
late and rearrange into new planes than larger micelles or bilayers do.
Consistent with this idea are the observations of Toister and Loyter
(1971, 1973) that high concentrations of calcium ions combined with
high pH can induce erythrocytes to fuse (see Section A,1). Under such
conditions divalent bridges may be expected to replace both monova-
lent cations (such as potassium and sodium) and protons as counterions
on the anionic sites of phospholipids, thereby effecting contraction of

lipid structures by water displacement and greater likelihood of membrane fusion (see Carvalho *et al.*, 1963; Sanui and Pace, 1967a,b).

If micellization is a reasonable description of the lipoidal state necessary for fusion, then here again lysosomes may be involved. Controlled release of lysosomal phospholipase on to the cell membrane (again, sublethal autolysis) could generate lysolecithin and encourage micellization.

One unsolved problem involving lysosomes is the mechanism enabling these organelles to fuse with endocytic vacuoles while not, for example, with mitochondria or nuclei. It is conceivable that the ability of various membranes to fuse is related to their differential susceptibility to phospholipase A activity. Sphingomyelin is a common constituent of plant and animal membranes, but it is not susceptible to the lipase activity. Membranes which are rich in sphingomyelin, or other resistent lipids, and poor in phosphoglycerides might therefore be less susceptible to phospholipase A activity and therefore less readily fused. Apropos of this idea, Bretscher (1972) proposed that the reason ruminant erythrocytes are resistant to phospholipase A when intact but susceptible when ruptured is that there exists an assymetric distribution of sphingomyelin in the two sides of their plasma membrane lipid bilayer: the outer side is rich in it and the inner poor. This idea would take on special relevance here if it were the case that membrane fusions consistently involved meetings of either inner *or* outer surfaces. While intercellular fusions of course entail the outer surfaces of two plasma membranes, intracellular fusions (e.g., vesicular fusion and sequestration) involve what corresponds to inner surfaces of plasma membranes. The latter type fusion is clearly more common *in vivo*, as would be predicted if Bretscher's proposal were correct and generalizable and if phospholipases were involved. What remains a curiosity is whether or not there are outer–inner surface fusions. The work of Kasper and Kubinski (1971) suggests that such fusions do not occur spontaneously; while microsomal particles (derived mostly from smooth and rough endoplasmic reticulum) fused spontaneously with each other, they did not fuse with erythrocyte ghosts. That inner × outer surface distinctions may be maintained has become credible through the work of Kornberg and McConnel (1971a,b) who showed that although lateral diffusion of lipid in synthetic vesicles is very fast, migration from one side of a bilayer to the other (a so-called "flip-flop") is slow. The half-life of flip-flops is measured in hours. Biomembrane stabilizers, like proteins, would further decrease the rate of flip-flopping.

From studies on surface thermodynamics, Dingle (1968) has argued that when two membranes of high surface tension come into contact with one another, fusion is probable unless there are stabilizing factors present. If one adds protein to suspensions of lipid from which a foam is to be made, the protein adds to the stability of the resulting foam in two ways: it lowers the surface tension of individual bubbles and it coats them with an elastic skin several molecules thick. This stability can be overcome either by removal of the protein or by loosening it from its association with lipid by surface active substances which are soluble in both the protein and lipid layers. Presumably another way of affecting surface tension is to raise or lower the number of lipid molecules packed per unit area membrane. However, it is not at all clear that thermodynamic principles derived from studies of bubbles and foams are directly useful in describing biological membranes.

F. A ROLE FOR ATP

It appears, then, that lipid structural rearrangement is also an insufficient prerequisite for reproducible cell fusion. As already indicated, lysolecithin yields a relatively small number of fused cells because of its tendency to disrupt membranes totally, beyond simply effecting fusion. There also seems to be a relatively narrow line between cell fusion and cell lysis when myxoviruses are used as the fusing agents (Yanovsky and Loyter, 1972). What probably is required is a mechanism to stabilize the membranes of the two cells during fusion so that their structural integrity is preserved. Such activity is undoubtedly energy-requiring.

As already noted in Chapter IV, the requirement for respiratory ATP in virus-induced fusion has been established. When one considers what enzyme systems might influence cell fusion and use ATP, several membrane-associated adenosinetriphosphatases (ATPases) come to mind. The best characterized of these are the Na, K-activated ATPase—the active transport enzyme described first by Skou (1971)—and the ATP-synthesizing enzyme of mitochondrial cristae membranes. (There are in addition several other less well described ATPases reported to be associated with bacterial and erythrocyte membranes.) While the mitochondrial enzyme clearly is involved, since it is the major source of cellular ATP, it is probable that its involvement is indirect. For one thing, it is part of an inaccessible membrane that is clearly not engaged in cell fusion. Since the other ATPases are associated with plasma membranes, they are more likely candidates for processes involving membrane coalescence. Of these, the Na, K-transport ATPase is probably uninvolved (Poste and Allison, 1971), unless, in the negative sense, its activity must be stopped for fusion to occur. Yanovsky and Loyter (1972) have presented evidence consistent with this idea. They showed that conditions which inhibit the activity of this enzyme, such as the presence of the drug ouabain, or high K–low Na concentrations, eliminate the need for calcium ions when Ehrlich ascites cells are fused with Sendai virus. Moreover, the effect of calcium on this magnesium-activated enzyme is inhibitory. Thus, conditions which inactivate this enzyme apparently promote fusion.

This conclusion is intriguing in light of the findings of Graham and Wallach (1971). They reported that in the presence of ATP and Mg^{2+}, the Na, K-ATPase promotes a distinct reordering of proteins in human erythrocyte ghosts. Some of them change from an α-helical or unordered (presumably globular) configuration to an antiparallel β-configuration. This recalls Kavanau's (1965) idea that membrane fusion requires a clumping of surface protein and that stabilization required reversion to the extended form (see Section B). Perhaps the structural changes which accompany Na, K-ATPase activity inhibit membrane fusion and promote membrane integrity. One of the several roles of calcium, then, may be to prevent these changes in membrane protein.

Among the less adequately described membrane-associated ATPases are others which are activated by calcium or by both calcium and magnesium. Some of these have been implicated in fusion in various ways. In virus-induced cell fusion, it may be recalled here, as mentioned in the previous chapter, that there is evidence for an ATPase in myxovirus envelopes. Murayama and Okada (1965; Okada and Murayama, 1966) believe that the role of calcium in Sendai virus-induced fusion is that of an ATPase activator. Poste and Allison (1971) suggest that fusion requires momentary removal of calcium and ATP from the membranes of fusing systems in order to provide a region of low rigidity which is susceptible to fusion. This may be effected by the calcium and ATP momentarily engaging a membrane ATPase. Similarly, Woodin and Wieneke (1964) showed that while calcium and ATP played no part in the release of enzymes from lysosomal granules when isolated from polymorphonuclear leukocytes, both were necessary for induced enzyme release from intact leukocytes. They suggested that the adherence of the granules to the plasma membrane depended upon the presence of calcium ion and ATP, but that the ultimate release of the enzymes from the cells involved removal of Ca^{2+} and ATP. This could be brought about in an ATPase reaction.

G. STABILIZATION

There is no evidence, however, which makes it obligatory that any ATPase involved in a positive way with fusion is *membrane*-associated.

Although it has been argued that the energy-requiring step in virus-induced cell fusion is the repair of a membrane site damaged by adsorbed virus (Okada, 1969; Okada *et al.*, 1966), it is conceivable that the repair of the molecular architecture of the lipid membrane per se is not what required energy. In the first place, coalescence of model lipid membranes of sufficient surface tension is energetically down hill. Then, too, there is the question of whether or not membrane coalescence suffices for the fusion of cells. Do fused 100-Å-thick bilayers or micellar arrays from two cells possess sufficient tensile strength to pull the separate cytoplasms together and stabilize them into a single unit? Polykaryons resulting from fusion of two or three cells are not formless, distended, or bilobed blobs of cytoplasm but are shapes characteristic of one of the parental lines or of something intermediate (see Chapter XI). When erythrocytes fuse with HeLa cells or myoblasts, the form is characteristic of the larger partner. When epithelial cells are fused with fibroblasts the resulting hybrids exhibit morphological characteristics of both. Moreover, there are occasions in nature when single-celled organisms join only partly, as during conjugation of *Paramecia*. Such joinings include at least the fusion of plasma membranes. Yet contrary to what might be expected if coalescence of plasma membranes were sufficient for cell fusion, these cells subsequently separate again without ever having become completely unified. It is likely, then, that the cell structures which determine cell shape, integrity, and movement are the principal agents for stabilizing fusion rather than tenuous membranes. Thus the ATPase activity called for during fusion may function in the operation of those structures and may not be membranous.

The shape-determining structures of cells are microtubules and microfilaments. Separately or together these fibrillar systems comprise the so-called "cytoskeleton" and in combination with membranes they often constitute a cell periphery which gives some modern significance to the old term, ectoplasm. As studies with colchicine and Colcemid have shown, microtubules determine cell shape and developmental changes in the morphology of lower organisms, plants, and animals. Holmes and Choppin (1968), in fibroblast homokaryons, and Gordon and Cohn (1970), in macrophage melanocyte heterokaryons, showed that the clustering of nuclei in the center of fused cells is dependent upon microtubular function. Recent studies with cytochalasin, which among other effects appears to disrupt microfilaments (Wessels *et al.*, 1971; Carter, 1972), show that the filaments are necessary in maintaining cell integrity. In the presence of the drug, nuclei are spontaneously extruded from cytoplasms of many cultured cells or may easily be spun free of their cytoplasm by low speed centrifugation (see Chapter

VIII). In either case, the plasma membrane is incapable of retaining the nucleus, but instead bulges, snips off, and fuses closed as the nucleus escapes, leaving both the anucleate cytoplasm and the isolated nucleus surrounded by plasma membrane.

In many if not all cells where they are found, either type of fibrillar structure may also be responsible for generating movement of various sorts. Microtubules produce meiotic, mitotic, ciliary, and flagellar movements, and actin- and myosin-like filaments account for cytoplasmic streaming, muscle contraction, cytokinesis in animal cells, and endocytosis. Moreover, both fibrillar systems are known to bind nucleoside triphosphates and to be associated with Ca-stimulated ATPase activity that functions during generation of movements (e.g., Gibbons, 1968; Bendall, 1969; Summers and Gibbons, 1971).

The location and ubiquity of actomyosin-like microfilaments make them good candidates for involvement in stabilization of newly fused cells. They are found in association with plasma membranes not only in cells where movement leads one to expect them [e.g., in amoeba (Marshall and Nachmias, 1965; Pollard and Ito, 1970); in slime molds (Nachmias *et al.*, 1970)] but also in as immotile cells as erythrocytes (Ohnishi, 1962; A. S. Rosenthal, *et al.*, 1970). In addition, their possession of ATPase activity which is calcium activated and which operates optimally at high pH's is consistent with the idea that they may be part of the process of fusion in which high calcium ion concentration and high pH are also optimal conditions. If, as seems likely, microfilaments are disrupted by cytochalasin, then the observations of Sanger *et al.*(1971), of Pasternak and Micklem (1973), and of Zurier *et al.* (1973) also become significant. The first group showed that cytochalasin B markedly affected the ability of cultured myoblasts to form myotubes. A majority of myotubes formed in the presence of the drug were binucleated and small, while control cultures produced myotubes with several hundreds of nuclei. The authors felt that the inhibition of fusion was similar to that of low calcium ion concentration. Similarly, the second group reported that Sendai-induced fusion of an ascites cell line was markedly inhibited by cytochalasin. These authors also reported that when challenged with particles which normally induce phagocytosis, cytochalasin B-treated leukocytes responded by releasing lysosomal enzymes directly into the surrounding medium without first ingesting the particles. The disrupted microfilaments apparently inhibited endocytosis without simultaneously preventing whatever mechanism allowed the normal lysosomal fusion with plasma membrane. If one is allowed to extrapolate from these observations, then, first, membrane coalescence is a phenomenon separable from cell fusion; second, the merging of cytoplasms is an activity involving microfilaments; and third, microfilamentous activity entails not only morphological considerations but more importantly cytoplasmic movement that opens nascent cytoplasmic bridges into major connections and ultimately creates a single polykaryotic unit.

Zurier *et al.* (1973) also observed that colchicine inhibited the cytochalasin-induced release of lysosomal enzymes. They interpreted these findings as showing the necessity for intact microtubules in the translocation of lysosomal granules to the plasma membrane. The experiments demonstrating involvement of microtubules in the aggregation of nuclei in the center of heterokaryons seem consistent with such an interpretation. But while this kind of role is easily envisioned for microtubules, there appears to be no evidence suggesting that they are directly engaged in fusion of two cytoplasms. It is even conceivable that conditions favoring fusion may inhibit microtubule integrity, for Weisenberg (1972) and Borisy and Olmsted (1972) observed that even low concentrations of calcium completely inhibited the *in vitro* repolymerization of tubulin into microtubules and even caused breakdown of intact microtubules.

It seems likely, then, that the actin- and myosin-like microfilaments subtending the plasma membranes of cells are the agents primarily responsible for effecting cytoplasmic union and stabilizing fusing systems, and that the energy expenditure involved in cell fusion is chiefly spent in the activity of these fibers (Fig. V-4D). Environmental conditions favoring their operation encourage fusion. Among these may be counted high calcium ion concentration and high pH, conditions which also encourage membrane coalescence. These are not, to be sure, conditions conducive to cell growth. Thus, when fused cells are returned to ideal culture media, normal membrane activities resume, and the cell surfaces are restored to states which promote growth and discourage further fusion.

VI

Fusion of Cells at Different Stages of the Cell Cycle

A. DNA SYNTHESIS IN POLYKARYONS

1. Synchronization

In the late 1960's, two groups (Yamanaka and Okada, 1966; Johnson and Harris, 1969a) showed that cells at all stages of the cell cycle fuse with each other when exposed to Sendai virus. Thus, in experiments with asynchronous cultures, polykaryons are formed from all combinations of G_1, S, G_2, and mitotic cells.* Polykaryons containing nuclei of more than one stage are said to be heterophasic, whereas those which by chance or by design contain nuclei from only one stage of the cell cycle are classified as homophasic (Fig. VI-1).

As a rule homokaryons and heterokaryons examined some time after cell fusion, show synchronization of DNA synthesis (Yamanaka and Okada, 1966; Johnson and Harris, 1969a,b; Westerveld and Freeke, 1971; Graves, 1972a). The degree of synchrony in homokaryons is usually greater than that in heterokaryons. The mechanisms responsible for synchronization have been analyzed by Rao and Johnson (1970,

---

* The designations G_1, S, G_2, and M (for mitotic cell) will be used to describe the phases of the cell cycle occupied by the individual nuclei within polykaryons. The numbers before the phase designation show the numbers of each type of nucleus. Thus, a polykaryon designated $3G_1$ contains 3 G_1 nuclei, a $2G_1/1S$ cell contains 2 G_1 nuclei and 1 S phase nucleus. A cell which is formed by fusion of one G_1 cell with two mitotic cells is referred to as a $(G_1)2M$ polykaryon.

64

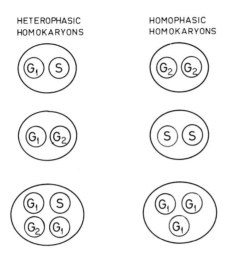

Fig. VI-1. Heterophasic and homophasic homokaryons (Ringertz, 1974).

1972a) by fusing cells in different phases of the cell growth cycle with each other. They found that in multinucleate cells containing only G_1 nuclei, all nuclei usually started DNA synthesis at the same time. In heterophasic G_1/S homokaryons of HeLa cells, DNA synthesis was induced in more than 50% of the G_1 nuclei within 2 h after fusion. The greater the proportion of S nuclei, the faster DNA synthesis was induced in the G_1 nuclei (Fig. VI-2). This dose-effect relationship

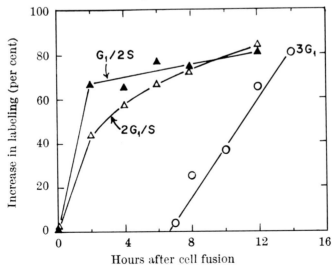

Fig. VI-2. Dosage effect on the induction of DNA synthesis in the G_1 nuclei of the heterophasic trinucleate cells of the G_1/S fusion (Rao and Johnson, 1970).

suggested that the triggering of DNA synthesis depended on the concentration of inducing factors present in the cytoplasms of the S cells,
and therefore that synchronization could result from a stimulatory effect
of S phase cytoplasm on G_1 nuclei. In none of their experiments was
there any indication that G_1 nuclei inhibited DNA synthesis in S nuclei.
Nor did the presence of G_2 nuclei seem to inhibit the initiation of DNA
synthesis in G_1 nuclei. Even in polykaryons with a high ratio of G_2 to G_1
nuclei, G_1 nuclei progressed into S phase. In G_2/S polykaryons, DNA
replication continued in the S phase nuclei, but DNA synthesis was
not induced in the G_2 nuclei. Whether this is an invariable rule for
other types of heterokaryons as well is not absolutely certain. In some
types of fusions, mononucleate hybrid cells frequently contain a tetraploid set of chromosomes from one parent and a diploid set from the
other parent (Jami et al., 1971). These particular hybrids could arise if
one nucleus of a "dikaryon" underwent two replication cycles (the second cycle is *endoreduplication*) while the other underwent just one
cycle before there was a synchronous mitosis. However, the possibility
that these hybrids arise from the fusion of tetraploid cells with diploid
cells or from fusion of three cells cannot be excluded.

2. Duration of G_1 and S Phases

Large polykaryons may remain in culture for a considerable time
without replicating or dividing. As is the case among nondividing
mononucleate cells, this almost certainly represents a prolongation of
the G_1 phase. Nonetheless, in those polykaryons which start DNA synthesis, the S phase may also be longer than normal (Heneen, 1971, Westerveld and Freeke, 1971). If, however, one considers only small polykaryons, the nuclei often retain the same cell cycle characteristics they
had as mononucleate cells. This was found to be true in a study by
Graves (1972a) where homokaryons and heterokaryons were prepared
by fusing synchronized HeLa cells with synchronized Chinese hamster
(DON) or mouse cells (NCTC 2472). These cells differ from each other in
the length of the phases of the cell cycle (Fig. VI-3). HeLa cells have a
longer G_1 phase than the other cell types while in NCTC cells the S
phase is longer than in either DON or HeLa cells. In homokaryons the
nuclei were found to retain the cell cycle time characteristic of the
parental cells. In heterokaryons, however, the time point at which DNA
synthesis was initiated was determined by the cell with the shorter G_1
phase. HeLa nuclei in heterokaryons started DNA synthesis earlier than
normal and at the same time as either DON or NCTC nuclei. There was
no indication that the time of initiation of DNA synthesis in DON or
NCTC nuclei was delayed by the HeLa component. On the other hand

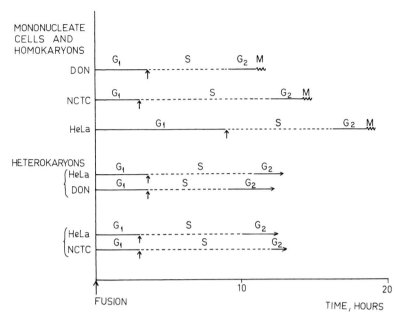

Fig. VI-3. Schematic summary of measurements of cell cycle stages in mononucleate cells, in homokaryons and heterokaryons of DON, NCTC, and HeLa cells based on Graves (1972a).

each nucleus retained the S phase duration typical of its mononucleate parental cell. Nuclei with a short S phase therefore completed their DNA replication before those with a longer S phase even though they all resided in a common cytoplasm.

3. Asynchronous DNA Synthesis in Some Heterokaryons

Although synchronization of DNA synthesis seems to be common in polykaryons and has been observed even when the parental cells come from widely different species, this coordination of nuclear events does not occur in all systems. There are cases of asynchrony (Sandberg *et al.*, 1966; Burns, 1971; Sheehy *et al.*, 1974) or even "negative synchrony" (Johnson and Harris, 1969c), and there is at least one case where DNA synthesis in the replicating nucleus seems to be inhibited by factors supplied by nonreplicating cells (Norwood *et al.*, 1974). These exceptions are important in helping to provide insight into the mechanisms controlling synchrony of DNA synthesis.

First, there are exceptions which imply that the state of the nucleus is important. One example of this is encountered where the two nuclei of a heterokaryon have widely different activities, as, for instance, those in a chick erythrocyte + HeLa cell heterokaryon. The chick erythrocyte nucleus does not normally synthesize DNA, but is induced to do so in the heterokaryon in concert with DNA synthesis in the HeLa nuclei (Bolund et al., 1969a; Johnson and Harris, 1969b; Johnson and Mullinger, 1975). Nonetheless, a significant number of chick erythrocyte nuclei may be found which are not making DNA even though the HeLa nuclei are. These nonsynthesizing erythrocyte nuclei are often smaller than actively replicating erythrocyte nuclei, which suggests that some nuclear swelling or other preparatory steps may be necessary before initiation of DNA synthesis. Such preparatory steps may in part account for the observation that among virus-induced heterophasic polykaryons, the degree of synchrony increases gradually with time after fusion (Yamanaka and Okada, 1966; Johnson and Harris, 1969a).

A striking exception to the rule of synchrony was found in HeLa + Ehrlich heterokaryons by Johnson and Harris (1969c)—indeed, a sort of "negative synchrony" was found. In these heterokaryons DNA synthesis in the HeLa nuclei was suppressed when Ehrlich nuclei were replicating. This phenomenon was particularly striking in heterokaryons containing a high ratio of Ehrlich to HeLa nuclei. A similar dominance of the Ehrlich nuclei was observed in Ehrlich–mouse fibroblast heterokaryons. It is not yet known if the suppression of DNA synthesis in nuclei from one of the parental cells is a property peculiar to heterokaryons made with Ehrlich cells or if this phenomenon is also found in other heterokaryons where one of the parental cells is an undifferentiated malignant tumor cell. The negative synchrony observed in HeLa–Ehrlich heterokaryons could be due to competition among nuclei for cytoplasmic initiation factors necessary for DNA synthesis, the Ehrlich nuclei being more competitive than other nuclei.

In one case the failure to synchronize nuclei in S phase appears to be due to suppression of DNA synthesis in replicating nuclei by factors contributed by cells that had "aged" in vitro. Aging (senescence) of cells in vitro is a characteristic of normal diploid cells (Hayflick, 1965). After primary cultures have been initiated from tissue explants, the cells grow rapidly for some time ("young cells"), then more slowly, and, finally, after a fixed number of cell cycles, they stop multiplying ("old cells"). At this stage DNA replication is also arrested. Transformed cells and tumor cells, by contrast, lack the senescence phenomenon and are capable of infinite growth in vitro. Norwood et al. (1974) fused senescent (old) fibroblasts with young fibroblasts. The old nuclei were not induced to synthesize DNA. Instead, DNA synthesis was inhibited in the young nuclei. Different results were obtained when old fibroblasts were fused with transformed cells (Norwood et al., 1975). In heterokaryons made with either HeLa cells or SV40 transformed human cells,

DNA synthesis was reinitiated in the old nuclei. The reason why trans-formed cells, but not young normal cells, induce DNA synthesis in old nuclei is not known.

4. Control of DNA Synthesis

The observations of the activities of heterophasic polykaryons re-counted in the foregoing sections lead rather directly to three hypothe-ses regarding DNA synthesis in eukaryotic cells:

(a) There are cytoplasmic substances in S phase cells which initiate nuclear DNA synthesis, while the inverse, that G_1 or G_2 cytoplasms contain inhibitors of DNA synthesis, appears less likely.

(b) Once DNA synthesis has been initiated nuclei and chromo-somes retain their own replicative rates and patterns, irrespective of their cytoplasmic environment.

(c) The state of the nonreplicating nucleus at the time of fusion is of some importance. Condensed inactive nuclei may require some prelim-inary swelling before the nucleus will respond to the cytoplasmic in-ducers with DNA synthesis.

The nature of the DNA synthesis-inducing substances is not known. However, successful initiation of DNA synthesis in mononucleate cells requires RNA and protein synthesis, suggesting that *de novo* synthesis of protein ("initiator protein") may be important (for a review see Klein and Bonhoefer, 1972). Furthermore, these proteins may have to pene-trate into the nucleus to exert their influence (Goldstein and Ron, 1969). In heterokaryons made with nucleated erythrocytes (Ringertz *et al.*, 1971), the activation of the erythrocyte nucleus is preceeded by a mas-sive influx of proteins from the cytoplasm into the nucleus (Chapter VII). It appears likely, therefore, that the initiation of DNA synthesis in G_1 nuclei of heterophasic G_1/S polykaryons depends on the migration of specific protein factors present in S phase cytoplasms into the G_1 nuclei.

These interpretations are consistent with results obtained in nuclear transplantation experiments in unicellular organisms (Goldstein and Prescott, 1967, 1968; Goldstein and Ron, 1969; Goldstein, 1972, 1974) and in experiments where somatic nuclei have been injected into frog eggs (Gurdon and Woodland, 1968; Merriam, 1969).

B. SYNCHRONIZATION OF MITOSIS

In spontaneously formed multinucleated cells, not only is DNA syn-thesis synchronous, but so, also, is mitosis (Fell and Hughes, 1949; Mazia, 1961; Oftebro and Wolf, 1967). In general, mitotic synchrony also seems to be the case among polykaryons formed by virus induced fusion. Rao and Johnson (1970) found almost perfect synchrony among nuclei in HeLa G_1/S or HeLa G_1/G_2 heterokaryons. Asynchronous divisions were observed in but five of 1000 fused cells. In S/G_2 fusions, however, 11% of the polykaryons showed mitotic asynchrony. The conclusion which can be drawn from these studies and which is supported by numerous other observations (Johnson and Rao, 1971; Rao *et al.*, 1975) is that mitotic synchrony is very common not only in homophasic but also in heterophasic polykaryons. It is worth noting, however, that there are a number of interesting exceptions to the rule of synchrony.

In some cases the lack of synchrony appears to be related to the size of the polykaryon. Thus, in some very large polykaryons, which were the results of fusion induced by measles virus, nuclei in different parts of the cells were seen to be out of phase (Cascardo and Karzon, 1965; Heneen *et al.*, 1970). In some of these cells there appeared to be "mitotic waves" suggesting that the concentration of mitotic triggers formed a gradient across the cell.

Other examples of mitotic asynchrony may be the consequence of asynchronous DNA replication. Thus, Johnson and Harris (1969c) found that HeLa–Ehrlich heterokaryons showed considerably less synchronized mitotic activity than HeLa homokaryons. Frequently, only the Ehrlich nuclei were found to be in mitosis while the HeLa nucleus was still in interphase. This asynchrony may be a consequence of the anomalous coordination of DNA synthesis referred to as negative synchrony. In HeLa + chick erythrocyte heterokaryons many erythrocyte nuclei fail to enter mitosis when the HeLa nucleus divides (Appels *et al.*, 1975b). Instead the erythrocyte nuclei are transferred intact to the daughter cells which result from the division of the heterokaryon.

Even in those cases where polykaryons undergo a normal bipolar division a certain degree of asynchrony may persist into the succeeding cell cycles of the mononucleate daughter cells. Thus differences in the degree of chromosome condensation have been found between the two parental chromosome sets of interspecific hybrid cells (see Chapter X).

The mechanisms by which synchrony of mitosis is established are not known but the following experiments suggest that more than one factor is involved.

(a) Late G_2 and mitotic cells seem to contain factors which induce G_1 or S nuclei to enter mitosis more quickly. These factors are discussed in the following section (Section C) where fusions between mitotic and interphase cells are considered.

(b) S phase cells contain factors which prevent or delay the entry of

G_2 nuclei into mitosis. These factors may be potent. Rao and Johnson (1970) found that in $1S/3G_2$ polykaryons the S component prevented $3G_2$ nuclei from entering mitosis.

(c) Mitosis itself may be prolonged so that lagging nuclei catch up with advanced nuclei during an extended metaphase (Oftebro and Wolf, 1967; Heneen *et al.*, 1970).

C. PREMATURE CHROMOSOME CONDENSATION (PCC)

Fusion of a mitotic cell with a cell in interphase results in a precocious attempt of the interphase nucleus to enter mitosis. The chromatin of the interphase nucleus condenses into chromosome-like structures, sometimes with a fragmented appearance, and the nuclear membrane disappears. In the following discussion we will refer to this phenomenon as *premature chromosome condensation** (PCC), a term introduced by Johnson and Rao (1970). The chromosome filaments which form as a result of this phenomenon will be called prematurely condensed chromosomes or "PC chromosomes."

This type of chromosome was first noted in polykaryons arising in cell cultures which had been infected with measles virus (Nichols, 1963; Nichols *et al.*, 1964, 1965; Aula, 1965; Norrby *et al.*, 1965; Heneen *et al.*, 1970), but have subsequently been observed in polykaryons induced by Sendai virus (Saksela *et al.*, 1965; Kato and Sandberg, 1968a,b; Aula and Saksela, 1966; Aula, 1970; Johnson and Rao, 1970) and other viruses as well (Cantell *et al.*, 1966; Stich *et al.*, 1964). (For more detailed information see Rao and Johnson, 1974.)

1. Different Types of PCC

Even before the correlation between PCC and cell fusion was fully understood, several workers noted that there is considerable variation from one polykaryon to another in the morphology and extent of PCC (Aula and Saksela, 1966; Nichols *et al.*, 1967; Kato and Sandberg, 1968b; Takagi *et al.*, 1969). In some polykaryons the PC-chromosomes appear totally fragmented ("pulverized"), whereas in other cells they are extended filaments with very few breaks. Takagi *et al.* (1969) and Stenman and Saksela (1969) made observations that suggested that the variable appearance of PC chromosomes could be attributed to the stage of the cell cycle at which the interphase cells fused with mitotic cells. This hy-

* The phenomenon of premature chromosome condensation has also been termed *chromosome pulverization* (e.g., Nichols *et al.*, 1964, 1965, 1967) and *prophasing* (e.g., Obara *et al.*, 1973a, 1974b).

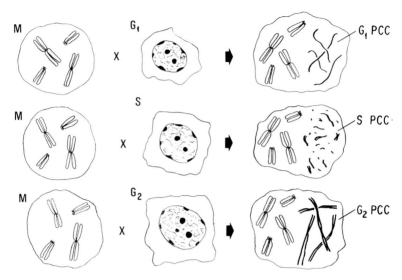

Fig. VI-4. Schematic illustration of the induction of premature chromatin condensation in G_1, S, and G_2 cells after fusion with mitotic cells (M) (Ringertz, 1974).

pothesis received strong support from a series of experiments carried out by Johnson and Rao (1970, 1971; Jonson et al., 1970; Rao and Johnson, 1970, 1971, 1972a,b; Sperling and Rao, 1974a,b). In these experiments one population of synchronized HeLa cells was fused with another at the same or a different stage in the cell cycle, and poly-karyons containing both mitotic and interphase nuclei were analyzed. The results obtained are summarized schematically in Fig. VI-4 and can be described as follows: G_1 interphase nuclei produced thin extended chromosome filaments whereas G_2 nuclei gave thicker filaments which were similar to normal mitotic chromosomes. Furthermore, chromo-some filaments derived from G_1 nuclei were single stranded while fila-ments of G_2 nuclei consisted of two strands (chromatids). S phase nuclei gave rise to irregular, fragmented chromatin masses. The different forms of PCC can be observed most easily in fusions of cells with a small number of chromosomes, as illustrated by studies by Röhme (1974) on Indian muntjac cells (7 chromosomes) (Figs. VI-5 through VI-7).

The single-stranded nature of G_1 PC chromosomes and the double-strandedness of G_2 PC chromosomes can easily be explained by the fact that G_1 represents the interphase period before DNA replication, while G_2 is the post replication period immediately preceding mitosis. The fragmented nature of S phase PC chromosomes is more difficult to un-derstand. Further studies of S phase PCC by Röhme (1974, 1975) show, however, that the S phase PC chromosomes may not be as fragmented as they appear in the light microscope. He labeled S phase cells with ³H-thymidine prior to fusion with metaphase cells and then performed autoradiography. In such preparations the chromosome fragments appear to be linked to each other by radioactively labeled filaments which are too thin to be visible in the light microscope, but which can be traced from the silver grains in the autoradiograms (Fig. VI-8a and b). Whereas the chromatin clumps seen in light microscopic preparations probably represent condensed chromatin which has already replicated or which has not yet started to replicate, the ³H-thymidine-labeled, "invisible" regions connecting these "fragments" are quite possibly greatly extended chromosome fibers which are in the process of DNA synthesis (Fig. VI-9). Although PC chromosomes have been studied in the electron microscope using sections (Sanbe *et al.*, 1970) the ultrastructure of these chromosomes requires further examination

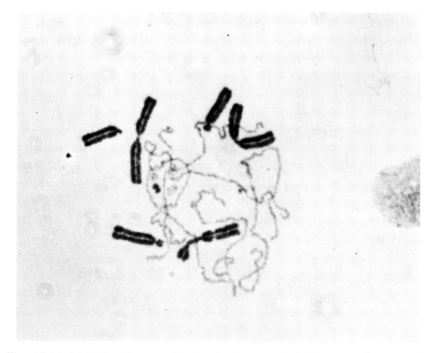

Figs. VI-5 to VI-7. Homokaryons showing the three morphological types of prematurely condensed chromosomes of the Indian muntjac. Mitotic inducers are muntjac metaphase cells. (Röhme, 1975.)
Fig. VI-5 G₁ type. The PC chromosomes consist of one chromatid.

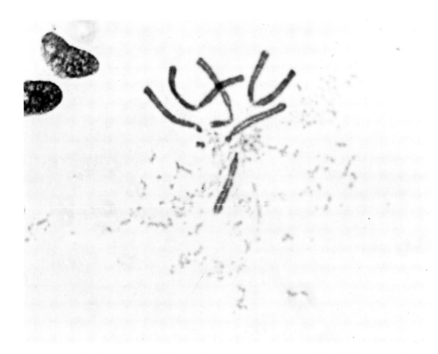

Fig. VI-6 S type. The PC chromosomes appear as heavily fragmented.

Fig. VI-7 G₂ type. The PC chromosomes consist of two chromatids.

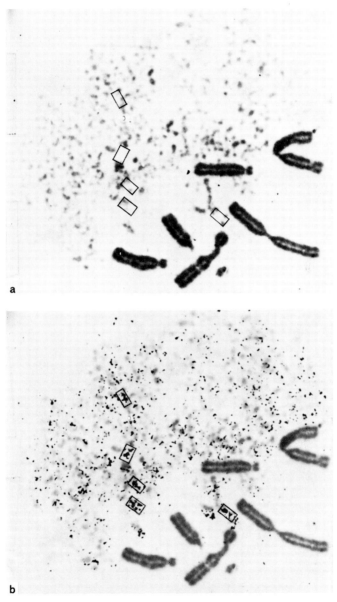

Fig. VI-8. S-type PCC of Indian muntjac cells. (a) Fragmented appearance of S-type PCC. (b) The same cell showing the distribution of ^{3}H-thymidine label after being processed for autoradiography (^{3}H-thymidine was given 15 min prior the induction of cell fusion). By comparing the two pictures it is evident that the label is predominantly located in the gaps between the visible fragments. Note the "hot spots" of label indicated by rectangulars. (Röhme, 1975).

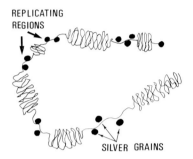

REPLICATING
REGIONS

SILVER GRAINS

Fig. VI-9. Schematic graph illustrating a likely explanation for the fragmented appearance of S phase PCC. Replicating regions can be detected by autoradiography after ³H-thymidine labeling but are too extended to be visible by light microscopy. Nonreplicating chromatin regions are more condensed and can therefore be observed by ordinary microscopy.

with spreading techniques in order to clarify the correlation between replication and chromatin structure.

2. Kinetics of PCC Induction

PCC occurs immediately after an interphase nucleus is confronted with the cytoplasm of a mitotic cell: with conventional cytological methods, it can be observed within 10 min of cell fusion (Aula and Saksela, 1966; Sandberg *et al.*, 1970). Analyses of cells containing varying proportions of mitotic and interphase nuclei have shown that the rate at which PCC is induced depends on the ratio of mitotic to interphase nuclei and the cell cycle stages of the interphase cells participating in the fusion. Thus, in the experiments of Johnson and Rao (1970), PCC was far more readily induced in G_1 nuclei (Fig. VI-10) than in either S or G_2 nuclei. In cells containing one M and three G_1 nuclei (1M/3G_1) the chromosomes of all G_1 nuclei had condensed 45 min after fusion. At this time 1M/3G_2 and 1M/3S cells failed to show any signs of PCC. With a higher proportion of M to G_2 or S nuclei (e.g., 2M/1S or 2M/1G_2), all of the G_2 and S nuclei could be induced to condense within 45 to 75 min after fusion.

3. PCC-Inducing Factors

Attempts have been made to identify the factors responsible for PCC in the hope of providing information about the control of normal mitosis and the way in which certain viruses damage chromosomes. Matsui *et al.* (1971) demonstrated that a factor necessary for premature chromosome condensation is synthesized 15–45 min before the onset of normal metaphase. The critical factor is not formed if protein synthesis

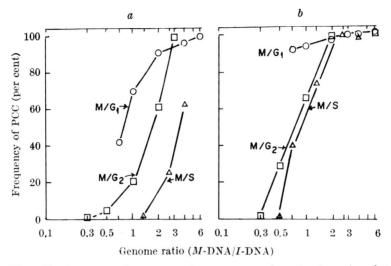

Fig. VI-10. The frequency of premature chromosome condensation in various hetero-phasic M/G$_1$, M/S, and M/G$_2$ cells, at (a) 15 and (b) 45 min after addition of virus, plotted as a function of the ratio of mitotic DNA to interphase DNA in the cells (genome ratio). The genome ratio is plotted on a log scale. Squares, M/G$_2$ cells; circles, M/G$_1$ cells; triangles, M/S cells (Johnson and Rao, 1970).

is blocked during the G$_2$ phase. The chemical nature of the PCC inducing protein(s) has not yet been elucidated. It is interesting to note, however, that the factors involved seem not to be species specific since PCC has been observed in interspecific heterokaryons (Ikeuchi and Sandberg, 1970; Johnson and Rao, 1970).

Rao and Johnson (1971) have studied factors which promote or inhibit the induction of PCC. Spermine, putrescine, and Mg^{2+} promoted PCC, whereas another positively charged compound, spermidine, inhibited it. These compounds can be expected to bind to DNA-phosphate groups in the chromatin. In doing so, it is likely that they influence the normal DNA-protein interaction thereby promoting or inhibiting the mitotic condensation of chromatin.

4. Comparison of PCC with Normal Mitotic Events

While premature chromatin condensation shares many features with normal mitosis, there are also some differences. In considering whether or not it is an appropriate model for studying the molecular mechanisms involved in normal mitosis, it is useful to spell out both the similarities and the differences between PCC and mitosis.

Just as in normal mitosis, the nuclear envelope breaks down as the interphase chromatin begins to condense (Aula, 1970). The prematurely

condensed G_1 and G_2 chromosomes are much more extended than
normal metaphase or prometaphase chromosomes. Moreover G_2 PC
chromosomes are obviously thicker than G_1 PC chromosomes.
Kinetochore-like structures have been identified on PC chromosomes
and there are indications that microtubular filaments orient themselves
towards these structures (Matsui *et al.*, 1972b).

Furthermore there are functional changes in the chromatin which resemble those of normal mitosis. RNA synthesis ceases (Stenman, 1971)
just as it does during mitosis (Taylor, 1960; Prescott and Bender, 1962).
There are also indications that RNA and protein associated with the interphase chromatin are released into the cytoplasm during PCC (Matsui
et al., 1972a). Similar releases occur during mitosis (for review see Goldstein, 1974).

The major difference between PCC and normal metaphase chromosomes is that G_1 and G_2 PC chromosomes seem unable to condense to
the extent seen in metaphase. Furthermore, as is discussed in the following section (Section A,5) unlike normal chromosomes, PC chromosomes are often lost during the first divisions of the fused cells.

5. Postmitotic Fate of PCC Material

The postmitotic fate of prematurely condensed chromosomes has
been examined by fusing ^{3}H-thymidine-labeled interphase cells with
unlabeled cells arrested in mitosis by a Colcemid block (Matsui *et al.*,
1972b; Rao and Johnson, 1972b). With this technique it has been possible to identify labeled PC chromosomes in the immediate fusion
products and their daughter cells during the first days after fusion.
Matsui *et al.* (1972b) found that during the first division, G_1 and S chromatin became randomly distributed among daughter cells as nuclear
fragments and micronuclei, while G_2 chromatin formed PC chromosomes which became integrated into the nuclei of the daughter cells and
emerged as normal chromosomes at the next division. Similar observations were made by Rao and Johnson (1972b) who observed that PC
chromosomes formed discrete entities which were not interspersed
with the metaphase chromosomes. But in these experiments too, some
PC chromosomes were incorporated into one or both groups of anaphase chromosomes. It is uncertain, however, whether PC chromosomes that are carried through one mitosis are normally replicated and
segregated during later divisions. Heterophasic fusions with two different glycine-requiring mutant cell lines of Chinese hamster indicated
that very few of the heterokaryons formed with mitotic cells gave rise to
viable synkaryons (Rao and Johnson, 1972b). In those cases where via-

ble hybrids were obtained it was difficult to prove that they resulted from a heterokaryon in which the interphase nucleus had undergone PCC. Even if the mitotic cell population is well synchronized it is likely to contain a small number of G_1 and G_2 interphase cells. In fusions with other cells these interphase "contaminant" cells may well be the ones which give rise to viable hybrids without undergoing PCC.

6. Practical Applications for PCC

The induction of PCC promises to be useful for a number of purposes that are not directly related to the study of mitosis, DNA replication, or the cell cycle. Among potential applications are its use to (a) visualize chromosomes in nondividing cells (Johnson and Rao, 1970; Yanishevsky and Carrano, 1975); (b) study the molecular architecture of interphase chromatin during G_1, S, and G_2 (Schor et al., 1975a); (c) facilitate a more detailed analysis of the structure of banding patterns after certain forms of chromosome staining (Unakul et al., 1973; Röhme, 1974; Sperling and Rao, 1974a,b); (d) detect chromosome damage arising from alkylating agents or radiation and to monitor the repair of such chromosome damage (Hittelman and Rao, 1974a,b, 1975; Waldren and Johnson, 1974; Schor et al., 1975a).

The first of these applications is illustrated by a study by Johnson et al. (1970) that showed that both nongrowing differentiated cells and rapidly multiplying cells of many different species can be induced to undergo PCC. Thus, fusion of a diverse collection of cells including horse lymphocytes, bovine spermatozoa, Chinese hamster ovary cells, embryonic chick fibroblasts and erythrocytes, *Xenopus* kidney, and mosquito cells with Colcemid blocked HeLa cells all gave rise to PC chromosomes in each of the interphase nuclei (Fig. VI-11). Using the same approach Yanishevsky and Carrano (1975) were able to induce G_1 PC chromosomes in senescent, nondividing WI 38 cells.

The use of the PCC phenomenon in the analysis of interphase chromatin is exemplified by investigations by Schor et al. (1975a). These authors studied in detail the morphology of PC chromosomes derived from chromatin of different stages of interphase. Their results show that PC chromosomes from early G_1 are more contracted than those from late G_1. The gradual attenuation which takes place during G_1 continues into S phase where parts of the chromosomes are so extended that they are invisible under the light microscope (Röhme, 1974). After DNA replication is complete, PC chromosomes become progressively more condensed. G_2 PC chromosomes are more contracted and thicker than G_1 chromosomes, and late G_2 PC chromosomes do not differ greatly from

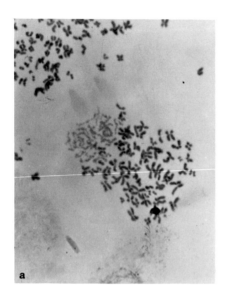
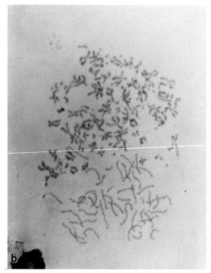
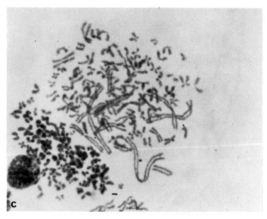

Fig. VI-11. Induction of PCC in nondividing cells (Johnson *et al.*, 1970). (a) HeLa/sperm heterokaryon showing the induction of PCC of the sperm nucleus. The chromosomes of the sperm nucleus are of the G_1 type. (b) G_1 PCC in *Xenopus* nucleus after fusion with mitotic HeLa cell. (c) G_2 PCC in Chinese hamster nucleus after fusion with mitotic HeLa cell.

normal metaphase chromosomes (Johnson and Rao, 1971). These observations indicate that the susceptibility of chromatin to condensation varies during interphase. Together with studies of the sensitivity of chromatin to proteolytic enzymes, and titrations of actinomycin binding sites (Pederson, 1972; Pederson and Robbins, 1972), the analysis of

PC chromosomes supports the hypothesis of cyclic changes in chromatin structure advanced by Mazia (1963). He suggested that chromosomes undergo a condensation–dispersion cycle, maximum condensation occurring at mitosis and maximum dispersion during DNA replication in interphase.

Because PC chromosomes from G_1 and G_2 nuclei are much more extended than normal metaphase chromosomes, they offer better possibilities for the identification of individual chromosomal regions (Röhme, 1974). When stained with quinacrine, Giemsa, or staining methods specific for centromeric heterochromatin (see Chapter X), the G_1 and G_2 PC chromosomes show bands similar to the Q-, G-, and C-bands of normal metaphase chromosomes (Unakul *et al.*, 1973; Röhme, 1974). These bands make it possible to identify individual chromosomes in species with relatively few chromosomes. Furthermore, PC chromosomes show more numerous and more distinct bands than metaphase chromosomes (Fig. VI-12). This suggests that some individual G-bands seen in mitotic chromosomes are in fact made up of several smaller bands. The greater resolution offered by the PCC banding patterns could be useful in gene mapping and analysis of chromosome rearrangements in species that have very few chromosomes. In cells with many chromosomes the intertwined PC chromosomes probably become too difficult to analyze.

The usefulness of being able to visualize interphase chromosomes by inducing PCC has also been appreciated in radiobiological work (Hittelman and Rao, 1973, 1974a,b; Waldren and Johnson, 1974). Damage to interphase chromatin caused by X irradiation, UV irradiation, or alkylating agents has been made apparent by fusing the interphase cells with synchronized mitotic cells. In an experiment with X irradiation, Hittelman and Rao (1974a) scored the radiation induced damage by analyzing G_2-type PC chromosomes for gaps, chromatid breaks, and chromatid exchanges. Comparison of these results with those obtained by analyzing metaphase chromosomes showed significantly higher incidence of aberration. The reasons the analysis of PC chromosomes detected more chromosomal damage appear to be that (1) it is easier to detect aberrations in extended PC chromosomes than among condensed chromosomes; (2) the induction of PCC allows discovery of chromosome aberrations immediately after X irradiation (i.e., before any significant repair of chromosome breaks has taken place); and (3) the method allows cells which are so damaged that they would never reach mitosis to be scored.

Chromosomal breakage induced by X irradiation can be measured quantitatively by fusing irradiated G_1 interphase cells with mitotic cells

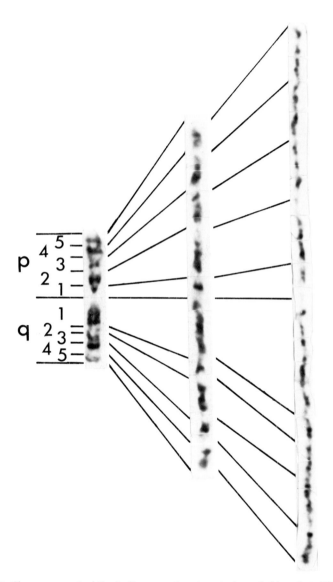

Fig. VI-12. Chromosome 1 of the Indian muntjac at metaphase (left) and at different degrees of premature chromosome condensation (middle and right). The chromosomes are subdivided into ten regions, p 1–5 in the short arm and q 1–5 in the long arm. The regions are defined on the metaphase chromosomes as the segments between middle points of successive, well recognizable, light bands starting from the centromere. Borders of corresponding regions in the different chromosomes are connected by lines (×2300) (Röhme, 1974).

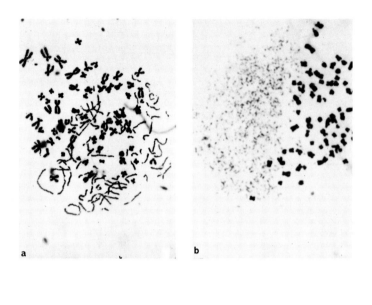

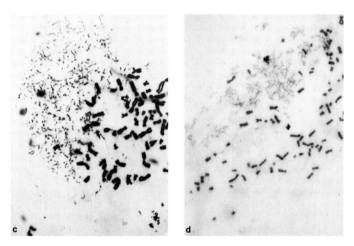

Fig. VI-13. Use of PCC to detect X-ray induced chromosome damage (Waldren and Johnson, 1974). (a) Unirradiated HeLa G_1 PCC with 69 chromosomes ($\times 1500$). (b) Massively fragmented G_1 PCC from a cell irradiated with 8000 rads and immediately fused with a mitotic cell to display the chromosomes. Very little continuity can be seen between the fragments ($\times 1300$). (c) G_1 PCC with about 230 fragments; the pieces are considerably longer than in (b). Fusion immediately after 800 rads of X irradiation ($\times 1500$). (d) G_1 PCC with about 150 fragments. The length of chromosome fragments is greater than in (c). Fusion immediately after 200 rads of X irradiation ($\times 1000$) (Waldren and Johnson, 1974).

(Fig. VI-13) and then counting PC chromosome fragments. Thus Waldren and Johnson (1974) found that the number of PC chromosome fragments (G_1-type) was linearly related to the dose of irradiation up to 1800 rads. Furthermore, for a given dose the number of breaks detectable in G_1 PC chromosomes was greater than the number scored in metaphase cells. If the irradiated cells were incubated for a few hours before fusion, the number of breaks detectable in PC chromosomes decreased markedly, suggesting that many breaks had been repaired during the prefusion incubation time.

Although UV irradiation did not produce any chromosome breaks detectable in G_1 PC chromosomes it did produce breaks detectable in metaphase chromosomes. It appears therefore that the UV-induced breaks require some time or specific stage of the cell cycle to manifest themselves. Alternatively, the UV-induced breaks develop during repair of the primary lesions of UV irradiation and become apparent only at mitosis. For a further discussion of the organization of chromosomes and their response to UV irradiation at different stages of the cell cycle see Schor et al. (1975a).

D. "TELOPHASING"

Occasionally fusions of mitotic cells with interphase cells fail to effect PCC. Instead, in many of these cases, a nuclear envelope forms around the metaphase chromosomes resulting in a telophase-like nucleus (Fig. VI-14). This phenomenon has been studied by Sandberg and co-workers (Ikeuchi et al., 1971; Obara et al., 1973, 1974a,b, 1975). They found that the incidence of "telophasing" was frequent among poly-karyons containing high proportions of interphase nuclei and depended on external factors such as pH and the concentration of divalent metal ions. Telophasing and PCC appeared to be mutually exclusive processes since conditions which favored PCC reduced the frequency of telophasing and vice versa. At pH's of 6.6–8.0, PCC predominated, while at pH 8.5, telophasing was more common (Obara et al., 1974b). Nuclei which had formed as a result of telophasing were able to return to interphase since they showed dispersion of chromatin, reformed nucleoli and resumed RNA synthesis (Obara et al., 1974a). The ability of interphase cells to induce nuclear envelope formation around chromosomes after fusion with mitotic cells varies with the cell cycle stage and is dependent upon proteins synthesized during interphase (Obara et al., 1975). Late G_2 cells are more potent than early G_1 cells in producing telophasing. It appears, therefore, that both PCC and telophasing re-

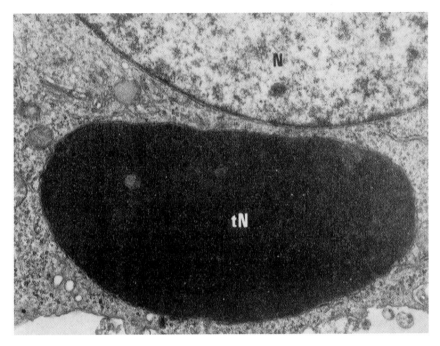

Fig. VI-14. Fused cell having interphase nucleus (N) and telophase-like nucleus (tN), harvested 40 min after incubation at 37°C with virus. Elliptical dense chromatin mass (tN) is continuously surrounded by the double membranes, most probably nuclear envelope. Ribosomes are seen on the cytoplasmic surface of the membranes (× 14 000) (Ikeuchi *et al.,* 1971).

flect mechanisms which are operating during normal mitosis. Cell fusion with synchronized cell populations appears to be a valuable technique which can be used to analyze these mechanisms.

E. CELL FUSION AND THE ANALYSIS OF THE CELL CYCLE

In summary, it is clear that the cell fusion technique has already contributed extensively to our knowledge of the cell cycle. Since S phase cells induce G_1 nuclei to initiate DNA synthesis, fusion with cells synchronized at these stages of the cell cycle can be used to identify factors that control the initiation of DNA synthesis. In a similar way fusion can also be used to identify factors that control mitosis. The PCC phenomenon has offered a new approach to the study of the organization of DNA replicating units and to the definition of factors influencing condensation of interphase chromatin into metaphase chromosomes. The examples presented also demonstrate that inducing PCC is a powerful technique for studying chromosome breakage brought about by radiation and chemicals. The PCC method makes it possible to score the extent of chromosome damage in cells which normally do not divide. In addition it is a more sensitive method than conventional chromosomal analysis because it reveals more lesions and permits a more detailed classification of the chromosome breaks. Moreover by varying the time and conditions of postirradiation incubation the PCC method also makes it possible to investigate repair and rejoining processes.

So far, most PCC studies have been carried out with cytochemical methods. But since synchronized cell populations can be fused with each other, there is hope that in the future, biochemical methods can also be used to study the mechanisms involved in DNA synthesis and mitosis.

VII

Heterokaryons

A. DEFINITIONS

As indicated in the introduction, the term heterokaryon is used to describe multinucleate cells formed by the fusion of different cell types. The parental cells may differ with respect to genotype or phenotype. In interspecific hybrids (e.g., mouse + man) there is clearly a genotypic difference. In fusions of liver cells and lymphocytes there is a distinct and stable difference between the phenotypes of the parental cells. Often, however, the difference between the parental cells may be small and difficult to define as in fusions of fibroblasts from two different mouse strains or in fusions of normal and neoplastic cells from the same tissue. In the following discussion multinucleate cells are classified as heterokaryons if there is a genotypic or stable phenotypic difference between the parental cells. The term homokaryon is reserved for poly-karyons formed by fusion of cells among which there is no known disparity in genotype or phenotype.

The identification of heterokaryons in a fusion mixture may sometimes be a problem. In order to distinguish heterokaryons from homo-karyons one may use parental cells differing in nuclear size (Harris, 1967), chromatin morphology (Le Douarin and Rival, 1975), nuclear staining properties (Moser *et al.*, 1975), or antigenicity (Carlsson *et al.*, 1974b). Alternatively one of the parental cells may be prelabeled with ^{3}H-thymidine after which the heterokaryons can be recognized as cells having both labeled and unlabeled nuclei.

In the previous chapter, the regulation of DNA synthesis and mitosis in heterokaryons and homokaryons is discussed. In this chapter other characteristics of heterokaryons are presented. The main focus is on the regulatory events which occur when two cells differing markedly in nuclear activity and/or phenotype are fused with each other.

B. REACTIVATION OF DORMANT CELL NUCLEI

Soon after Harris and Watkins (1965) began using Sendai virus to produce multinucleated heterokaryons from cells with active and inactive nuclei, they noted that after fusion, inactive nuclei, among them chick erythrocyte nuclei and mouse macrophage nuclei, underwent marked reactivation and showed accelerated synthesis of RNA (Harris *et al.*, 1965, 1966). The reactivation of chick erythrocyte nuclei was further studied by Harris (1967) and by our group in Stockholm (for reviews see Ringertz and Bolund, 1974a; Sidebottom, 1974; Appels and Ringertz, 1975). Macrophage heterokaryons were investigated by Gordon and Cohn (1970, 1971a,b; Gordon, 1973). These studies, together with investigations on heterokaryons involving gametes, have provided considerable information about nucleocytoplasmic interaction and its role in the control of nuclear activity. The salient findings are reviewed in the following pages.

1. Chick Erythrocyte Heterokaryons

Until quite recently it was generally believed that inactivation of erythrocyte nuclei during erythropoiesis was an irreversible process (for a review see Ringertz and Bolund, 1974b). This situation changed when Harris (1965, 1967) demonstrated that chick erythrocyte nuclei are reactivated in heterokaryons formed with mouse fibroblasts, human tumor cells, or other active cells. One special feature of erythrocyte fusions is that these cells often undergo virus-induced lysis and lose their cytoplasmic contents before the fusion takes place (Schneeberger and Harris, 1966; Dupuy-Coin *et al.*, 1976). The erythrocyte ghosts which result from this lysis consist of small condensed nuclei surrounded by plasma membranes. In the presence of Sendai virus, the ghosts agglutinate to each other and to other cells present. Electron microscopic studies of ghosts and HeLa cells agglutinated by Sendai virus (Schneeberger and Harris, 1966) show that the plasma membranes are broken down at some points of contact. Consequently, HeLa cytoplasms gradually fill the "empty bags" of the erythrocytes and the interconnections widen. The ultrastructural studies (Schneeberger and Harris, 1966; Dupuy-Coin *et al.*, 1976) suggest that the erythrocyte nuclei are transplanted into HeLa cytoplasm virtually free of chick cytoplasm. Despite that generalization, however, immunological techniques detect chick hemoglobin in many newly formed heterokaryons, suggesting that at least among some erythrocytes, lysis is not complete by the time of fusion (Carlsson *et al.*, 1974b). To some extent the degree of lysis can be controlled by modifying the ionic composition of the fusion medium (Zakai *et al.*, 1974a,b). At low Ca^{2+} concentration erythrocytes tend to lyse before they fuse, whereas at high calcium concentration they fuse without lysis.

Once it has been incorporated into the cytoplasm of an actively growing cell, the erythrocyte nucleus responds to its new environment by undergoing a reactivation process which involves a series of morphological and chemical changes. The main events that occur during reactivation in heterokaryons made with human tumor cells (HeLa) or other actively growing cells can be summarized as follows:

1. Early changes in the physicochemical properties of deoxyribonucleoprotein (Bolund *et al.*, 1969a; Ringertz, 1969)

2. Increase in nuclear volume and dry mass and a dispersion of the condensed chromatin (Harris, 1967; Bolund *et al.*, 1969a)

3. Migration of *human* nucleospecific proteins into the chick nucleus (Ringertz *et al.*, 1971; Ege *et al.*, 1971; Goto and Ringertz, 1974; Appels *et al.*, 1974a,b)

4. Initiation of HnRNA synthesis (Harris, 1965; Harris *et al.*, 1969b) and increase in nucleoplasmic RNA polymerase activity (Carlsson *et al.*, 1973)

5. Formation of a nucleolus (Harris, 1967; Sidebottom and Harris, 1969), increase in nucleolar RNA polymerase activity, and initiation of ribosomal RNA synthesis (Harris *et al.*, 1969b; Carlsson *et al.*, 1973)

6. Initiation of chick specific protein synthesis including chick nucleolar antigens, surface receptors and antigens, and enzymes, etc. (Harris and Cook, 1969; Harris *et al.*, 1969b; Cook, 1970; Ringertz *et al.*, 1971; Dendy and Harris, 1973; Kit *et al.*, 1974)

7. DNA synthesis (Harris, 1965; Bolund *et al.*, 1969a; Johnson and Harris, 1969b; Johnson and Mullinger, 1975)

8. Formation of mononucleate synkaryons which rapidly eliminate most of the chick chromosomes (Schwartz *et al.*, 1971; Kao, 1973; Trisler and Coon, 1973; Boyd and Harris, 1973; Kit *et al.*, 1974; Leung *et al.*, 1975)

In the following sections some of these steps are considered in more detail since they illustrate how heterokaryons can be used to analyze nucleocytoplasmic interactions in the control of nuclear activities.

a. CHROMATIN DISPERSION AND NUCLEAR GROWTH

At an early stage of the reactivation of a chick erythrocyte nucleus in HeLa cells, the DNA protein complex undergoes drastic physicochemical changes which manifest themselves in an increased binding of basic dyes and an increased susceptibility to thermal denaturation and acid hydrolysis (Bolund *et al.*, 1969a). These changes are paralleled by a marked dispersion of the tightly condensed erythrocyte chromatin (Harris, 1967; Dupuy-Coin *et al.*, 1976; Lindberg, 1974) which can easily be observed by UV microphotography (Fig. VII-1) or electron microscopy (Fig. VII-2). At the same time there is an increase in nuclear size which is associated with an increase in nuclear dry mass. These changes occur even if the initiation of RNA synthesis in the chick nucleus is inhibited by UV irradiation of the erythrocytes before fusion. This indicates that the growth of the chick nucleus postfusion is a passive "swelling" process which is independent of chick RNA synthesis (Bolund *et al.*, 1969b). Most of the increase in dry mass appears to be due to an accumulation of protein. Immunological (Ringertz *et al.*, 1971), cytochemical (Bolund *et al.*, 1969b; Appels *et al.*, 1974a), and biochemical (Goto and Ringertz, 1974; Appels *et al.*, 1974b) data obtained with HeLa–chick erythrocyte heterokaryons suggest that a considerable proportion, if not all, of the proteins initially accumulating in the erythrocyte nucleus are human proteins. The migration of human macromoleucles into the chick erythrocyte nucleus during the reactivation process has been followed by means of indirect immune fluorescence studies using species-specific antibodies to human nuclear antigens (Ringertz *et al.*, 1971). Immediately after fusion, chick erythrocyte nuclei in heterokaryons do not bind antibodies directed against human nuclei. As the chick nuclei enlarge, however, human antigens appear in their nucleoplasms, nuclear envelopes, and nucleoli.

An important point about the accumulation of human antigens in the growing erythrocyte nucleus is that the antigens are exclusively nuclear. Antigens characteristic of human cytoplasms do not enter. Therefore, the growth of chick erythrocyte nuclei in heterokaryons made with HeLa cells is not due to a random uptake of material from the surrounding cytoplasm but to a *selective concentration of human nucleospecific macromolecules.* The individual human antigens show exactly the same distribution in the fully reactivated erythrocyte nucleus as they do in the HeLa nucleus. Human nuclear membrane, nucleoplasmic and

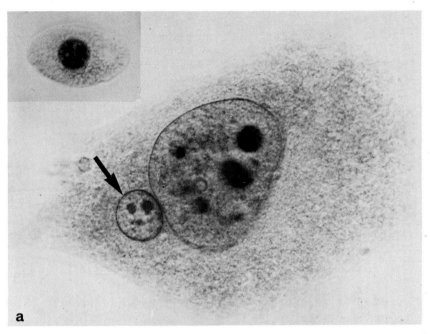

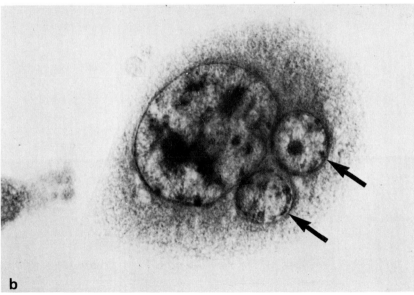

Fig. VII-1. UV microphotographs (265 nm) of two heterokaryons (a and b) formed by fusing chick erythrocytes with human HeLa cells. The heterokaryons were photographed 73 h postfusion. At this stage the erythrocyte nuclei (arrows) have undergone reactivation. Insert shows unfused erythrocyte with condensed, inactive nucleus (Ege *et al.*, 1975a).

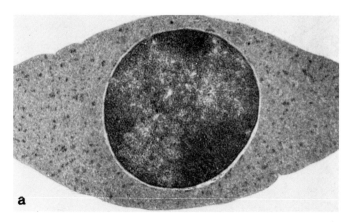

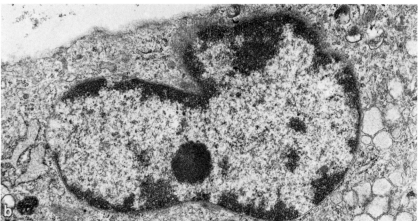

Fig. VII-2. Electron micrographs of (a) a 13-day embryonic erythrocyte with a condensed inactive nucleus and (b) a reactivated chick erythrocyte nucleus containing a well developed nucleolus (arrow). The heterokaryon was fixed 48 h postfusion (Dupuy-Coin *et al.*, 1976).

nucleolar antigens, thus, occur in the equivalent structures of the chick nuclei. Using heterokaryons made by fusing chick erythrocytes with SV40 virus-transformed hamster cells, Rosenqvist *et al.* (1975) found that an SV40 specified nuclear antigen, the T antigen, is also concentrated by erythrocyte nuclei as they enlarge (Fig. VII-3a). This observation is of special interest since some information exists about the chemical nature of this antigen (Gilden *et al.*, 1965; DelVillano and Defendi, 1973) and because the gene specifying this antigen is believed to be on a small piece of virus nucleic acid integrated into the hamster chromosomes.

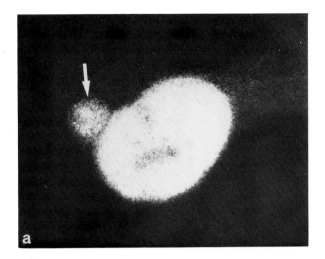

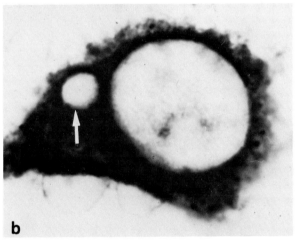

Fig. VII-3. (a) Migration of SV40 T antigen into a chick erythrocyte nucleus (arrow) reactivated in a heterokaryon made by fusing an SV40 transformed hamster cell with a chick erythrocyte (Rosenqvist *et al.,* 1975). (b) Cytochemical assay for lactate dehydrogenase (LDH) showing a strong cytoplasmic reaction but no nuclear reactions in a HeLa + chick erythrocyte heterokaryon. Arrow indicates the position of a reactivated chick erythrocyte nucleus (Appels *et al.,* 1975a).

Cytochemical measurements, using staining methods for total protein or for histone show that most of the dry mass increase in the reactivating erythrocyte nuclei is due to protein accumulation. While there is some increase in the histone content, the major class of proteins accumulating in the reactivating chick nuclei is the nonhistone type (Appels *et al.*, 1974a, 1975a). Attempts have been made to define the types of nonhistone proteins that increase during the reactivation process. With cytochemical methods for RNA polymerase, Carlsson *et al.* (1973) found that both nucleoplasmic and nucleolar types of RNA polymerase increase during the reactivation of chick erythrocyte nuclei in rat epithelial cells. Data obtained by Darzynkiewicz and Chelmicka-Szorc (1972) and Paterson *et al.* (1974b) show that *human* DNA repair enzymes appear in erythrocyte nuclei reactivated in human + chick heterokaryons. Moreover, studies by Dubbs and Kit (1976), using temperature sensitive mutants of Chinese hamster cells, demonstrate that the initiation of DNA synthesis in the erythrocyte nuclei depends on factors supplied by the hamster cells. All this indicates that during the first days of reactivation many important functions in the chick nucleus are controlled by mammalian nonhistone proteins and enzymes which migrate via the cytoplasm of the heterokaryon into the chick nucleus.

Cytochemical studies on the distribution of several enzymes with an exclusively cytoplasmic localization support the immunological data which show that the uptake of protein by the erythrocyte nuclei is a selective process. Neither mammalian nuclei nor reactivated erythrocyte nuclei showed any lactate, malate, or succinic dehydrogenase activity (Fig. VII-3b), whereas the cytoplasms of the heterokaryons were strongly positive for the enzymes (Appels *et al.*, 1975b). These enzymes, therefore, appear to be selectively excluded from the nuclear compartments.

It has been possible to analyze the reactivation of erythrocyte nuclei biochemically by isolating nuclei from mammalian + chick erythrocyte heterokaryons (Fig. VII-4) and then separating the chick from the mammalian nuclei by sucrose gradient centrifugation (Fig. VII-5) (Goto and Ringertz, 1974). Using this approach, it was found that prelabeled mammalian protein accumulated in the erythrocyte nuclei, whereas mammalian RNA made before fusion was excluded (Goto and Ringertz, 1974). A more detailed analysis (Appels *et al.*, 1974b) showed that erythrocyte nuclei preferentially took up mammalian nonhistone protein relative to mammalian histones. Moreover, SDS-gel analysis revealed that nonhistone proteins with molecular weights less than 50,000 Daltons selectively accumulated relative to larger nonhistone proteins. Despite the preference for nonhistone proteins at least one mammalian histone also entered the erythrocyte nuclei: the human F1 histone. At the same time as F1 histone concentration increased, the f2c fraction, specific for avian erythrocytes, was lost.

The results obtained by immunological, cytochemical, and biochemical analysis of the reactivating erythrocyte nuclei are summarized in

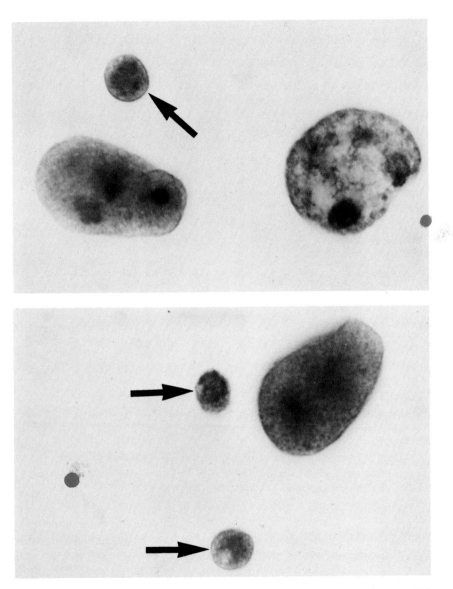

Fig. VII-4. HeLa and chick erythrocyte nuclei (arrows) isolated from heterokaryons 15 h postfusion (Appels *et al.*, 1975a).

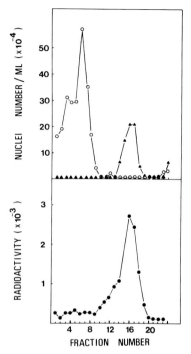

Fig. VII-5. Gradient centrifugation to separate L-cell nuclei and chick erythrocyte nuclei isolated from heterokaryons 15 h postfusion. Top graph shows the first centrifuge run. The right peak is the erythrocyte peak, the left peak contains L-cell nuclei. Lower graph shows the result of a recentrifugation of the erythrocyte nuclei on a second gradient (Goto and Ringertz, 1974).

Table VII-1. The general conclusions that can be drawn from these data are that (1) the erythrocyte nucleus grows by selectively taking up and concentrating mammalian "nucleospecific" proteins from the surrounding cytoplasm; (2) enzymes and antigens characteristic of the mammalian cytoplasm and mammalian cytoplasmic RNA are selectively excluded from the enlarging erythrocyte nuclei, and (3) some proteins of the chick erythrocyte chromatin are lost.

Nuclear growth and changes in nuclear nonhistone proteins have been observed in many other biological systems where relatively inactive nuclei are transformed into a more active state (Baserga and Stein, 1971; Auer and Zetterberg, 1972; Baserga, 1974). At least in some of these cases, most, if not all, of the initial increase in nuclear dry mass increase must be due to an influx of preformed macromolecules from the cytoplasm. Nuclei from differentiated *Xenopus* cells, such as brain cells and erythrocytes, rapidly swell up to 60 times their original size

TABLE VII-1

Uptake of Mammalian Macromolecules and Loss of Chick Nuclear Components during Reactivation of Chick Erythrocyte Nuclei in Mammalian + Chick Heterokaryons

	References
(a) Taken up and selectively concentrated	
Antigens	
Human nucleoplasmic antigen	Ringertz *et al.*, 1971
Human nucleolar antigen	Ringertz *et al.*, 1971
Virus SV40 T antigen	Rosenqvist *et al.*, 1975
Proteins	
RNA polymerase[a]	Carlsson *et al.*, 1973
DNA repair enzymes (human)	Darzynkiewicz, 1971; Darzynkiewicz and Chelmicka-Szorc, 1972; Paterson *et al.*, 1974b; Paterson and Lohman, 1975
Many classes of human nonhistone nuclear protein	Appels *et al.*, 1974a,b
f1 histone (human)	Appels *et al.*, 1974b
(b) Macromolecules excluded	
Antigens	
Human cytoplasmic antigen	Ringertz *et al.*, 1971
Proteins	
Lactate dehydrogenase	Appels *et al.*, 1975a
Malate dehydrogenase	Appels *et al.*, 1975a
Succinic dehydrogenase	Appels *et al.*, 1975a
RNA	
Human cytoplasmic RNA	Goto and Ringertz, 1974
(c) Macromolecules lost from chick nucleus	
f2c histone	Appels *et al.*, 1974b

[a] Species of origin not determined.

and take up a spectrum of proteins from the cytoplasm when transplanted into normal or enucleated *Xenopus* eggs (Graham *et al.*, 1966; Arms, 1968; Merriam, 1969; Gurdon, 1974). Nuclear concentration of certain types of proteins has also been observed in microinjection experiments with oocytes (Bonner, 1975a,b) and other types of cells (Paine and Feldherr, 1972).

b. RNA SYNTHESIS

Shortly after fusion of chick erythrocytes with HeLa cells, the erythrocyte nucleus resumes RNA synthesis (Harris, 1967). As it grows in size and its chromatin becomes more dispersed, the rate of labeling with ³H-uridine increases. Autoradiographic data suggest that the rate of RNA synthesis increases in direct relation to nuclear growth (Harris, 1967, Carlsson *et al.*, 1973). The RNA produced in the erythrocyte nucleus during the first days after fusion has a high molecular weight and is polydisperse on sedimentation (Harris et al., 1969b). This RNA is not transported to the cytoplasm and appears to be similar to the HnRNA found in other cell systems. Ribosomal RNA (rRNA) cannot be detected until about the third day after fusion. The appearance of the rRNA fractions coincides with the development of nucleoli. Transport to the cytoplasm of RNA synthesized in the chick nucleus and new synthesis of chick proteins (e.g., surface antigens) cannot be detected before nucleoli become visible (Harris et al., 1969b; Sidebottom and Harris, 1969) and rRNA is synthesized. The continued synthesis of chick proteins in heterokaryons containing a fully reactivated erythrocyte nucleus also appears to be dependent upon nucleolar function. If the single nucleolus of an erythrocyte nucleus within a heterokaryon is irradiated with a UV microbeam, the synthesis of chick proteins ceases even though nucleoplasmic RNA synthesis continues in the erythrocyte nucleus (Deak et al., 1972). The late activation of nucleolar RNA synthesis observed by Harris and co-workers is paralleled by a late appearance of nucleolar RNA polymerase activity (Carlsson et al., 1973). Whether the RNA polymerases transcribing the chick DNA are endogenous chick enzymes or mammalian polymerases taken up from the cytoplasm is not known. Both alternatives seem plausible since chick erythrocyte nuclei have been shown to contain some endogenous RNA polymerases (Scheintaub and Fiel, 1973).

c. DNA SYNTHESIS

Some 10–15 h after fusion the majority of the erythrocyte nuclei in HeLa + chick erythrocyte heterokaryons have increased markedly in volume and dry mass but still contain the G_1 amount of DNA. At about this time some of the chick erythrocyte nuclei begin to incorporate thymidine (Harris, 1965) and by 48 h after fusion many erythrocyte nuclei contain increased amounts of DNA. Microspectrophotometric measurements of Feulgen stained cells show that at this stage many of the nuclei contain the premitotic amount of DNA (Bolund et al., 1969a). Studies by Johnson and Mullinger (1975) show, however, that a great number of erythrocyte nuclei are lagging behind the HeLa nuclei with respect to their DNA replication cycle. As a result, many erythrocyte nuclei have not completed their DNA synthesis when the HeLa nucleus enters mitosis and consequently they undergo PCC and show the fragmented appearance of S phase PC chromosomes.

There is no close correlation between the extent of nuclear growth and the initiation of DNA synthesis, but at least some enlargement seems to be necessary before DNA synthesis can be initiated in the erythrocyte nuclei. It is not known whether the DNA polymerase is of chick or mammalian origin, nor whether it is present in the erythrocyte nucleus or taken up from the surrounding cytoplasm. In either case, data obtained by Dubbs and Kit (1976) indicate that the erythrocyte nucleus requires factors from the surrounding cytoplasm for the initiation of DNA synthesis. In heterokaryons where the mammalian cell was a Chinese hamster mutant cell which was defective in the initiation of DNA synthesis at a certain temperature (nonpermissive or restrictive temperature; see Chapter IX), the erythrocyte nuclei enlarged but failed to synthesize DNA. When the temperature was then shifted to one where the Chinese hamster nucleus synthesized DNA (permissive temperature) the erythrocyte nucleus also replicated. These results indicate that DNA synthesis in the erythrocyte nucleus depends on the state of the surrounding cytoplasm, but they do not provide any specific information about the chemical nature of the factors required. Such information will be evident, though, if the genetic defect of the temperature sensitive mutant cells can be identified.

2. Macrophage Heterokaryons

Macrophages obtained from the peritoneal fluid of rabbits, mice, or rats can be separated from other cells by exploiting their ability to attach rapidly and firmly to glass surfaces. Macrophages do not normally synthesize DNA, but appear to be arrested in the G_1 phase as are many other differentiated cells. Although their kidney-shaped cell nuclei are smaller than those of more active nuclei from other tissues, they nevertheless do have small nucleoli and synthesize RNA. When macrophages are fused with more active cells such as HeLa cells (Harris *et al.*, 1966) or melanoma cells (Gordon and Cohn, 1970) the macrophage nuclei become swollen, (Fig. VII-6) the nucleoli become more prominent, and RNA synthesis accelerates 4 to 10 times. The macrophage nuclei are also triggered to resume DNA synthesis, and some heterokaryons subsequently enter mitosis.

The reactivation of macrophage nuclei in macrophage + melanocyte heterokaryons appears to be a more rapid process than the reactivation of chick erythrocyte nuclei in heterokaryons made with HeLa tumor cells or melanoma cells (Gordon and Cohn, 1971a). Macrophage nuclei swell and show increased RNA synthesis within an hour, and a wave of DNA synthesis starts 3 h after fusion. The explanation for the more rapid reactivation may well be that macrophage nuclei are less condensed and less completely inactivated than are erythrocyte nuclei.

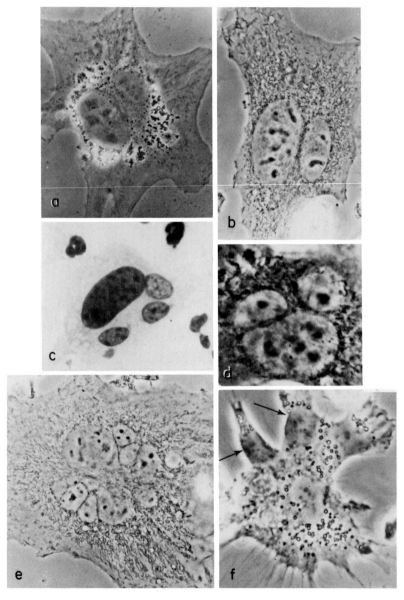

Fig. VII-6. Heterokaryons obtained by fusing mouse macrophages with undifferentiated melanoma cells. (a) A 1:1 heterokaryon containing a large melanoma nucleus and a small macrophage nucleus. (b) A 1:1 heterokaryon nucleus 2 days after fusion. The macrophage nucleus on the right is now enlarged and contains a prominent nucleolus. (c) A 3:1 heterokaryon. (d) A 2:1 heterokaryon 1 day after fusion. The macrophage nuclei have devel-

Gordon and Cohn (1971a,b) examined the reactivation of macrophage nuclei to see if it is dependent upon RNA or protein synthesis specified by the macrophage or the melanoma nuclei. Their results, using inhibitors of RNA and protein synthesis, showed that the reactivation of DNA synthesis in the macrophage nuclei is dependent upon RNA synthesized by the *melanoma* nucleus. Inhibition of RNA synthesis in the macrophage nucleus did not prevent the initiation of macrophage DNA synthesis. Treatment of the melanoma parent with inhibitors of protein synthesis before fusion also effectively prevented the initiation of DNA synthesis in the macrophage nuclei after fusion. And again, prevention of protein synthesis in the macrophage parent played no role in the initiation of macrophage DNA synthesis. These results suggest that the melanoma cells provide the macrophage nuclei with products which trigger their DNA synthesis.

3. Gamete Heterokaryons

Sendai virus has been used to fuse spermatozoa (Brackett *et al.*, 1971; Sawicki and Koprowski, 1971; Gledhill *et al.*, 1972; Gabara *et al.*, 1973; Bendich *et al.*, 1974) and eggs (Graham, 1969; Baranska and Koprowski, 1970; Lin *et al.*, 1973) with somatic cells to generate heterokaryons. In another case the virus has been used to overcome barriers to fertilization *in vitro* (Ericsson *et al.*, 1971).

Following spontaneous (Bendich *et al.*, 1974) or Sendai virus-induced penetration of sperm into somatic cells, the sperm heads tend to retain their characteristic form and to remain in an inactive state in the cytoplasm (Fig. VII-7), perhaps because they are confined to phagosomes. Some sperm heads remain visible in heterokaryons for as long as 1 month after fusion (Sawicki and Koprowski, 1971). In other cases, however, the chromatin of the spermatozoa has been dispersed (Zelenin *et al.*, 1974) and DNA synthesis has been initiated (Gledhill *et al.*, 1972). Furthermore, there are indications that some heterokaryons formed by the fusion of mouse spermatozoa and diploid Chinese hamster cells express mouse specific antigens (Bendich *et al.*, 1974). Convincing proof that sperm DNA is transcribed and sperm genes are expressed is, however, lacking. On the contrary, results obtained by Phillips *et al.*

---

oped nucleolei. (e) A large heterokaryon containing one melanoma nucleus and many macrophage nuclei. Nuclei accumulate in the center of the giant cell. (f) A macrophage homokaryon after 5 h exposure to Colcemid. The nuclei (arrows) have failed to accumulate in the center of the cell and are located in the cell periphery in stubby pseudopods. Also the postfusion reorganization of the lipid droplets and other cytoplasmic organelles is inhibited by Colcemid (Gordon and Cohn, 1970).

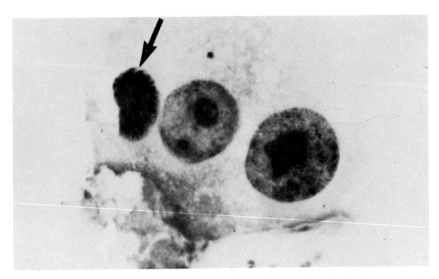

Fig. VII-7. Polykaryon formed by the fusion of two HeLa cells and a bovine spermato-
zoan. The sperm nucleus (arrow) has undergone some swelling but retains its characteristic form (Johnson *et al.*, 1970).

(1976) indicate that hybrids arising in mixtures of sperm and mutant
mammalian cells originate from the fusion of somatic cells contaminating the sperm preparations. In fusions of synchronized populations
of mitotic somatic cells with sperm, Johnson *et al.* (1970) found that
some sperm nuclei undergo premature chromosome condensation
(PCC). The frequency at which sperm heads are activated or induced to
undergo PCC is very low if one considers the relative ease with which
large numbers of spermatozoa can be introduced into somatic cells.
There are indications, however, that haploid nuclei at a somewhat earlier stage of spermiogenesis may be easier to reactivate. Thus, Nyormoi
et al. (1973b) found that proliferating hybrid cells could be produced by
fusing rat spermatids with cells of an established mouse cell line (see
Chapter X).

A somewhat different application of the virus-induced fusion technique is the promotion of *in vitro* fertilization. In order to fertilize eggs
in vitro, sperm has to undergo a physiological change known as *capaci-
tation.* While the exact nature of the induced change is not known, this
phenomenon is normally induced in mammals by storing sperm within
the female genital tract. Ericsson *et al.* (1971) have found, however, that
if Sendai virus is present uncapacitated sperm can be used for *in vitro*
fertilization. Eggs thus fertilized form polar bodies and undergo cleavage.

Attempts have been made to fuse unfertilized and fertilized mouse eggs with somatic cells (Graham, 1969; Baranska and Koprowski, 1970; Lin *et al.*, 1973). At first this type of fusion met with limited success both because many eggs lysed when exposed to UV-inactivated Sendai virus and because the technique of culturing eggs was poorly developed. Some of these problems have been overcome by using chemically inactivated virus and improved culture conditions (Graham, 1971, 1974). Baranska and Koprowski (1970) noted that when unfertilized mouse eggs were fused with a variety of mouse, hamster, monkey, or human cells, development proceeded to the early morula stage.

Fertilized eggs at different cleavage stages have also been fused with somatic cells. Thus, mouse blastomeres from 2–8 cell embryos have been fused with established mouse cell lines (Bernstein and Mukherjee, 1972). At the 2-cell stage mouse A9 nuclei in A9 + blastomere heterokaryons showed a reduction in the rate of RNA synthesis. On the other hand the blastomere nuclei, which normally show a low level of transcription at this stage, had an accelerated RNA synthesis. In heterokaryons using blastomeres from the 4-cell stage, a stage where blastomere nuclei normally synthesize RNA 5–10 times more rapidly than at the 2-cell stage, no depression in A9 RNA synthesis could be detected. It appears, therefore, that the RNA synthesis of the A9 nuclei is controlled by the state of the blastomere cytoplasm.

Another interesting result was obtained when cleaving eggs were exposed to Sendai virus. The blastomeres fused with each other (Graham, 1971; Bernstein and Mukherjee, 1972), and continued to cleave until they produced tetraploid blastocysts. On transfer to foster mothers these blastocysts gave rise to large trophoblast masses.

C. REGULATION OF NUCLEIC ACID SYNTHESIS IN HETEROKARYONS

Some of the most interesting studies using heterokaryons have examined changes in DNA and RNA synthesis among the participating nuclei. Table VII-2 summarizes results obtained with several different types of heterokaryons. Three major conclusions can be drawn from this work:

1. The active partner in the fusion almost always stimulates the nucleus of the inactive partner to synthesize more RNA and/or DNA. Reported exceptions to this rule are (a) fusions of somatic cells with 2-cell blastomeres (i.e., a situation where the inactive partner contributes 50 times more cytoplasm than does the active partner); (b) heterokaryons containing a very high ratio of inactive to active nuclei (in these cells

TABLE VII-2

RNA and DNA Synthesis in Some Heterokaryons and Their Parental Cells[a]

Parental cell type	RNA	DNA
HeLa (human)	+	+
A9 "fibroblast" (mouse)	+	+
Established fibroblast (monkey)	+	+
Melanocyte (mouse)	+	+
Myoblast (rat)	+	+
Myotube (rat)	+	−
Macrophage (rabbit)	+	−
Lymphocyte (rat)	+	−
Neuron (mouse)	+	−
Erythrocyte (chick)	−	−

Heterokaryon		RNA synthesis Nucleus		DNA synthesis Nucleus		
Cell I	Cell II	I	II	I	II	References
HeLa	Macrophage	+	*	+	*	Harris, 1965
HeLa	Lymphocyte	+	*	+	*	Harris, 1965
HeLa	Erythrocyte	+	*	+	*	Harris, 1965; Harris *et al.*, 1966
A9	Erythrocyte	+	*	+	*	Harris *et al.*, 1969b
Macrophage	Lymphocyte	+	+	−	−	Harris *et al.*, 1966
Macrophage	Erythrocyte	+	*	−	−	Harris *et al.*, 1966
Myoblast	Erythrocyte	+	*	+	*	Carlsson *et al.*, 1970
Myotube	Erythrocyte	+	*	−	−	Carlsson *et al.*, 1970
Melanocyte	Macrophage	+	*	+	*	Gordon and Cohn, 1971a,b
Fibroblast	Neuron	?	?	+	*	Jacobsson, 1968

[a] Explanation of symbols: +, synthesis; −, no synthesis; *, induced or accelerated synthesis; ?, not tested.

there may be competition among the nuclei for macromolecules specifically involved in transcription); (c) fusions involving senescent cells as one of the parental cells (in this case the senescent cell may contribute inhibitory factors).

2. The inactive nucleus is stimulated to attain the same level of activity as the active nucleus. Thus, where the active cell is synthesizing RNA, but not DNA (macrophages, myotubes), the inactive nucleus is only stimulated to make RNA. If, on the other hand, the active partner is synthesizing both types of nucleic acid, then the inactive nucleus is also stimulated to synthesize both types of nucleic acid.

3. The signals which trigger RNA and DNA synthesis are not species specific. Active cells of mouse origin can stimulate inactive human, rabbit, or chick as well as mouse nuclei.

The nature of the signals which regulate nuclear activities are not yet known, but studies of erythrocyte nuclei undergoing reactivation in heterokaryons provide some general clues, as already noted.

The loss of the erythrocyte specific f2c histone may be an important step, but its precise relationship to the reactivation process is difficult to evaluate. The mechanism by which this histone is lost could be by diffusion out of the erythrocyte nucleus or by degradation. In this respect it is interesting that Darzynkiewicz *et al.* (1974a,b) found that treating HeLa–erythrocyte heterokaryons with specific inhibitors of proteolytic enzymes inhibits the reactivation reaction, as measured by morphology and transcription from chick DNA. It is conceivable that synthesis of some chick proteins early in the reactivation could be a necessary prerequisite for later steps in the reactivation process. This does not seem to be the case, though, because experiments with UV-irradiated erythrocyte nuclei showed that large reductions in the transcription from chick DNA do not affect dispersion of chromatin, nucleoprotein changes, and protein accumulation (Bolund *et al.*, 1969b).

Migration of proteins from the cytoplasm into the erythrocyte nucleus, however, seems to be a prerequisite for the induction of RNA synthesis and possibly also for the induction of appropriate chromatin changes. Since an increase in nonhistone nuclear proteins occurs very rapidly after fusion, it seems likely that some of these proteins function as positive signals in the reactivation. They may do so by contributing important enzyme functions in which the erythrocyte nucleus is deficient or by making chromatin available for transcription. Whatever the chemical nature of these proteins may turn out to be they appear to be in relatively short supply in the cytoplasm surrounding the erythrocyte nuclei. The observation that the rate of reactivation as measured by the rate of transcription, nuclear enlargement, uptake of mammalian nucleospecific antigens, or increase in RNA polymerase activity is slower when more than one chick nucleus is present (Carlsson *et al.*, 1973; Toister and Loyter, 1973) indicates that rate limiting factors for the reactivation process exist in the cytoplasm of the mammalian cell and that all nuclei in the polykaryon compete for the same pool of molecules.

The fact that mammalian proteins seem to function well in avian nuclei may at first seem somewhat remarkable in view of the phylogenetic disparity involved. It is, however, consistent with several other lines of evidence. Both histones (Ohlson and Busch, 1974) and nonhistone nuclear proteins (Elgin and Bonner, 1970; Jost *et al.*, 1975) are structurally similar over a wide range of species and therefore seem to have undergone little change during evolution. It is logical therefore that nuclear proteins of one species can substitute functionally for those of another species. Direct experimental evidence for this is provided by the rapidly growing interspecific hybrid cells discussed in later chapters of this book and biochemical observations which show that RNA polymerases of one species are capable of transcribing DNA of a variety of other species. The possibility that some regulatory proteins may be species-specific should, however, be kept in mind.

D. PHENOTYPIC EXPRESSION IN HETEROKARYONS

1. Gene Expression Patterns

So far only a few attempts have been made to analyze phenotypic expression in heterokaryons (Table VII-3). In the following paragraphs some of the patterns of gene expression which have been observed in heterokaryons are discussed. For a more detailed discussion of gene regulation in hybrid cells in general, Chapter XI should be consulted.

Many heterokaryons express properties characteristic of both parental cells (*coexpression* or *codominance*). Thus, Watkins and Grace (1967) found surface antigens characteristic of both parental cells in heterokaryons made by fusing human HeLa cells with Ehrlich cells. Coexpression was also observed by Carlsson *et al.* (1974a) in studies of hybrid myotubes; both rat and chick myosin were synthesized in myotubes formed by spontaneous fusion of rat and chick myoblasts. But in other types of heterokaryons, properties characteristic of one parental cell are *extinguished* in the heterokaryons. Extinction of differentiated properties was observed by Gordon and Cohn (1970, 1971c) in macrophage + melanoma heterokaryons, where the macrophage marker properties disappeared. In heterokaryons formed by fusing rat hepatoma cells with rat epithelial cells, Thompson and Gelehrter (1971) noted that the inducibility of tyrosine aminotransferase (TAT), a property characteristic of liver differentiation, disappeared in the heterokaryon. These examples illustrate the two most important patterns of phenotypic expression, coexpression and extinction of properties characteristic of the parental cells.

Two types of heterokaryons have been examined in much greater detail than others: chick erythrocyte and macrophage heterokaryons. Because the pattern of phenotypic expression has been best characterized in these heterokaryons, they are presented here in more detail.

TABLE VII-3

Phenotypic Expression in Heterokaryons

Marker	Parental cells	References
Coexpression		
Antigens, mouse + human	Ehrlich (mouse) + HeLa (human)	Watkins and Grace, 1967
Myosin, chick + rat	Myoblast (chick) + myoblast (rat)	Carlsson *et al.*, 1974a
Hemoglobin, frog + tadpole	Erythroblast (frog) + erythroblast (tadpole)	Rosenberg, 1972
Extinction		
Phagocytosis	Macrophage (mouse) + melanoma (mouse)	Gordon and Cohn, 1970, 1971a,b,c
Lysosomal enzymes	Macrophage (mouse) + melanoma (mouse)	Gordon and Cohn, 1970, 1971a,b,c
Membrane ATPase	Macrophage (mouse) + melanoma (mouse)	Gordon and Cohn, 1970, 1971a,b,c
Hemoglobin	Erythroblast (chick) + A9 (mouse)	Harris, 1970b
Inducibility of tyrosine aminotransferase (TAT)	Hepatoma (rat) + epithelial cells (mouse)	Thompson and Gelehrter, 1971
Failure to reactivate synthesis		
Hemoglobin (chick)	Erythrocyte (chick) + A9 (mouse)	Harris, 1970b
Myosin (chick)	Erythrocyte (chick) + myotube (rat)	Carlsson, *et al.*, 1974a
Serum albumin (chick)	Erythrocyte (chick) + hepatoma (rat)	Szpirer, 1974

2. Erythrocyte Heterokaryons

Heterokaryons containing fully reactivated chick erythrocyte nuclei with well developed nucleoli express many different chick properties. The spectrum of chick macromolecules identified in HeLa or mouse L cell heterokaryons includes surface antigens (Harris *et al.*, 1969b), surface receptors (Dendy and Harris, 1973), nucleolus specific antigens (Ringertz *et al.*, 1971), and at least three different enzymes involved in nucleotide synthesis: hypoxanthine guanine phosphoribosyltransferase (Harris and Cook, 1969), adenosine phosphoribosyltransferase (Clements, 1972), and cytoplasmic thymidine kinase (Kit *et al.*, 1974). Hemoglobin synthesis has not been detected in heterokaryons made with mature erythrocytes of late embryonic stages, but, if immature erythroid chick cells are used in fusions with undifferentiated mouse cells (Harris, 1970b; Davis and Harris, 1975), hamster cells (Davis and Harris, 1975), or avian fibroblasts (Alter and Ingram, 1975), hemoglobin synthesis continues at a declining rate for 1–2 days after fusion. Heterokaryons produced by fusing two types of immature, nucleated erythrocytes, namely tadpole and frog, synthesize both tadpole and frog hemoglobins (Rosenberg, 1972). These results suggest that non-erythroid cytoplasm may contain factors which switch off hemoglobin synthesis at the transcriptional or translational level.

Attempts have been made to study how the cytoplasm influences the phenotype expressed by chick erythrocyte nuclei by reactivating such nuclei in the cytoplasms of rat and mouse cells representing different lines of cell differentiation. Chick erythrocyte nuclei were introduced into rat myoblasts and rat myotubes (Fig. VII-8) at times when rat myosin synthesis was going on (Carlsson *et al.*, 1970). No significant chick myosin synthesis could be demonstrated in myotubes although the chick erythrocyte nuclei developed nucleoli, produced RNA, and expressed chick antigens. A small number of what may have been rat + chick synkaryons did, however, react with antibodies to chick myosin, but the very small number of these cells present did not allow their hybrid nature to be conclusively established. In control experiments, heterokaryons made by spontaneous fusion of chick myoblasts with rat myoblasts synthesized both chick and rat myosin. It appears, therefore, that although practically all chick erythrocyte nuclei underwent reactivation in rat myoblasts or myotubes very few if any were reprogrammed to specify the synthesis of chick myosin (Carlsson *et al.*, 1974a). Similarly Szpirer (1974) failed to detect chick serum albumin synthesis in heterokaryons obtained by fusing albumin producing rat hepatoma cells with chick erythrocytes.

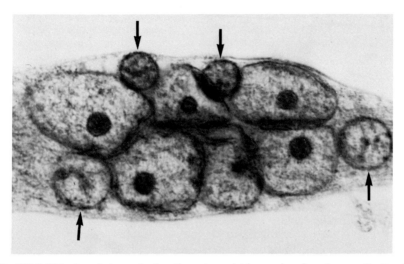

Fig. VII-8. UV microphotograph of erythrocyte nuclei (arrows) undergoing reactivation in hybrid myotubes formed by virus induced fusion of chick erythrocytes with rat myoblasts (S.-A. Carlsson, R. E. Savage, and N. R. Ringertz, unpublished work).

3. Macrophage Heterokaryons

Macrophages possess several well-defined marker properties which make them useful as parental cells for cell fusion experiments that study mechanisms regulating phenotypic expression. Gordon and Cohn (1970, 1971a,b,c; Gordon, 1973) have studied the following three markers of macrophage differentiation in heterokaryons: (1) the ability to bind and ingest antibody-coated red cells, dextran sulfate, and other macromolecules by phagocytosis; (2) the presence of a Mg^{2+}-activated membrane bound ATPase; (3) the presence of lysosomes and lysosomal enzymes (acid phosphatase).

Heterokaryons formed by fusing macrophages with melanoma cells lost their phagocytic activity after a few days (Fig. VII-9). The rate at which this property was lost was a function of the proportion of macrophage to melanoma nuclei in individual heterokaryons: the higher the proportion of melanoma nuclei, the more rapid the loss of phagocytic activity. The loss appeared to be due to a masking of receptors on the cell surface. Macrophages normally carry membrane receptors which bind antibody-coated particles. They can be demonstrated by mixing macrophages with sheep red cells coated with 7 S antibody. Such re-

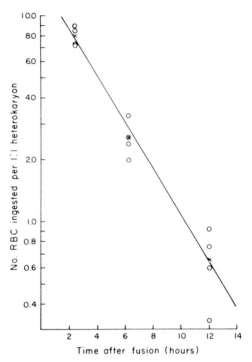

Fig. VII-9. Extinction of phagocytic function in 1:1 macrophage + melanoma cell hetero-karyons. Phagocytosis was measured by allowing the heterokaryons to ingest sensitized sheep red cells (Gordon and Cohn, 1970).

ceptors, present on the heterokaryons immediately after fusion, disap-peared after a short time (Gordon and Cohn, 1970) but could be exposed again by treating the heterokaryons with trypsin. If subjected to further cultivation in the absence of trypsin, the receptors again became masked. Inactivation of melanoma nuclei before fusion preserved macrophage receptors on the surface of the heterokaryons, whereas in-activation of macrophage nuclei was without effect.

Moreover the extinction of phagocytic function in macrophage het-erokaryons was clearly related to the state of activity of the nonmacro-phage parent (Gordon and Cohn, 1971c). When this cell was a rapidly growing Ehrlich tumor cell the receptors disappeared much faster than when it was a normal primary chick fibroblast. In macrophage-chick erythrocyte heterokaryons, the macrophage receptors could be detected for at least 7 days. This suggests that extinction of the macrophage markers only occurs when the nonmacrophage parent is an active cell.

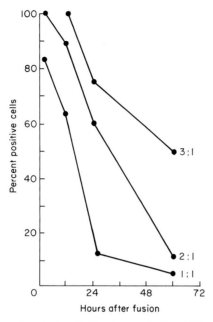

Fig. VII-10. The loss of cytochemically demonstrable ATPase activity in macrophage + melanoma heterokaryons containing 1, 2, or 3 macrophage nuclei and one melanoma nucleus (Gordon and Cohn, 1970).

   Initially macrophages had about 30 times more surface ATPase activity than melanocytes. After fusion, the ATPase activity was lost in heterokaryons at a rate determined by the ratio of macrophage to melanocyte nuclei although in the same experiments homokaryons containing only macrophage nuclei retained their ATPase activity (Fig. VII-10). Other macrophage markers such as lysosomes and lipid droplets also disappeared from heterokaryons but were retained by macrophage homokaryons.

4. Viral Studies with Heterokaryons

   The preceding sections illustrate the use of heterokaryons as experimental systems for studying nucleocytoplasmic interactions controlling nuclear activity and phenotypic expression. In addition to these applications heterokaryons have been used to analyze the interplay between nucleus and cytoplasm in virus-infected or virus-transformed cells. Synkaryons have been used for these studies as well, so for expediency both types of hybrid cells are considered together in Chapter XV.

E. GENE COMPLEMENTATION ANALYSIS WITH HETEROKARYONS

Studies of heterokaryons made with fibroblasts from patients suffering from xeroderma pigmentosum (de Weerd-Kastelein *et al.*, 1972, 1973; Bootsma, 1974; Gianelli *et al.*, 1973; Gianelli and Croll, 1971; Gianelli and Pawsey, 1974; Paterson *et al.*, 1974a), Tay-Sachs' and Sandhoff's disease (Galjaard *et al.*, 1974a; Migeon *et al.*, 1974) GM_1-gangliosidosis (Galjaard *et al.*, 1975) maple syrup disease (Lyons *et al.*, 1973) and cobalamin deficiency (Gravel *et al.*, 1975) show clearly the value of a cell fusion as a general system for gene complementation analysis of genetically defective cells (Table VII-4).

Xeroderma pigmentosum (XP) is an autosomal recessive disease in which the skin is extremely sensitive to sunlight (for a review see Robbins *et al.*, 1974). This is due to genetic defects which affect DNA repair enzymes. Biochemical studies on bacteria suggest that this form of repair involves the excision of UV-induced thymine dimers by combined action of endo- and exonucleases, synthesis of new polynucleotide segments by a special DNA polymerase and linking of the newly synthesized DNA to the preexisting DNA by special ligases (Fig. VII-11). In principle DNA repair can be impaired if any one or a combination of these functions is impaired by the presence of a defective enzyme. The fact that the disease varies in severity from one patient to another suggests that xeroderma pigmentosum is a genetically heterogenous group of afflictions. In the classical form only skin is affected, whereas in the De Sanctis–Cacchione syndrome severe neurological abnormalities are also present. The genetic heterogeneity of xeroderma cases has also been demonstrated by examining the extent to which damaged (UV-irradiated) herpes simplex virus can grow on normal fibroblasts and fibroblasts from different xeroderma patients. The damaged viral nucleic acid appears to be repaired by host cell repair enzymes and different xeroderma cell lines differ with respect to their abilities to support growth of UV-damaged virus.

DNA repair can be examined in cultivated cells by irradiating them with UV light and then incubating them with ^{3}H-thymidine for 1–2 h. After fixation the cells are examined by autoradiography. In normal cultures only S phase cells incorporate thymidine, but after UV irradiation, cells in the nonreplicating stages (G_1 and G_2) will also incorporate thymidine (Fig. VII-12a). Xeroderma cells, by contrast, only show incorporation during S phase and UV-irradiated G_1 and G_2 cells remain unlabeled since they cannot perform DNA repair synthesis. (Fig. VII-12b).

De Weerd-Kastelein *et al.* (1972, 1973) found that in homokaryons containing two fibroblast nuclei from a single xeroderma pigmentosum

TABLE VII-4

Analysis of Genetic Disease in Man by Gene Complementation Studies in Fibroblast Heterokaryons

Genetic defect affecting	Disease	Abbreviation[a]	Type of fusions	Markers studied	Technique	References
DNA repair	Xeroderma pigmentosum (and De Sanctis-Cacchione)	XP	XP + XP XP + N[a]	DNA repair replication	Cytochemical	de Weerd-Kastelein et al., 1972, 1973; Gianelli et al., 1973; Gianelli and Pawsey, 1974
				DNA repair replication; single strand excision	Biochemical	Paterson et al., 1974a
Hexosaminidase A activity	Tay-Sachs disease	TS	TS + S	Hexosaminidase A,B,C	Biochemical	Galjaard et al., 1974a Thomas et al., 1974
Hexosaminidase A and B activity	Sandhoff's disease	S				
Degradation of branched chain amino-acids[b]	Maple syrup urine disease	MSUD	MSUD + MSUD	BCKA decarboxylase	Biochemical	Lyons et al., 1973
β-Galactosidase of lysosomes	GM₁-Gangliosidosis	GM	GM₁ type 1 + GM₁ type 4	β-Galactosidase	Microchemical	Galjaard et al., 1975

[a] N, normal cell.
[b] Valine, leucine, and isoleucine.

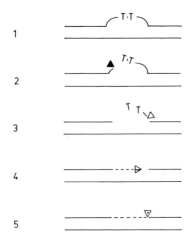

Fig. VII-11. Schematic representation of repair mechanisms involved in the removal of thymine dimers induced by UV irradiation. The scheme presented is based on studies with bacteria, but the repair mechanism of eukaryotic cells is believed to be similar. (1) Thymine dimer in one of the strands of a double stranded DNA molecule; (2) a specific endonuclease (▼) breaks the damaged DNA strand near the dimer; (3) a small region containing the dimer is excised by an exonuclease (▽); (4) a DNA polymerase (▷) synthesizes a new segment of DNA with the intact strand as a template; (5) a specific ligase (▽) joins the end of the newly synthesized DNA to the old DNA.

patient, both nuclei were unable to repair their DNA. However, in heterokaryons containing one nucleus from a normal person and one from a patient suffering from the classic type of disease (XP4), both nuclei showed thymidine incorporation after UV irradiation (Fig. VII-13). A similar result was obtained in fusions of cells from classic xeroderma cases with cells from a De Sanctis-Cacchione case (XP14). This suggests that different genes are involved in the basic defects of the De Sanctis-Cacchione and the classic syndromes. The binucleate XP12/XP4 heterokaryons in Fig. VII-14 show intergenic complementation and therefore are able to synthesize all factors needed for the repair synthesis. In later experiments complementation has also been observed in fusions of cells from different patients suffering from the classic form of the disease. At the moment, the xeroderma syndrome appears to fall into 4–6 different *complementation groups* (Kraemer *et al.*, 1973, 1975; de Weerd-Kastelein *et al.*, 1974) but further analysis of more cases may reveal additional groups. Whether these results mean that at least four different enzymes are involved in DNA repair synthesis is uncertain. A more detailed dissection of the genetic defect in xeroderma pigmentosum will undoubtedly require biochemical analysis of whole heterokaryon cultures. Such investigations have been initiated by Paterson *et al.* (1974a) and their

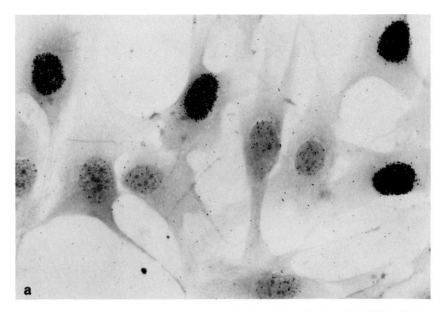

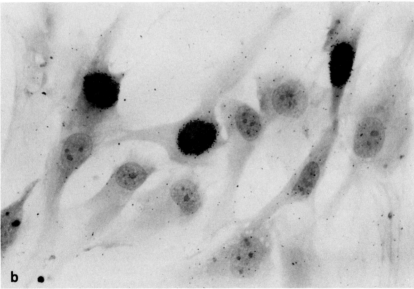

Fig. VII-12. Cytochemical detection of UV-induced DNA repair synthesis (Step 4 in Fig. VII-11) involves irradiation of cell cultures with ultraviolet light, pulse labeling with ^{3}H-thymidine of high specific activity and autoradiography. By this method it is possible to detect repair synthesis (unscheduled DNA synthesis) in G_1 and G_2 cells. (a) Autoradiogram of a fibroblast culture from a skin biopsy of a healthy donor. Four nuclei are in S phase and therefore heavily labeled. Six interphase nuclei are weakly labeled because of DNA repair synthesis. (b) Autoradiogram from comparable culture but derived from a patient suffering from xeroderma pigmentosum. Three nuclei are in S phase and therefore strongly labeled. The other nuclei are unlabeled because of the inability of XP cells to carry out repair synthesis. (Courtesy of Drs. E. A. de Weerd-Kastelein and W. Keijzer.)

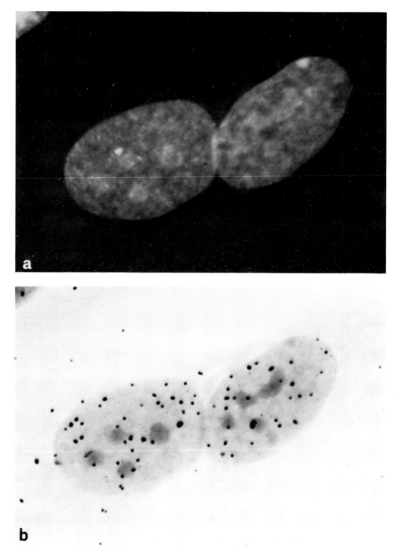

Fig. VII-13. Binuclear cell formed after fusion of a normal female cell with a male XP cell. (a) Microphotograph showing fluorescence after atebrin staining. The Barr body in the normal female nucleus (right) is located at the periphery while the Y-body in the male xeroderma nucleus (left) is in the center of the nucleus. (b) Autoradiogram of the same cell as shown in (a). Both nuclei show DNA repair synthesis in the heterokaryon. (Courtesy of Drs. E. A. de Weerd-Kastelein and W. Keijzer.)

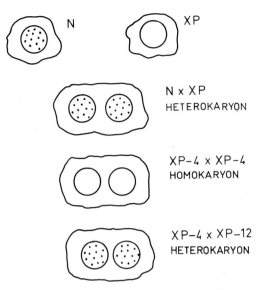

Fig. VII-14. Schematic representation of complementation analysis by fusing cells from normal persons (N) and from patients suffering from xeroderma pigmentosum (XP), a disease which is associated with defective repair mechanisms. When cells from different xeroderma patients (XP-4 and XP-12) are fused into heterokaryons both nuclei sometimes show DNA repair synthesis (dots in nuclei represent silver grains in autoradiograms following ^{3}H-thymidine labeling). By pairwise fusions of cells from different XP patients it is possible to identify those who belong to the same or different complementation groups.

results show that complementation in xeroderma heterokaryons can be monitored by measuring individual steps in the repair process. The number of cases analyzed to date is too small to permit a conclusion about which biochemical defects correspond to which gene complementation group.

Biochemical analysis of gene complementation in heterokaryons has also been used to analyze the genetic defects in Tay-Sachs disease and Sandhoff's disease (Galjaard *et al.*, 1974a). Both conditions are lysosomal storage diseases with characteristic enzyme deficiencies. Normal human cells contain three different lysosomal isoenzymes of hexosaminidase. Tay-Sachs cells lack hexosaminidase A (Hex A), whereas Sandhoff's cells are defective with respect to both Hex A and Hex B. By analyzing homokaryons and heterokaryons of normal and diseased cells it was established that two different gene mutations are involved in these two diseases.

Similarly heterokaryons produced by fusing cells from different patients suffering from maple syrup urine disease (Lyons *et al.*, 1973) and GM_1 gangliosidosis (Galjaard *et al.*, 1975) showed complementation.

These findings again suggest that different cases of the same disease carry different gene mutations and that cell hybridization may help to elucidate the genetic and biochemical defects in genetic diseases. When the complementation analysis can be based on cytochemical analysis (Galjaard *et al.*, 1974b) the number of cells required for a test is very small. This may allow cells obtained by amniocentesis to be used in cell fusion experiments.

Moreover, at least in principle, the analysis of heterokaryons should be applicable to a number of other syndromes, as, for example, glycogen storage diseases, or disorders related to amino acid, carbohydrate, lipid, purine and pyrimidine metabolism. The main requirements for an analysis of these conditions in heterokaryons are that the genetic defects manifest themselves in cultured cells (see Table IX-3) and that methods suitable for single cell analysis are available. The latter requirement is not even an absolute one. Biochemical methods can often be used where the frequency of fusion is high enough to ensure a large number of heterokaryons.

F. COMMENTS ON STUDIES USING HETEROKARYONS

The main disadvantages of using heterokaryons for the analysis of gene regulatory processes and gene complementation are (a) their limited life span, which may be too short for some regulatory interactions to manifest themselves, (b) the fact that the interacting genomes are separated from each other by membrane barriers, and (c) the technical problems in analyzing heterokaryons in the presence of homokaryons and unfused parental cells.

On the other hand, the advantages include the fact that phenotypic expression can be studied from the time of fusion, *before* any selection or chromosomal losses have occurred. Furthermore, it is often possible to examine the role of gene dosage in phenotypic expression since virus-induced fusion gives rise to heterokaryons containing varying numbers and proportions of the parental nuclei. Since many investigations only require examination of single cells, one can use culture conditions which favor cell differentiation even if this means that cell multiplication is inhibited.

VIII

Fusion with Cell Fragments

A. RECONSTRUCTION OF CELLS FROM CELL FRAGMENTS

An interesting development of great potential is the generation of viable cells from the fusion of anucleate and nucleate cell fragments. Fragments suitable for this type of experiment are prepared by centrifuging cells in the presence of the drug cytochalasin B. In this way it is possible to prepare anucleate cells that consist of most of the cytoplasm of the intact cell surrounded by an intact plasma membrane and minicells* that are the extruded nuclei surrounded by a small amount of cytoplasm and plasma membrane. The viability of both these cell fragments is limited to 2 days or less, but they can be fused together to give reconstituted cells which survive for longer periods and divide. Anucleate cells can also be fused with intact cells to give cytoplasmic hybrids (cybrids).

If the enucleation technique is applied to cells in which micronucleation has previously been induced by agents such as colchicine, then it is possible to prepare microcells which contain a micronucleus surrounded by a small amount of cytoplasm and a cell membrane. Since microcells can also be fused with intact cells and since micronuclei can contain as few as one chromosome, this provides a technique by which

---

* The terminology used is that introduced by Ege and Ringertz (1974; Ege *et al.*, 1974a). A different terminology has been suggested by Veomett *et al.* (1974): anucleate cells are referred to as cytoplasts and minicells as karyoplasts.

a small part of the genome of one cell can be introduced into another cell. It provides a less drastic means for transferring genes from cell to cell than the technique of fractionating cells, isolating metaphase chromosomes, and inserting them into a new cell.

Other recently developed methods make use of lysed erythrocyte ghosts or artificial membrane vesicles (liposomes) to introduce substances into cells. This chapter describes the methods used in preparing these various cell fragments and discusses some of their properties and their uses in fusion experiments.

B. CYTOCHALASIN-INDUCED ENUCLEATION

Cytochalasin B (Fig. VIII-1) is a metabolite produced by the fungus *Helminthosporium dematioideum*. In 1967, Carter observed that cytochalasin B induced enucleation of mouse L cells growing in tissue culture. Since then the biological effects of cytochalasin have been extensively studied, but the mechanism of its actions has not yet been fully elucidated. At low doses (1–5 µg/ml) cell attachment, endocytosis, and membrane permeability are affected, and there is inhibition of cell motility and cytokinesis following mitosis. At higher doses (> 10 µg/ml) a small proportion of the cells undergo enucleation (for a review see Poste, 1973). Early in the extrusion the nucleus is found in a protrusion over the central part of the cell body. The protrusion with its enclosed nucleus extends from the cell body until it is connected with the main cytoplasm only by a narrow stalk (Shay *et al.*, 1975). Enucleation presum-

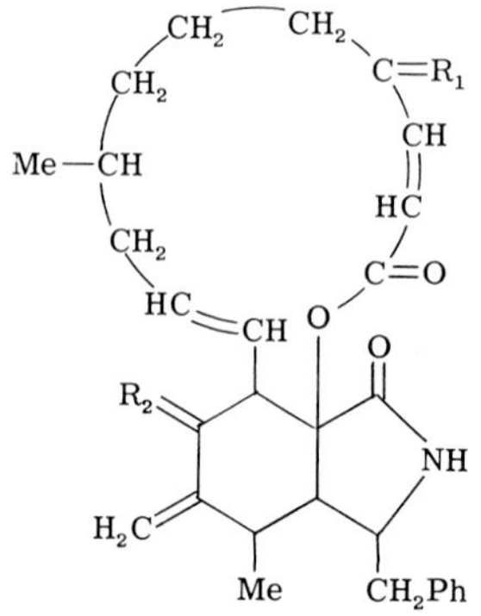

Fig. VIII-1. Chemical structure of cytochalasin B.

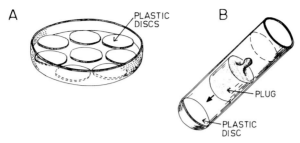

Fig. VIII-2. Enucleation of cell monolayers according to Prescott *et al.* (1972) is achieved by growing cells on round plastic discs (A) and then centrifuging these discs, cell-side down, in a medium containing cytochalasin. A plastic plug fixes the position of the disc during centrifugation (B).

ably results from the pinching off of such stalks. In spite of these rather drastic effects cytochalasin does not significantly inhibit protein, RNA, or DNA synthesis (see Poste, 1973). Furthermore, the effects on cell motility, attachment, cell form, and permeability are readily reversible if cytochalasin is removed.

In 1972, Prescott *et al.* and Wright and Hayflick (1972) developed a method of increasing the frequency of cytochalasin-induced enucleation through centrifugation. Plastic or glass discs carrying monolayers of cells were placed cell-side down in centrifuge tubes containing cytochalasin solution (Fig. VIII-2). During centrifugation the nuclei were pulled out of the cells (Fig. VIII-3), leaving the cytoplasms attached to the plastic discs. The nuclei could then be collected from the bottoms of the centrifuge tubes. The efficiency of this enucleation varies from one cell type to another and depends on cell density, dose of cytochalasin, speed of centrifugation, and temperature (Prescott *et al.*, 1972; Prescott and Kirkpatrick, 1973; Wright, 1973; Poste, 1973; Goldman and Pollack, 1974; Ege *et al.*, 1974a). Under optimal conditions, close to 100% of the cells undergo enucleation. Nevertheless, some cell types adhere so poorly that they fall off the discs without undergoing enucleation. This difficulty may sometimes be avoided by precentrifuging without cytochalasin so as to remove loosely attached cells or by pretreating the surfaces of the discs with collagen (Goldman and Pollack, 1974), concanavalin A (Gopalakrishnan and Thompson, 1975), or polylysine (T. Ege, unpublished observation). Alternatively, enucleation can be carried out with cells in suspension by centrifuging in Ficoll (Wigler and Weinstein, 1975) or colloidal silica (Bossart *et al.*, 1975) density gradients containing cytochalasin.

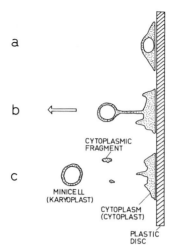

Fig. VIII-3. Schematic illustration of the enucleation of cells adhering to plastic discs. At first the cells are flat (a). During centrifugation the nuclei are pulled out of the cells into protrusions (b) by the combined action of cytochalasin and the centrifugal force (arrow). The stalks connecting the nuclei and the cytoplasms then break (c). The pellet material on the bottom of the tube at the end of the centrifugation consists mainly of nuclei surrounded by a rim of cytoplasm and a plasma membrane (minicells or karyoplasts), small cytoplasmic fragments, and a few intact cells detached from the plastic discs. The main part of the cytoplasm remains on the disc in the form of an anucleate cell (cytoplast) (Ringertz *et al.*, 1975).

C. PROPERTIES OF CELL FRAGMENTS

1. Anucleate Cells

At first the anucleate cell which remains attached to a disc after a minicell has been extruded has an altered morphology: most of the cytoplasm is concentrated in a central bulge from which long slender strands of cytoplasm extend. Within 15–20 min after the removal of cytochalasin, the cytoplasm flattens out again and assumes the shape of the original cell (Fig. VIII-4). From a technical point of view it is of interest that the anucleate cells can be detached from glass or plastic surfaces by trypsinization or EDTA treatment and subcultivated in new dishes (Goldman and Pollack, 1974; Goldman *et al.*, 1975). The reseeded cells attach and spread out in a normal way. The fact that anucleate cells can be suspended also facilitates counting, fusion and fractionation experiments.

Electron microscopic studies of the ultrastructure of anucleate cells reveal a surprisingly normal cytoplasm (Fig. VIII-5) with mitochondria, Golgi apparatus, lysosomes, ribosomes, endoplasmic reticulum, micro-

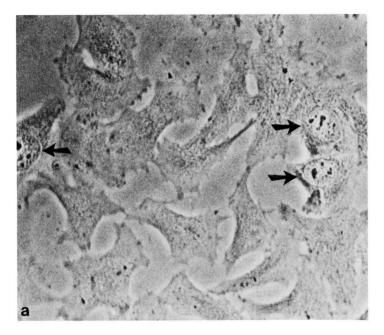

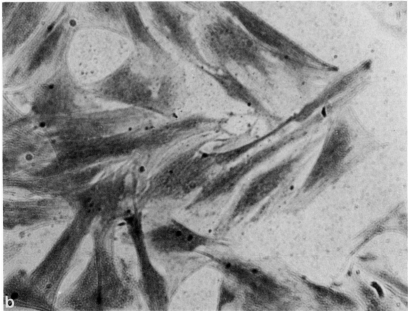

Fig. VIII-4. Partially enucleated monolayers of (a) HeLa cells (Ege *et al.*, 1973) and (b) rat L6 myoblasts (Ege *et al.*, 1974a). Arrows indicate remaining intact cells.

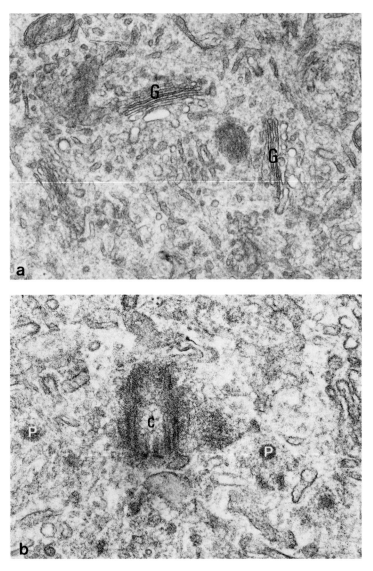

Fig. VIII-5. Ultrastructure of enucleated cells (a–c) 30 min after enucleation. Immediately after enucleation the cytoplasm contains normal cell organelles and has a normal appearance. After some time the cytoplasm becomes vacuolated and the cisternae of the Golgi apparatus become dilated (Wise and Prescott, 1973). (a) G, golgi apparatus; (b) C, centrioles; P, paracentriolar satellites; (c) F, microfilaments; and M, microtubules.

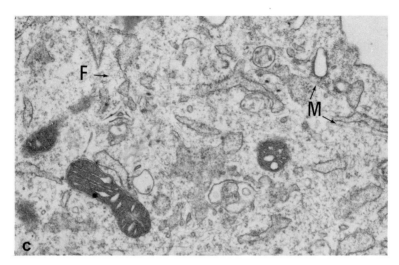

Fig. VIII-5c

filaments, microtubules, and centrioles (Wise and Prescott, 1973; Goldman and Pollack, 1974; Shay *et al.*, 1974).

Cytochemical measurements (Table VIII-1) on enucleated mouse L cells and rat L6 myoblasts indicate that their dry mass is equivalent to approximately 60% of that of the intact cells (Ege *et al.*, 1973, 1974a). Some cytoplasm is lost with the minicells and some as very small cytoplasmic fragments, the largest of which are found as contaminants in the minicell pellet after centrifugation.

The anucleate cells retain several functional characteristics of intact cells, such as membrane ruffling, active cell movement and endocytosis

TABLE VIII-1

Dry Mass of Cell Fragments Isolated from L6 Myoblasts[a]

	Dry mass	
	Relative units	In % of intact cells
Intact cells	15.47 ± 0.83	100
Enucleated cells	8.72 ± 0.66	56
Detergent isolated nuclei	3.09 ± 0.12	20
Minicells	4.34 ± 0.41	28
Cytoplasmic fragments	0.49 ± 0.06	3

[a] From Ege *et al.*, 1974a.

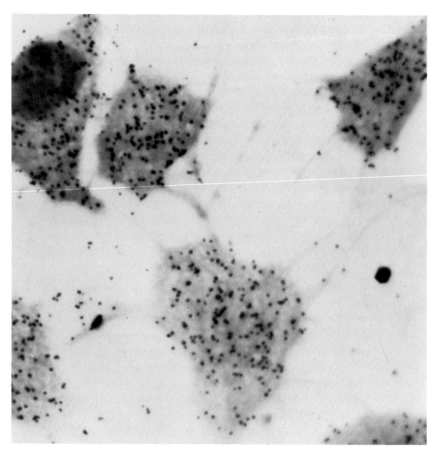

Fig. VIII-6. Autoradiographs of L cells incubated in 10μCi/ml of ^{3}H-leucine for 60 min beginning 30 min after centrifugation-cytochalasin treatment. According to grain counts the rate of protein synthesis in enucleated cells is the same as in nucleated cells (Prescott *et al.*, 1972).

(Goldman *et al.*, 1973; Goldman and Pollack, 1974). While no RNA or DNA synthesis has been detected protein synthesis continues. Immediately after enucleation the rate of ^{3}H-amino acid incorporation is similar to that of the intact cell (Fig. VIII-6) if the smaller size of the anucleate cell is taken into account (Prescott *et al.*, 1972; Goldman and Pollack, 1974). However, there is a gradual decrease in the rate of protein synthesis (Poste, 1972; Poste and Reeve, 1972; Prescott *et al.*, 1972; Goldman and Pollack, 1974) so that after 12–18 h many anucleate cells fail to show amino acid incorporation at a level which can be detected by autoradiography. At this stage the cytoplasm becomes vacuolated and the cisternae of the Golgi apparatus are abnormally dilated (Wise and Prescott, 1973). Most anucleate cells die between 16–30 h after enucleation (Poste, 1972; Prescott *et al.*, 1972; Wright and Hayflick, 1972).

2. Minicells

Nuclei drawn out of cell monolayers during enucleation by the cytochalasin method differ from detergent isolated nuclei in that they retain a plasma membrane and a rim of cytoplasm equivalent to 10–20% of the original cytoplasm of the intact cell (Figs. VIII-7 and VIII-8). Dry mass measurements on individual minicells show that their average weight is about one third of that of the average intact cell (Fig. VIII-9).

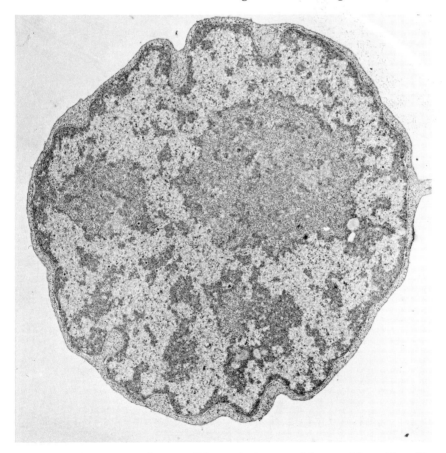

Fig. VIII-7. Ultrastructure of a minicell (karyoplast) prepared from rat L6 myoblasts (Ege *et al.*, 1974a).

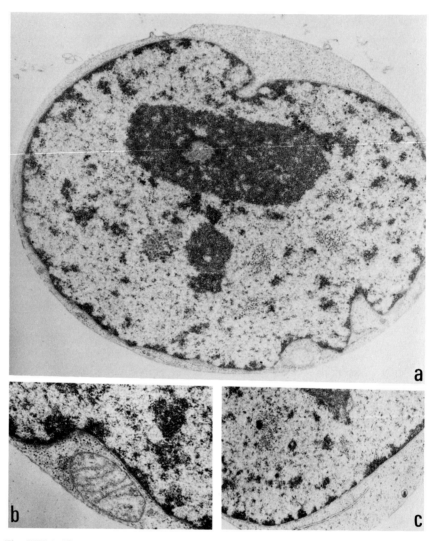

Fig. VIII-8. Electron micrographs of minicells prepared from mouse A9 cells. (a) The nuclei of minicells have a normal appearance. The narrow rim of cytoplasm which surrounds the nucleus sometimes contains mitochondria (b) and parts of the endoplasmic reticulum (c). (From T. Ege, H. Dalen, and N. R. Ringertz, unpublished work.)

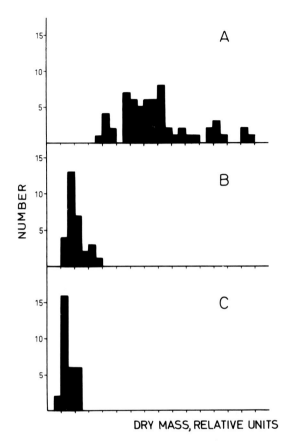

Fig. VIII-9. Histograms of microinterferometric dry mass determinations on (A) intact L6 rat myoblasts; (B) minicells prepared from L6 cells; and (C) detergent isolated nuclei from L6 cells (Ege *et al.*, 1974a).

As already noted, the minicells can be collected from the bottom of the centrifuge tubes after enucleation. Usually the pellet also contains intact cells, which have been stripped off the plastic discs during centrifugation, and small cytoplasmic fragments. The number of intact cells can be drastically reduced by precentrifugation in the absence of cytochalasin.

Although some minicells collected in this way are damaged the majority exclude dyes in viability tests and synthesize RNA and DNA for some time after enucleation (Prescott and Kirkpatrick, 1973; Ege *et al.*, 1974a). Nonetheless, when cultured under conditions in which intact cells thrive, minicells fail to regenerate a cytoplasm and multiply. Instead they undergo a progressive decrease in dry mass, lose their

ability to synthesize RNA and DNA, become permeable to dyes in viability tests, and ultimately lyse. Most minicells die within 24 h of their
preparation, and under normal tissue culture conditions, all appear to
be dead after 40 h. Using a somewhat different method of enucleation
and incubating at a higher cell density, Lucas *et al.* (1976) have found
that 0.1–1.4% of the minicells survive and give rise to clones of cells.

3. Ghosts and Membrane Vesicles

Okada and collaborators (Furusawa *et al.*, 1974) have developed a
technique for intracellular microinjection which uses erythrocyte ghosts
as "syringes." The first step in this procedure is to induce a gradual hemolysis of erythrocytes during which most of the hemoglobin is replaced by the macromolecules that one wishes to introduce into
"target" cells (e.g., fluorescein isothiocyanate or proteins). Under
appropriate ionic conditions the erythrocyte membranes spontaneously
reseal. The resulting erythrocyte ghosts, filled with the appropriate solution, can then be fused to cells with Sendai virus.

Artificially generated lipid vesicles (liposomes) may be used in a similar manner (Papahadjopoulos *et al.*, 1974; Pagano *et al.*, 1974; Weissmann *et al.*, 1975). These vesicles are prepared by dispersion and
sonication of highly purified lipids (phosphatidylserine, phosphatidylcholine, etc.) in calcium- and magnesium-free solutions. The resulting vesicles consist of a single lipid bilayer enclosing an aqueous
space. Sometimes vesicles have multilamellar walls. Both types of vesicles fuse spontaneously with living cells and can also be used as an
agent which fuses cells into polykaryons (Papahadjopoulus *et al.*, 1973).

4. Microcells

Microcells are even smaller than minicells and differ from these by
having only a fraction of the genome of the intact cell (Ege and Ringertz,
1974). The first step in generating microcells is to induce micronucleation. Several mitotic inhibitors (colchicine, Colcemid, vinblastine, and
griseofulvin) and X irradiation are known to produce large numbers of
micronucleated cells (Fig. VIII-10). The exact mechanism by which micronuclei arise is not known and may be different for different agents.
There is evidence that mitotic poisons effect micronucleation by preventing the assembly of microtubules (for a review see Margulis, 1973).
This causes divergent anaphase movements and as a result scattered
individual chromosomes and small groups of chromosomes function as
foci for the reassembly of the nuclear membrane (Fig. VIII-11) and,
thus, form micronuclei (Levan, 1954; Mazia, 1961; Das, 1962; Stubble-
field, 1964; Phillips and Phillips, 1969). In those cells that show max-

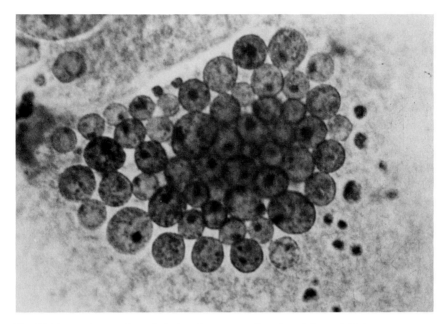

Fig. VIII-10. Micronucleated Syrian hamster cell induced by colchicine treatment. (Ringertz *et al.*, 1975.)

imum micronucleation, each individual chromosome appears to form its own micronucleus. Thus, the maximum number of micronuclei observed in Chinese hamster cells (11 pairs of chromosomes) is very close to 22 (Stubblefield, 1964; Phillips and Phillips, 1969). In other types of cells there also exists a rough correlation between the modal chromosome number and the maximum number of micronuclei which can be obtained (Ege *et al.*, 1976). The frequency of cells which undergo micronucleation varies depending on the mitotic inhibitor used, the dose

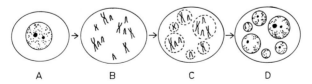

A   B   C   D

Fig. VIII-11. Schematic illustration of a hypothetical mechanism for the formation of micronuclei. As an interphase nucleus (A) is blocked in metaphase by colchicine the chromosomes will condense. (B) Because of microtubular disturbances caused by the drug the chromosomes will be widely scattered. Micronuclei (C) form as the reappearing nuclear envelope encloses individual chromosomes or groups of chromosomes. Within these micronuclei the chromosomes decondense and return to the interphase state (D).

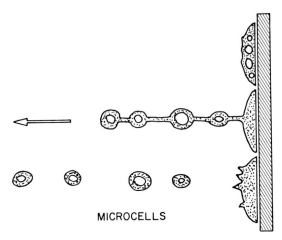

MICROCELLS

Fig. VIII-12. Schematic illustration (top to bottom) of the formation of microcells during centrifugation of micronucleated cells in the presence of cytochalasin. The arrow indicates the direction of the centrifugal force.

of inhibitor, the time of exposure, the growth rate of the cells, etc. Cells from different species vary in their sensitivity to mitotic inhibitors. Thus, it is necessary to work out optimal micronucleation conditions for each cell type examined. With a number of rodent cells it is possible to induce micronucleation in 70–90% of the cells but a minority of mono-nucleate cells always remains.

Formation of micronuclei is associated with a decondensation of mi-totic chromosomes into interphase chromatin of a dispersed type. Nu-cleoli reform in many but not all of the micronuclei (Phillips and Phillips, 1969). Parallel to these changes RNA synthesis is resumed and there is evidence that DNA replication takes place (Das, 1962; Stubble-field, 1964). Some micronucleated cells enter a second mitosis but the exact fate of these cells is not known.

Cytochalasin treatment and centrifugation by the enucleation method developed by Prescott *et al.* (1972) results in the formation of microcells (Ege and Ringertz, 1974) (Fig. VIII-12). The appearance (Fig. VIII-13) and DNA content (Fig. VIII-14) of individual microcells varies consid-erably. Feulgen microspectrophotometry indicates that the smallest of the microcells prepared from L6 rat myoblasts have a DNA content equivalent to a single chromosome (Ege and Ringertz, 1974). The largest cells in the microcell preparations, on the other hand contain as much DNA as do normal G_1 nuclei. Presumably these are minicells formed from cells which failed to undergo micronucleation.

Schor *et al.* (1975b) obtained microcells by arresting HeLa cells in metaphase by means of nitrous oxide block at 5 atm and then expos-

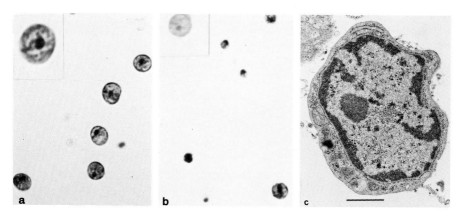

Fig. VIII-13. Giemsa-stained minicells (a) and microcells (b) prepared from L6 rat myoblasts. Inserts show one minicell and one microcell at a higher magnification. (Ege and Ringertz, 1974), (c) Electron micrograph of a microcell from a Py 6 mouse cell. Bar = 1 μm (H. Hamberg, T. Ege, and N. R. Ringertz, unpublished observation).

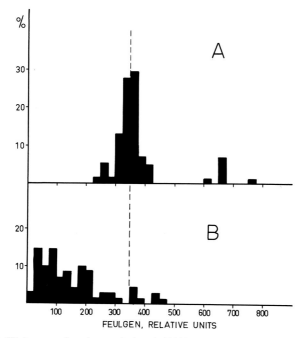

Fig. VIII-14. Histogram showing variation in DNA content in populations of (A) minicells; (B) microcells prepared from rat L6 myoblasts. The majority of the microcells have a subdiploid DNA content. The smallest of the microcells contain DNA equivalent to 1–2 chromosomes. Broken line indicates diploid G_1 DNA value (Ege and Ringertz, 1974).

ing them to a 9 h cold shock at $+4°C$. This treatment causes abnormal cleavage when the cells are returned to 37°C. A larger number of bud-like protuberances form and are pinched off to give microcells.

D. CYTOPLASMIC HYBRIDS (CYBRIDS)

When an anucleate cell is fused with a nucleated cell the resulting cell is endowed with a hybrid cytoplasm. Bunn *et al.* (1974) suggested the word "cybrid" as a convenient term for this kind of cytoplasmic hybrid cell. The first cybrids produced by cytochalasin enucleation and virus induced fusion were reported by Poste and Reeve (1971, 1972). In these experiments the anucleate cells were obtained by separating "spontaneously" enucleated cells from the remaining whole cells by density gradient centrifugation on a Ficoll gradient. After fusion of the anucleate cell fractions with normal nucleated cells, the cybrids were identified as nucleated cells which had acquired markers characteristic of the cytoplasmic donor. Fusion of enucleated macrophages with L cells produced cybrids which carried macrophage receptors for sensitized red blood cells on their surface. These cells also resembled macrophages in that they remained attached to their substratum after trypsinization while L cells and L cell homokaryons were detached. The expression of these macrophage markers, however, was found to decrease with time after fusion. In other experiments where anucleate mouse L cells were fused with human HEp-2 cells, the resulting cybrids were identified as nucleated cells carrying murine surface antigens.

Croce and Koprowski (1973) and Poste *et al.* (1974) fused monkey cells with enucleated SV40 transformed mouse cells to see whether reactivation of SV40 virus production depended upon cytoplasmic and/or nuclear factors in the permissive monkey cells. It was found that SV40 virus production was activated, as when the transformed mouse cells fused with the permissive monkey cytoplasms. The nucleus of the permissive cell therefore was not needed for SV40 activation. These experiments illustrate nicely how fusion experiments with anucleate cells may be used to analyze important biological problems such as the relationship between tumor viruses and their host cells. These aspects are discussed in more detail in Chapters XIV and XV.

In another instance the potential value of cybrids in studying cytoplasmic heredity is well illustrated by the study of Bunn *et al.* (1974). These authors fused anucleate mouse L cells, carrying a mitochondrial gene marker in the form of chloramphenicol (CAP) resistance, with a chloramphenicol sensitive but BUdR resistant subline of mouse L cells (Fig. VIII-15). The BUdR resistance of the latter cell type is due to nu-

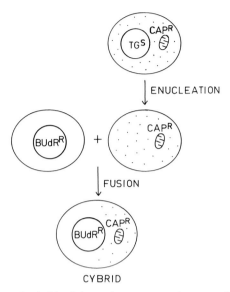

CYBRID

Fig. VIII-15. Isolation of cybrids. Schematic summary of an experiment by Bunn *et al.* (1974) in which a cell with a nuclear drug resistance marker (BUdR[R]) is fused with an enucleated cell carrying a mitochondrial drug resistance marker (CAP[R]). This results in a cybrid (cytoplasmic hybrid) resistant to both BUdR and CAP. In a medium containing both drugs the parental cells will die whereas the cybrids survive and multiply.

clear gene mutation(s) which results in cells deficient in the enzyme thymidine kinase (see Chapter IX). Fusion of the anucleate, CAP resistant cells with the BUdR resistant (but CAP sensitive) L cells gave cybrids which survived and multiplied in a selective medium containing both CAP and BUdR. Those L cells which did not fuse with anucleate cells died because of their sensitivity to CAP.

The importance of these experiments is that they show that mitochondrially specified genes can be propagated in multiplying cells where the nucleus and part of the cytoplasm is derived from another cell. The interrelationship between nuclear genes and mitochondrial functions is considered in more detail in Chapter XII where cell organelles in hybrid cells are considered.

Wright and Hayflick (1975a) have developed another selection method for the isolation of cytoplasmic hybrids*. This system is based on the ability of anucleate cells containing active enzymes to rescue cells poisoned with iodoacetate. The same principle can probably be used with other enzyme inhibitors. The intact cells are treated with the

---

* Cytoplasmic hybrids are referred to as "heteroplasmons" by Wright and Hayflick (1975a).

inhibitor at doses which kill all cells and then fused with anucleate but otherwise healthy cells. In this way the intact cell, or rather its nucleus, is rescued and can multiply to give a colony of cytoplasmic hybrids. One may expect, however, that problems will exist in defining the lethal dose and proving that the multiplying cells are in fact cytoplasmic hybrids formed by fusion and not intact cells which have been rescued by crossfeeding of nutrients released by dying anucleate cells. In spite of these difficulties the selection method of Wright and Hayflick may be an extremely useful first step. If possible the technique should combine the use of a nuclear gene marker in the intact cell (e.g., resistance to azaguanine or bromodeoxyuridine) and a cytoplasmic gene marker (e.g., chloramphenicol or erythromycin resistance) in the anucleate cell. It would then be possible to test at a later stage if the cells isolated are resistant to both drugs and, thus, are true cytoplasmic hybrids.

Wright and Hayflick (1975b) have used the iodoacetate selection method to test if young anucleate cells can rescue senescent cells which would normally stop dividing within a few cell generations. The negative results suggest that senescence is under nuclear rather than cytoplasmic control.

E. RECONSTITUTED CELLS

While cytoplasmic hybrids offer many new possibilities for cell genetic studies there are also experimental situations in which it would be more useful to combine the cytoplasm of one cell with a cytoplasm-free nucleus from another cell.

Although nuclei and cytoplasms have been successfully recombined in experiments with protozoa (Goldstein and Prescott, 1967; Goldstein, 1974) and amphibian oocytes (for review see Gurdon, 1974), there has been virtually no success with mammalian cells. Isolated nuclei, free of cytoplasm can apparently be taken up by endocytosis by both nucleated and enucleated mammalian cells (Poste, 1973) and addition of Sendai virus increases the uptake, probably by stimulating endocytosis. Unfortunately such nuclei undergo fragmentation and rapid degeneration (Poste, 1973). It appears therefore that in order to survive transfer to new cytoplasms, nuclei may need protective coats of cytoplasm. This conclusion is supported by observations made in nuclear transplantation experiments with protozoa (Jeon and Danielli, 1971) and oocytes (Gurdon and Woodland, 1970). Furthermore, in order to fuse with anucleate cells in the presence of Sendai virus the nuclei probably have to be surrounded by a cell membrane carrying the appropriate virus receptors.

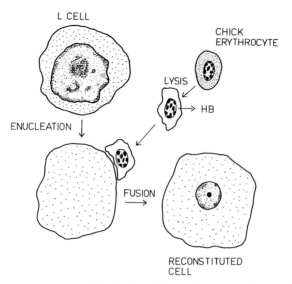

Fig. VIII-16. Reconstitution of cells by fusing chick erythrocytes with anucleate mouse L cells (Ringertz *et al.*, 1975).

Attempts to reconstitute cells by fusing nucleated avian erythrocytes with enucleated mammalian cells have been made by several groups (Ladda and Estensen, 1970; Poste and Reeve, 1972; Ege *et al.*, 1973, 1974b, 1975). The reason these fusions approximate reconstitution is that erythrocytes lyse when exposed to Sendai virus (Fig. VIII-16). Electron microscopical studies (Schneeberger and Harris, 1966; Dupuy-Coin *et al.*, 1976) have shown that during fusion the erythrocytes are essentially ghosts consisting of little more than condensed nuclei and plasma membrane. Even though there are indications that some erythrocytes fuse without previous lysis (Zakai *et al.*, 1974a,b; Carlsson *et al.*, 1974a; Appels and Ringertz, 1975) and that chick hemoglobin is introduced into heterokaryons formed by the fusion of erythrocytes with mammalian cells, the amount of cytoplasm contributed by the chick erythrocytes is likely to be small when compared with the volume of the anucleate mammalian cell. One advantage with this type of reconstituted cell is that the avian nucleus is inactive before fusion (see Chapter VII). Reactivation of nucleic acid synthesis after fusion can therefore be used as an indication that the erythrocyte nucleus has been properly integrated into the cytoplasm and is responding to regulatory signals.

Ladda and Estensen (1970) observed nuclear swelling and ^{3}H-uridine incorporation in erythrocyte nuclei introduced into enucleated L cells,

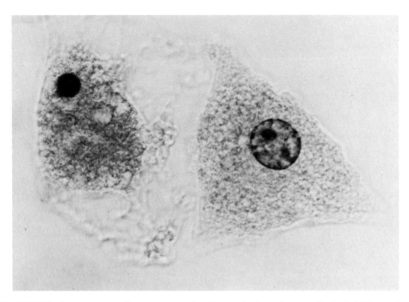

Fig. VIII-17. Appearance of reconstituted cells consisting of a reactivated chick erythro-cyte nucleus and a mouse cytoplasm 73 h after fusion. The cell to the left is dying (Ege *et al.* 1975a).

but Poste and Reeve (1972) failed to find any signs of reactivation in chick erythrocyte nuclei fused into anucleate macrophages or L cells. Subsequently Ege *et al.* (1973, 1975) found that chick erythrocyte nuclei introduced into anucleate L cells or HeLa cells underwent an abortive reactivation. At first the erythrocyte nuclei grew in size as fast as during reactivation in heterokaryons (see Chapter VII), RNA synthesis began, and nucleoli developed (Fig. VIII-17). A small percentage of the nuclei even synthesized DNA. But after approximately 12 h, the rate of ^{3}H-uridine incorporation began to decrease and at 48 h RNA synthesis was negligible. Nucleolar growth stopped, never reaching the size and level of organization it does in heterokaryons (Dupuy-Coin *et al.*, 1976). The introduction of chick erythrocyte nuclei did not succeed in retarding the decay of protein synthesis in the anucleate cytoplasm. Although the re-constituted cells remained well attached to the substrate somewhat longer than did anucleate cells, nonetheless after 3 days practically all reconstituted cells had died. In more recent experiments (P. Elias and N. R. Ringertz, unpublished observations) cells reconstituted by fusing 8 day embryonic erythrocytes with enucleated chick cytoplasms have been found to survive for up to 8 days. It appears therefore that the erythrocyte nuclei, although they become active, are not capable of rescuing the cytoplasms so as to form cells capable of multiplication. Perhaps the reactivation in these systems is simply not fast enough for this to occur.

Because minicells are already active in RNA and DNA synthesis and because they are surrounded by a narrow protective layer of cytoplasm they doubtless have a greater chance of producing reconstituted mammalian cells capable of long term survival and cell multiplication. Retention of a plasma membrane containing virus receptors makes them susceptible to Sendai-induced fusion.

Recently, two groups independently reported reconstitution of mammalian cells after fusion of minicells with anucleate cells (Veomett *et al.*, 1974; Ege *et al.*, 1974b; Ege and Ringertz, 1975). Veomett *et al.* (1974) fused minicells (karyoplasts) prepared from L cells with anucleate L cells (cytoplasts). In order to distinguish the reconstituted cells from cybrids formed by the fusion of intact cells present as contaminants in the minicell preparations, an ingenious labeling procedure was used. The cells were allowed to ingest latex particles of specific sizes and these then acted as markers for the cytoplasms. The cytoplasm of the nuclear donor was labeled with *small* latex spheres and with ^{3}H-thymidine in the nucleus. The cytoplasmic donor, on the other hand, was labeled only in the cytoplasm with *large* latex spheres. Reconstituted cells were then identified as cells which had ^{3}H-thymidine labeled nuclei and cytoplasm containing only large latex particles. Some of the reconstituted

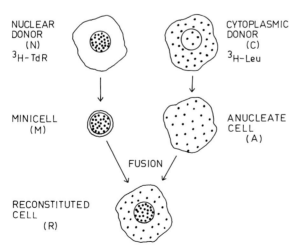

Fig. VIII-18. Labeling procedure used by Ege *et al.* (1974b) in order to identify reconstituted cells and cybrids and at the same time testing for the viability of these cells (^{3}H-hypoxanthine incorporation). Further explanations are given in the text.

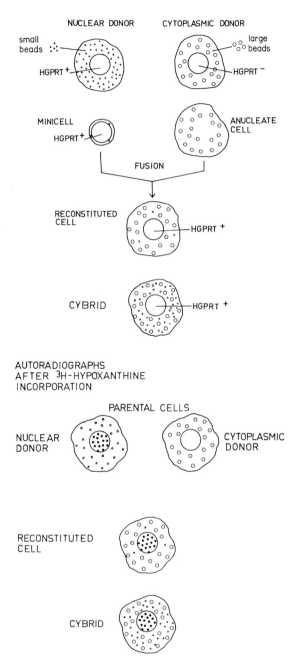

Fig. VIII-19. Labeling procedure used by Ege and Ringertz (1975) in order to identify re-constituted cells and cybrids and at the same time testing for the viability of these cells.

cells were found in different stages of mitosis suggesting that these cells were capable of proliferation.

Ege *et al.* (1974b) used a different labeling procedure in experiments where minicells and anucleate cells prepared from rat L6 myoblasts were fused to form reconstituted cells (Fig. VIII-18). The minicells were prepared from myoblasts whose nuclei were heavily labeled with ³H-thymidine but whose cytoplasms were unlabeled. Quantitative cytochemical methods showed that the minicells contained less than 10% of the cytoplasm of the whole cell (Ege *et al.,* 1974a). The cytoplasmic donor was prepared from cells weakly labeled with ³H-leucine in both nuclei and cytoplasms. Reconstituted cells were then identified as those having strongly labeled nuclei and weakly labeled cytoplasm. This made it possible to distinguish these cells from the few contaminating intact cytoplasmic donors (weakly ³H-leucine labeled nuclei and cytoplasms) and intact nuclear donors (strongly ³H-thymidine labeled nuclei, unlabeled cytoplasms).

The identification of viable reconstituted cells and cybrids is facilitated if enzyme deficient cells are used as nuclear or cytoplasmic donors (Fig. VIII-19). Ege and Ringertz (1975) labeled HGPRT⁻ rat myoblasts with large latex beads according to Prescott and co-workers (Veomett *et al.,* 1974). After enucleation the cytoplasms of these cells were fused with minicells prepared from normal L6 rat myoblasts. Before enucleation these cells had been labeled with a large number of small latex beads. Some of these remained in the minicells and served as a measure of the amount of cytoplasm introduced by the minicells into the reconstituted cells. At different time points after fusion the cells were exposed to ³H-hypoxanthine, a metabolite which can only be used by HGPRT⁺ cells. In autoradiograms of the fused preparations reconstituted cells were identified as cells having ³H-hypoxanthine labeled nuclei and large latex beads in the cytoplasm. Cybrids were radioactively labeled and had both large and small latex beads. As seen in Fig. VIII-20

The nuclear donor cell is labeled in the cytoplasm with small latex beads. The cytoplasmic donor cells are enzyme-deficient cells (HGPRT⁻) which are unable to incorporate ³H-hypoxanthine. Before preparing anucleate cells the cytoplasmic donor is labeled in the cytoplasm with large latex beads. The anucleate cells are then fused with minicells prepared from the nuclear donor cells. At varying times after fusion the fused preparations are exposed to ³H-hypoxanthine and then subjected to autoradiography. Reconstituted cells have silver grains over their nuclei and large beads in their cytoplasms. Cybrids are cells which contain both large and small beads and have labeled nuclei. This labeling procedure also makes it possible to identify contaminating nuclear and cytoplasmic donor cells which fail to undergo enucleation. The appearance of the different cell types is illustrated in Fig. VIII-20.

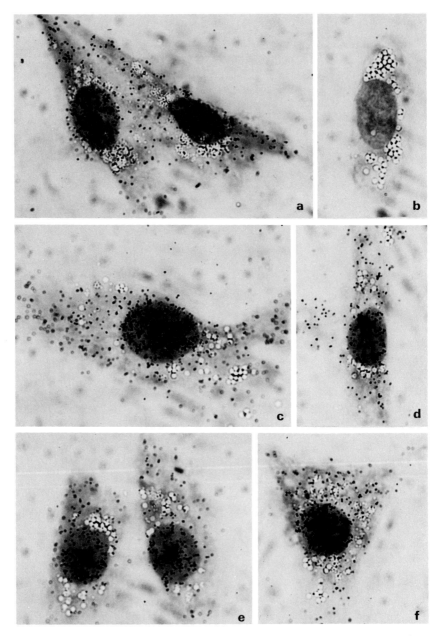

Fig. VIII-20. Appearance of reconstituted cells, cybrids, and intact cytoplasmic and nuclear donor cells in experiments where HGPRT⁺ minicells from L6 myoblasts were fused

the reconstituted cells and cybrids were also easily distinguished from contaminating intact nuclear and cytoplasmic donor cells.

Very little is now known about the properties and viability of reconstituted cells. The evidence which exists can be summarized as follows:

1. The nucleus of reconstituted cells appears to be normally integrated into the cytoplasm. Electron microscopic studies of chick erythrocyte nuclei introduced into anucleate rat cells show that the outer layer of the nuclear envelope is in direct contact with the rat cytoplasm and that the Golgi apparatus assumes its normal position relative to the nucleus (Dupuy-Coin *et al.*, 1976).

2. Normal nucleocytoplasmic relations are established with respect to the exchange of macromolecules between the nuclear and the cytoplasmic compartments. In reconstituted cells containing a chick erythrocyte nucleus and a rat cytoplasm the erythrocyte nucleus takes up and concentrates rat nuclear antigens (Ege *et al.*, 1973). In cells reconstituted from L6 minicells and L6 cytoplasms (Ege and Ringertz, 1975) newly synthesized RNA appears to be transported from the nucleus into the cytoplasm.

3. The nucleus responds to regulatory signals from the surrounding cytoplasm. Inactive chick erythrocyte nuclei introduced into the cytoplasm of active cells resume RNA synthesis, form nucleoli, and in a few cases also synthesize DNA (Ege *et al.*, 1973, 1975).

4. The reconstituted cells are viable in the sense that they synthesize macromolecules and survive longer than the cell fragments from which they were made (Ege and Ringertz, 1975).

5. Fusion of rat L6 myoblast minicells with mouse A9 cytoplasms has recently been found to produce cells that multiply and form clones (U. Krondahl, N. Bols, T. Ege, S. Linder, and N. R. Ringertz, in press).

The results obtained so far are encouraging since they show that reconstituted cells are capable of short-term survival. Unfortunately, it is much more difficult to obtain information about the long-term survival of such cells. The main problem is to identify the reconstituted cells at late time points when the properties of the cytoplasm have been modified by the influence of the new nucleus. In order to be able to identify

---

with enucleated cytoplasms from HGPRT⁻ L6 mutant cells. The nuclear donor cell (a) was labeled with small latex beads and the cytoplasmic donor (b) with large beads. At different time points after fusion the cells were exposed to ³H-hypoxanthine. Reconstituted cells (c and d) were then identified as radioactively labeled cells with large beads in their cytoplasm. At late time points (48 h) after fusion many reconstituted cells are found in pairs (e) suggesting mitotic divisions. Cybrids (f) were ³H-labeled and had both large and small beads (Ege and Ringertz, 1975).

reconstituted cells at this stage the cytoplasm must retain some form of markers. As indicated in the previous section on cybrids, mitochondrial gene mutations for chloramphenicol resistance are the only form of cytoplasmic gene markers known so far. Since chloramphenicol resistance has been successfully used in the selection of multiplying cybrids, there are reasons to believe that the same principle could be used to isolate the progeny of reconstituted cells. Thus, minicells containing a nuclear drug resistance marker could be fused with a cytoplasm carrying a mitochondrial drug resistance. The reconstituted cells should then be resistant to both drugs while contaminating intact nuclear and cytoplasmic donors should be killed by one of the two drugs.

Studies of reconstituted cells, cybrids, and their progenies may be useful in the analysis of a number of important cell biological problems. The main areas of application are likely to be in experiments designed to analyze:

1. Nucleocytoplasmic interactions, in particular the role of the cytoplasm in the control of nuclear activity
2. Gene regulation and cell differentiation
3. Dependence of mitochondria on nuclear genes
4. Dependence of virus replication on nuclear and cytoplasmic factors

F. MICROCELL HETEROKARYONS

Just as it has been possible to introduce isolated nuclei into other cells by microsurgery or endocytosis, so has it also been possible to introduce isolated chromosomes or chromosome fragments (McBride and Ozer, 1973; Shani *et al.*, 1974; Sekiguchi *et al.*, 1973, 1975b; Willecke and Ruddle, 1975; Burch and McBride, 1975; Wullems *et al.*, 1975). Unfortunately, however, it has been difficult to exploit the full potential of these operations since in most cases the gene material introduced appears to be rapidly eliminated. Microcells may prove to be the vehicles, which allow the introduction of small amounts of chromosomal material into cells without subsequent destruction. Ege and Ringertz (1974) have shown that microcells can be fused with intact cells using Sendai virus (Fig. VIII-21). Furthermore the micronuclei show RNA synthesis after they have been incorporated into intact cells (Ege *et al.*, 1974b). It is not unreasonable to hope, therefore, that genetic information encoded in the micronuclei will be expressed in heterokaryons. At the moment the long-term fate of microcell fusion products is unknown, but observations by Schor *et al.* (1975b) show that microcell heterokaryons divide. If

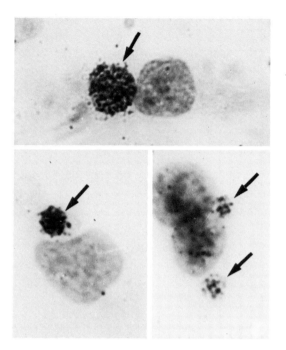

Fig. VIII-21. Heterokaryon formed by Sendai virus-induced fusion of ^{3}H-thymidine labeled microcells prepared from rat L6 myoblasts with intact L6 myoblasts. Micronuclei (arrows) are covered with silver grains (Ege and Ringertz, 1974).

the chromosomes in the micronuclei continue to duplicate and attach in a normal way to the host cell's mitotic spindle there is a possibility that their genetic information will be carried into daughter cells. If this proves to be the case the microcells will be useful in chromosome mapping since it should be possible to generate in one step, cell hybrids in which only one or a few chromosomes from one of the parental genomes is present. Exactly what type of hybrid is obtained may be controlled by using appropriate mouse mutant cells and human microcells. If the products of such fusions are grown on appropriate selective media (see Chapter IX) a specific type of hybrid will be obtained, namely one in which the chromosomes introduced with the human microcell complement the defect in the mouse genome (see also Chapter XIII). It may also be possible to use the microcells in studies of regulation in which one examines whether or not the introduction of a few chromosomes from a cell of one epigenotype will change the differentiation of a cell of another epigenotype. Moreover, exper-

iments in which microcells are used to analyze the interaction between tumor viruses and their host cells should be possible.

G. MICROINJECTION EXPERIMENTS WITH GHOSTS
AND MEMBRANE VESICLES

Ghost and membrane vesicles represent a type of artificial or semiar-
tificial cell fragment that can be used to modify the cytoplasmic composition of target cells. Experiments by Furusawa et al. (1974) demonstrate
that erythrocyte ghosts filled with protein or fluorescein isothiocyanate
can be fused to intact living cells with Sendai virus. As a result of such
fusions membrane components of the erythrocyte ghosts coalesce with
the membranes of the target cells and the contents of the ghosts are injected into the cytoplasms. This technique was used by Schlegel and
Rechsteiner (1975) to inject thymidine kinase and serum albumin into
mammalian cells. Human erythrocyte ghosts charged with small particles have been used by Loyter et al. (1975) to introduce ferritin and
latex spheres (diameter 0.1 μm) into animal cells.

Membrane vesicles fuse spontaneously with living cells without any
immediate cytotoxic effects or are taken up by endocytosis. In the same
way as with ghosts, the membrane components become part of the cell
membrane and the contents of the aqueous space are injected into the
recipient cell (Papahadjopoulos et al., 1974; Pagano et al., 1974).

Both ghosts and membrane vesicles are potentially interesting tools
in cell biological experiments. So far, however, there are very few practical results with these techniques. If they can be used to inject substances into living cells in vitro it is also possible that there may be practical results of medical interest. Membrane vesicles containing drugs
could then perhaps be used in chemotherapy to introduce drugs into
the cells of patients. Alternatively, such vesicles could be used to introduce enzymes into enzyme-deficient cells (see Weissmann et al., 1975).

IX

Isolation of Growing Hybrid Cells

A. FORMATION OF MONONUCLEATE HYBRID CELLS (SYNKARYONS)

The majority of mononucleate hybrid cells are probably formed by mitotic division of binucleate heterokaryons. Though coalescence of several different nuclei into one could also result in mononucleate hybrid cells, there is relatively little evidence to support this route of formation in animal cell hybrids (Heneen *et al.*, 1970). On the contrary, time-lapse studies of polykaryons show that nuclei retain their individuality during interphase (Fell and Hughes, 1949). At mitosis, however, chromosomes from two different nuclei may be included into one nucleus thereby achieving nuclear fusion (Yamanaka and Okada, 1968). In spite of the fact that heterokaryons must possess supernumerary centrioles they often undergo bipolar divisions and form two mononucleate daughter cells. Many of these mononucleate hybrids continue to multiply by well-balanced and regular mitosis, thus, giving rise to subsequent generations of hybrid cells. A large number of heterokaryons, however, never reach mitosis or fail to complete a normal mitosis. Because of abnormalities in the organization of the spindle and abnormal chromosome movements the first divisions of the heterokaryons and their daughter cells often fail. In some cells the chromosomes never reach the equatorial plate while in others anaphase separation of daughter chromosomes is disorganized. Multipolar divisions are common, presumably because of supernumerary centrioles and the formation of several spindles. Gordon and Cohn (1970) found that 30–80%

147

of binucleate macrophage + melanocyte heterokaryons divided to produce pairs of descendants within 2–4 days after fusion. Eventually practically all heterokaryons divided. Fewer cells underwent a second division and cells undergoing a third division were extremely rare. Thus,
very few heterokaryons produced growing clones of hybrids. Also in
other fusions the efficiency of hybrid cell formation from heterokaryons
has been found to be low (Harris *et al.*, 1965; Murayama and Okada,
1970a). The "effective mating rate" as measured by the number of hybrid clones capable of long term multiplication has been studied for
several cell combinations (Davidson, 1969; Coon and Weiss, 1969; Davidson and Ephrussi, 1970; Murayama and Okada, 1970a; Klebe *et al.*,
1970). It appears to be generally higher when one or both of the parental
cells are present in a monolayer than when a mixture of cells is fused in
suspension (Klebe *et al.*, 1970; Hitchcock, 1971; Hosaka, 1970). The
reason for this difference may be that the relative proportion of binucleate heterokaryons is high in monolayer fusions whereas suspension
fusion favors the formation of giant heterokaryons which die soon after
fusion. Other factors which affect the fusion yield are discussed in
Chapters IV, V, and X.

B. DIFFERENT ISOLATION METHODS

The first hybrid cells were discovered and isolated because they proliferated more rapidly than did either of the two parental cell types
(Barski *et al.*, 1961). This phenomenon, sometimes referred to as hybrid
vigour, is unusual for in most cases hybrids do not overgrow the
parental cells. Hybrid cells which have no selective advantage over the
parental cells on normal media can, however, be isolated from a cell
mixture either by nonselective isolation methods such as single cell
cloning (see below) or selective methods. Usually the experimenter is
faced with the problem of isolating a small number of slow-growing hybrid cells from a large number of rapidly proliferating parental cells. By
far the easiest solution to this problem is to use a *selective medium*
which favors the growth of the hybrid cells, while killing or inhibiting
the parental cells. On the other hand differentiated cells (e.g., nerve
cells, lymphocytes, spermatozoa) which grow poorly or not at all *in
vitro*, form proliferating hybrids when fused with more active cells (like
tumor cells or cells of established lines). Such hybrids can be isolated
by semiselective media in which the selection is directed only against
the rapidly proliferating parent (Davidson and Ephrussi, 1965; Harris *et
al.*, 1969a). Irrespective of the method used for their isolation it is
usually desirable that different hybrid lines are derived from indepen-

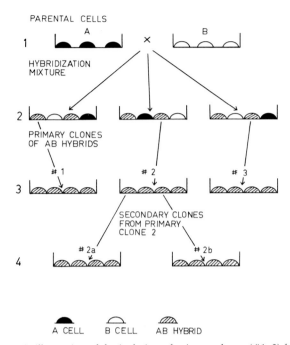

Fig. IX-1. Schematic illustration of the isolation of primary clones (#1–3) from a hybridization mixture followed by isolation of two secondary clones (#2 a and b) by subcloning one of the primary clones (#2).

dent fusion events, i.e., from different heterokaryons. For this reason the initial fusion mixture is immediately divided up and only one clone is then propagated from each culture dish (Fig. IX-1). Such clones are sometimes referred to as *primary clones,* as opposed to *secondary clones* that are obtained by subcloning primary clones. Secondary clones may be useful in the analysis of genetic variation arising during long term growth of primary clones.

C. NONSELECTIVE ISOLATION

If in a given culture the proportion of hybrid cells is great, relative to the number of parental cells, pure colonies of hybrid cells can sometimes be isolated by selecting clones with a hybrid karyotype from an array of clones grown up from randomly isolated single cells (Scaletta and Ephrussi, 1965). In practice, this is difficult because of the amount of work involved, and one may choose to study only those colonies which differ from the parental cells in their morphology. It is important to remember then that the technique is no longer completely "nonselective" since the experimenter may ignore those hybrid clones which resemble one of the parental cell types with respect to colony morphology.

D. SELECTIVE ISOLATION

1. Nonspecific Selection

Animal cells derived from different sources often vary markedly with respect to their ability to grow *in vitro*. The differences manifest themselves in different temperature optima for growth, differences in nutritional requirements, and ability to grow in suspension. Cells also vary with respect to their adhesiveness and need for a solid substratum ("anchorage dependence"), sensitivity to contact inhibition, and ability to persist in culture (aging or senescence phenomena). Although the molecular background for such differences is unknown they can be used in the isolation of hybrid cells. In most cases, however, it is an advantage to select hybrid cells on the basis of properties which are easier to define. The following sections discuss the specific selection of hybrids from parental cells which carry genetic markers in the form of drug resistance, nutritional requirements, or temperature sensitivity.

2. Selection of Hybrids Made from Drug Resistant Cells

Szybalski *et al.* (1962) demonstrated that it is possible to obtain mutant cells defective in specific enzymes by subjecting a normal cell population to selection with drugs. On the basis of these observations Littlefield (1964a, 1966) developed a general method for the isolation of hybrid cells. This isolation method uses cells resistant to azaguanine (AG^R) and bromodeoxyuridine (BUdR^R), and since it has been widely used in cell hybrid work a brief description will be given of the methods used in generating and isolating these forms of drug resistant cells. More detailed information can be found in reviews by Thompson and Baker (1973), Clements (1975), and Siminovitch (1976).

Drug resistant cells arise spontaneously in cell cultures, but X irradiation or chemical mutagens are usually applied in order to increase the mutation rate. After exposure to the mutagen, the drug is included in the normal tissue culture medium. Azaguanine and thioguanine (TG) kill normal cells because the drugs are metabolized and interfere with nucleotide and nucleic acid synthesis. This effect is mediated by the enzyme hypoxanthine guanine phosphoribosyl transferase (HGPRT) which converts the drugs to "abnormal" nucleotides. Mutant cells

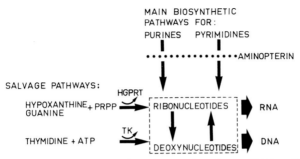

Fig. IX-2. Biosynthetic pathways for nucleotides. On the selective HAT medium the main biosynthetic pathways are blocked by the folic acid analogue aminopterin. Normal cells survive by utilizing hypoxanthine and thymidine. Mutant cells lacking HGPRT or TK enzymes die on the selective medium.

which lack the enzyme (HGPRT⁻) are resistant to the drugs (AG^R or TG^R) and therefore survive treatment.

Bromodeoxyuridine resistant cells are obtained by a similar procedure. In normal cells this drug will first be phosphorylated by thymidine kinase (TK) and then incorporated into DNA. This normally results in the death of the cell. The exact mechanism of killing is not known but there is evidence that chromosome breaks and mutations are induced and that cell growth and differentiation are affected. Mutant cells deficient in thymidine kinase (TK⁻) fail to phosphorylate and incorporate BUdR into DNA and, therefore, are drug resistant.

In the selection system developed by Littlefield (1964a, 1966) hybrid cells are isolated after fusion of azaguanine resistant (AG^R) cells that lack HGPRT with cells that are resistant to bromodeoxyuridine (BUdR^R) and lack TK. The genetic defects are of little importance during growth on normal tissue culture media, since these enzymes are only involved in salvage pathways for nucleotide synthesis. But when the main biosynthetic pathways for purine and pyrimidine nucleotides are blocked by the folic acid analogue aminopterin (Fig. IX-2), normal cells (HGPRT⁺, TK⁺) can survive if supplied with exogenous hypoxanthine and thymidine whereas the mutant cells die because of their inability to synthesize nucleotides from hypoxanthine (HGPRT⁻ cells) or from thymidine (TK⁻ cells).

Littlefield demonstrated that hybrid cells formed from spontaneous fusion of HGPRT⁻ and TK⁻ cells could be isolated by selection in a medium containing *h*ypoxanthine, *a*minopterin, and *t*hymidine, the so called HAT medium developed by Szybalski *et al.* (1962). Usually this selective medium is also supplemented with glycine since aminopterin blocks the synthesis of this amino acid and makes the cells dependent upon exogenous glycine.

In Littlefield's selective system, the hybrid cells, containing one chromosome set which was HGPRT⁻ but TK⁺ and one set that was HGPRT⁺ and TK⁻, were able to produce enough of the HGPRT and the TK enzyme to survive on the HAT medium. Thus, when combined in one cell the two parental genomes complement each other.

With the development of the HAT medium and its use in the isolation of hybrid cells, a technique long used in microbial genetics was for the first time applied to eukaryotic systems. Its impact on the development of somatic cell genetics has been great. The technique has already been utilized in the isolation of many different types of somatic cell hybrids and is now extensively used in connection with chromosome mapping (see Chapter XIII). In addition to the HAT system there are now other selection methods based on the fusion of drug resistant cells and selection in special media (Albrecht *et al.*, 1971; Kusano *et al.*, 1971; Sobel *et al.*, 1971; Chan *et al.*, 1975). The system of Kusano *et al.* (1971), for example, uses instead of HGPRT⁻ cells, cells deficient in adenine phosphoribosyltransferase (APRT⁻). These cells are obtained after selection with fluoroadenine. A new system developed by Chan *et al.* (1975) uses double mutants which are deficient in deoxycytidine kinase as well as deoxycytidine deaminase (dCD⁻). In a medium containing hypoxanthine, aminopterin, and 5-methyl deoxycytidine (HAM medium) the main biosynthetic pathways for pyrimidine synthesis will be blocked by the aminopterin. Under these conditions dCD⁻ cells are killed while dCD⁺ cells survive because they can convert 5-methyl deoxycytidine to thymidine which is then phosphorylated to thymidylate by thymidine kinase. The HAM selective system is analogous to the HAT system, the only differences being that (a) 5-methyl deoxycytidine replaces thymidine in the medium and (b) one of the parental cells is dCD⁻ instead of TK⁻.

Chan *et al.* (1975) used the HAM-medium to select hybrids from a fusion of human cells (dCD⁺) with a mouse double mutant that was dCD⁻ and also deficient in deoxycytidine kinase (dCK⁻). Although in these experiments the human cells were HGPRT⁺ and were selected against by their slower growth, the HAM system should be useful in the isolation of hybrids from mixtures where one parental cell is HGPRT⁻ dCD⁺ and the other HGPRT⁺ dCD⁻.

A new selection system suggested by Baker *et al.* (1974) and Jha and Ozer (1976) has the advantage that only one of the parental cells carries drug resistance markers. In this system the "marked" cell is a double mutant and is both thioguanine resistant (i.e. HGPRT⁻) and ouabain

TABLE IX-1

Examples of Drug Resistant Mutant Cells

Drug	Alteration in resistant cell line	References
Azaguanine (AG), thioguanine (TG)	Hypoxanthine guanine phosphoribosyltransferase deficiency (HGPRT⁻)	Szybalski et al., 1962; Littlefield, 1964a,b; Gillin et al., 1972; Shin, 1974
Fluoroadenine	Adenine phosphoribosyltransferase deficiency (APRT⁻)	Kusano et al., 1971
5-Bromodeoxyuridine (BUdR)	Thymidine kinase deficiency (TK⁻)	Djordjevic and Szybalski, 1960; Hsu and Sommers, 1962; Dubbs and Kit, 1964; Littlefield, 1965; Kit et al., 1963
Cytosine arabinoside (ara C)	Deoxycytidine kinase deficiency (dCK⁻)	Chu and Fisher, 1965; de Saint-Vincent and Buttin, 1973; DeChamps et al., 1974
5-Bromodeoxycytidine (BCdR)	Deoxycytidine deaminase deficiency (dCD⁻)	Chan et al., 1975
α-Amanitin	Resistant RNA polymerase II (nucleoplasmic)	Chan et al., 1972; Lobban and Siminovitch, 1975
Aminopterin	Over production of tetrahydrofolate reductase	Littlefield, 1969
Ouabain	Resistant Na⁺ K⁺-activated ATPase	Baker et al., 1974
Colchicine	Permeability barrier	Minor and Roscoe, 1975

resistant (Ou^R). When this mutant is fused with normal, ouabain-sensitive cells (HGPRT$^+$, Ou^S) the resulting hybrids will be able to grow on HAT medium containing ouabain whereas the parental cells die.

Additional selective systems probably will soon be developed since many other types of drug resistant cells have been reported (Table IX-1). Among these are cells which are deficient in specific enzymes while others seem to owe their drug resistance to over production of enzymes or to permeability changes. Although characterizing the different mutants requires much work, the necessary time is well spent. Knowing the nature of the genetic lesion it is possible not only to design appropriate selection systems for the hybrids but also to obtain important new biological information about biochemical pathways and cell physiological processes. These aspects are considered in Chapter XIII which discusses gene complementation analysis with hybrid cells.

3. Hybrids Made from Auxotrophic Mutants

The techniques for producing cells requiring special nutrients (auxotrophs) from nonrequiring cells (prototrophs) were first developed for bacteria. Through the work of Puck and Kao (Kao and Puck, 1967, 1968, 1970, 1972a,b; Kao *et al.*, 1969a,b) and Chu *et al.* (1969, 1972), similar methods have been developed for eukaryotic cells and found to be very useful for cell hybridization work. Thus, mutants that require amino acids, carbohydrates, purines, or pyrimidines have been isolated (Table IX-2). Double or triple auxotrophs, that is, cells requiring the addition of 2 or 3 different substances, have also been reported.

Auxotrophic mutants can be isolated by several different methods. Puck and Kao (1967) used a technique where the majority of normal cells were allowed to multiply rapidly on a minimal medium from which one amino acid had been excluded. Under these conditions mu-

TABLE IX-2

Examples of Auxotrophic Mutants

Substance required	Abbreviations	References
Glutamine	Glu$^-$	De Mars and Hooper, 1960; Chu *et al.*, 1969
Serine	Ser$^-$	Jones and Puck, 1973
Proline	Pro$^-$	Kao and Puck, 1967
Glycine	Gly$^-$	Kao and Puck, 1968; Kao *et al.*, 1969b
Adenine	Ade$^-$	Kao and Puck, 1972a; Puck, 1974; Patterson, 1975; Patterson *et al.*, 1974
Inositol	Ino$^-$	Kao and Puck 1968

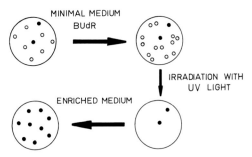

Fig. IX-3. Selection for auxotrophic mutants is achieved by growing cells on a minimal medium where a specific nutrient has been excluded. The wild-type cells (empty circles) will incorporate BUdR into DNA as they multiply, while auxotropic cells (filled circles) requiring the nutrient fail to grow and therefore do not incorporate BUdR. When the cultures are irradiated the wild-type cells will be killed because of the photosensitizing effect of BUdR while the nongrowing auxotrophic cells survive. The minimal medium is replaced with one containing the nutrient required by the auxotrophic cells, these cells then resume multiplication.

tant cells requiring this amino acid do not proliferate. The cultures were then supplied with BUdR that was incorporated into DNA by the growing DNA-synthesizing cells. DNA which has incorporated BUdR is more sensitive to light-induced chromosome breakage than is normal DNA. When the cultures were irradiated with UV light, the rapidly proliferating normal cells were therefore selectively killed, whereas the mutant cells survived (Fig. IX-3). When the cells were then shifted to an enriched medium, the mutant cells proliferated. An alternative approach to using BUdR and light is to kill all DNA synthesizing cells with highly radioactive thymidine (Siminovitch, 1974).

Another method used in the isolation of auxotrophic cells (De Mars and Hooper, 1960) is the thymineless death method, a technique originally developed in microbial genetics.

The use of auxotrophic mutants as parental cells in fusion experiments greatly facilitates the isolation of hybrids. To illustrate this, consider for example, an experiment in which an auxotrophic mutant requiring glycine is mixed with another mutant requiring proline. By means of gene complementation, hybrids arising from this cell combination will be prototrophic and grow on a minimal medium, deficient in both glycine and proline. At the same time both the parental cell types will die because of their defects. Sometimes two different cell lines require the same substance (e.g., adenine) but differ with respect to their gene mutations. For instance, the mutations may affect two different enzymes in the same metabolic pathway or two polypeptide chains on one and the same enzyme. If this is the case, fusion of two auxotrophic cell lines requiring the same substance will also give hybrids which are prototrophic.

4. Hybrids Made from Temperature Sensitive (ts) Mutants

Cultured mammalian cells are capable of multiplying in a temperature range from 32°C to about 40°C, the optimum temperature being 37°C. Using methods similar to those employed in the isolation of auxotrophic mutants it is possible to isolate temperature sensitive (ts) mutants that differ from normal (wild-type) cells in that they are unable to grow at 38°–39°C (nonpermissive temperature) whereas they grow like normal cells at 33°–34°C (permissive temperature) (Naha, 1969; Thompson et al., 1970, 1971, 1975; Siminovitch, 1976). Also "cold-sensitive" mutants can be and have been isolated. Both types of ts mutants offer interesting possibilities for genetic studies of cell functions by means of complementation analysis in hybrid cells.

In order to isolate ts mutants, cells are first cultured and exposed to a chemical mutagen at the permissive temperature (Fig. IX-4). After removal of the mutagen the cells are grown under normal conditions for some time in order to permit the recovery of the cells and fixation of the mutations. The cultures are then shifted to the nonpermissive temperature which is usually taken at or slightly above 37°C. Normal cells, but not the ts mutants, will grow at this temperature. After allowing some time for the mutant phenotypes to express themselves a selective agent (e.g., BUdR or ^{3}H-thymidine) is added in order to kill all cells capable of growing at the nonpermissive temperature. The time of treatment must be long enough to kill normal ("wild-type") cells but not so long that the nonmultiplying ts mutants are irreversibly damaged. The mutant cells can then be recovered by shifting the temperature to the permissive level where they will multiply. The biochemical background of ts mutants in eukaryotic cells is not known. Presumably the ts mutations

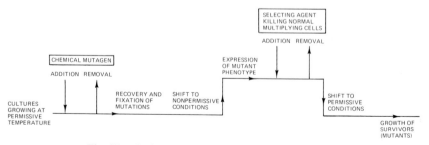

Fig. IX-4. Isolation of temperature sensitive mutants.

involve conformational changes in proteins or RNA which cause them to denature or become inactive when the temperature is changed from the permissive to the nonpermissive level.

Hybrid cells have been isolated after fusing two different ts mutants and from crosses of ts mutants with drug resistant and auxotrophic mutants (Meiss and Basilico, 1972; Thompson and Baker, 1973; Basilico, 1974). The observation that ts mutants, when crossed with each other, frequently produce hybrids which grow at the nonpermissive temperature indicates that the ts mutants may involve many different types of genetic lesions. In spite of the fact that the number of complementation groups and the biochemical defects responsible for the ts phenotype are known only in a few instances (review see Siminovitch,1976), these mutants are very valuable in cell genetic work (see also Chapter XIII).

Temperature can also be used as a selection pressure for the isolation of interspecific hybrids from crosses of nonmutant cells. In these cases the parental cells have been derived from species differing in body temperature. Thus, Goldstein and Lin (1972b) used temperature in order to select hybrid cells from a fusion mixture of Galapagos tortoise cells and golden hamster cells. The experiments of Zepp *et al.* (1971), in which hybrids between insect cells and human cells were isolated, illustrate the combined use of temperature and the composition of the medium for the selection of hybrid cells. In these experiments the hybrid cells grew at 37°C but not at room temperature, whereas the opposite was true for the insect cell line. At 37°C, therefore, the hybrid cells increased in number while the insect parent was gradually eliminated. The human parental cells were selected against by using a culture medium for insect cells.

5. Hybrids Made from Mutant Human Cells

A large number of inborn errors of metabolism are known in man and many of these conditions can be detected *in vitro* in fibroblasts growing out from skin biopsies (Krooth and Sell, 1970; Hsia, 1970). One advantage of using mutant cells from patients is that their genetics is often well documented on the basis of pedigree analysis. Table IX-3 lists some syndromes that may be of interest in future cell fusion work and that involve defects in carbohydrate, lipid, amino acid, and nucleotide metabolism. For five of these diseases (galactosemia, orotic aciduria, citrullinemia, arginosuccinyluria, and Lesch-Nyhan syndrome) there are selective media which prevent the proliferation of normal cells but permit growth of the homozygous mutant cells. For example, patients with the Lesch-Nyhan syndrome lack the enzyme HGPRT and their cells can therefore grow in 8-azaguanine or thioguanine-containing media.

TABLE IX-3

Some Genetic Defects in Human Cells of Potential Use in Somatic Cell Hybridization[a]

Syndrome	Molecules affected
Arginosuccinic aciduria[b]	Arginosuccinase
Citrullinemia[b]	Arginosuccinate synthetase
GM$_1$ gangliosidosis type I[c]	Absence of β-galactosidase A, B, and C
GM$_1$ gangliosidosis type II[c]	Absence of β-galactosidase B and C
GM$_2$ gangliosidosis type I (Tay-Sachs disease)	Absence of hexosaminidase A
GM$_2$ gangliosidosis type II (Sandhoff's disease)	Absence of hexosaminidase A and B
G6PD deficiency[d]	Glucose-6-phosphate dehydrogenase
Galactosemia[b]	UDP-galactose transferase
Gaucher's disease	Glucocerebrosidase
Hunter's syndrome[d]	α-L-iduronidase
Hurler's syndrome	Sulfiduronate sulfatase
Maple syrup urine disease	Branched chain ketoacid decarboxylase
Metachromatic leucodystrophy[c]	Arylsulfatase A
Lesch-Nyhan syndrome[b,d]	Hypoxanthine-guanine phosphoribosyl-transferase
Orotic aciduria[b]	Orotidine-5'-phosphate pyrophosphorylase and Orotidine-5-phosphate decarboxylase
Xeroderma pigmentosa	DNA repair enzymes

[a] Modified from Krooth and Sell (1970).
[b] Selective culture media exist.
[c] Lysosomal enzyme defect.
[d] X-linked.

Hybrids between TK$^-$ cells and cells from these patients can be isolated in Littlefield's selective HAT system (Migeon and Miller, 1968; Croce and Koprowski, 1974c).

E. SOME PROBLEMS IN USING MUTANT CELLS FOR HYBRIDIZATION

The work discussed in the preceding pages illustrates how the use of mutants can facilitate the isolation of hybrid cells. To complete the picture, however, it is necessary to discuss some of the complications which may arise in selecting hybrids from mixtures of mutant parental cells. Probably the most common difficulty is that the cells isolated turn out to be altered parental cells and not hybrid cells. The reason for the failure of the isolation procedure may be that the parental cells were phenotypic variants or they were mutant cells which underwent a reverse mutation that enabled them to grow on the selective medium. Another problem which makes cell hybrid data difficult to interpret is the fact that different types of mutations may give rise to the same phenotype (see the discussion of *xeroderma* cell hybrids in Chapter VII). Thus, the absence of a certain enzyme activity may be due to many different types of structural gene modifications or to loss or modification of regulatory genes. The fact that mutants are heterogeneous with respect to their genetic lesions makes it possible to use them in the analysis of metabolic pathways, enzyme structure, and complex cell functions but it may also be a problem in the isolation of hybrid cells. The following sections therefore discuss the criteria used to define mutant cells, the heterogeneity of mutants, and some factors which complicate the isolation of hybrid cells.

1. Definition of Mutant Cells

Ideally, a series of criteria should be satisfied before a variant cell is classified as a mutant. The most important of these criteria are that the altered phenotype (a) occurs randomly and with a frequency that is increased by mutagens; (b) is inherited by daughter cells and persists after mutagens and selective pressures have been removed; (c) is due to changes in specific gene products, e.g., proteins; (d) can be assigned to a certain chromosome or be mapped to a specific chromosome region; and (e) can be reversed and the reversion frequency increased by the addition of mutagens.

Unfortunately very few "mutant" cells have been characterized with respect to all these criteria. Among those which have been used in cell hybrid work, the majority do, however, satisfy criteria (a) and (b). Usually information concerning points (c)–(e) is either lacking or incomplete. Some cell lines which satisfy criteria (a)–(d) do not meet the last criterion (e) since they fail to show revertants. In these cases it is probable that the altered phenotype is due to the loss of a specific chromosome region thus making the change irreversible (deletion mutants).

Interpretation of the results of cell fusion studies may also require knowledge of whether (a) the mutant gene shows dominant or recessive expression and (b) the mutant gene or chromosome is present in one, two, or more copies. In this context it is important to remember that animal cells, with the notable exceptions of some haploid frog cell lines (Freed and Mezger-Freed, 1970; Mezger-Freed, 1972) are either *diploid, tetraploid, octoploid,* or *heteroploid.* In diploid cells where all autosomes occur in pairs there are *homozygous* (AA or aa) and *heterozygous* Aa or aA) states for any given gene depending on whether or not the two homologous chromosomes involved carry the same or dif-

ferent alleles. In polyploid and heteroploid cells gene dosage relationships may be very complex and difficult to determine. Even if chromosome banding techniques (see Chapter X) make it possible to establish
that a certain chromosome occurs in two or more copies the cells may be
functionally *hemizygous* (haploid) for individual genes or groups of
genes on this chromosome pair. The discrepancy between functional
gene frequency and ploidy has given rise to a great deal of controversy
in studies of mutation rates in cells of different ploidies, causing some
workers to doubt that mammalian cell mutants are in fact true mutants.
For a recent discussion of these problems see Siminovitch (1976).

2. Heterogeneity of Mutants with a Common Phenotype

As already indicated there is evidence that many different types of
structural or regulatory gene mutations may result in the same phenotype. Clearly genetic heterogeneity within a given class of mutants may
influence the spectrum of phenotypes that can be observed in hybrids
(for a discussion of these problems see Siminovitch, 1976). This is well
illustrated by the much used azaguanine resistant (AG^R and $HGPRT^-$)
mutants. Thus, by selecting at different AG concentrations it is possible
to obtain cell lines with different levels of resistance (Littlefield, 1963,
1964b). The most likely explanation for this is that different cell lines
vary with respect to the degree of enzyme deficiency (number of
$HGPRT^+$ genes) or that AG resistance may be caused by mechanisms
other than HGPRT deficiency. A second point which illustrates the heterogeneity of AG^R mutants is that a few cell lines are AG^R but $HGPRT^+$
(Szybalski *et al.*, 1962; Gillin *et al.*, 1972). In these cases it seems likely
that the drug resistance is caused by permeability changes rather than
enzyme deficiency.

Cell fusion experiments can be used to examine the heterogeneity
among mutants of a certain type. Sekiguchi and co-workers (Sekiguchi
and Sekiguchi, 1973; Sekiguchi *et al.*, 1974, 1975a) examined independently derived $HGPRT^-$ mutants by fusing them pairwise in different
combinations and then testing for cells capable of growing on HAT
medium. As a result a series of $HGPRT^+$ hybrid cells were obtained.
This indicates that some combinations of $HGPRT^-$ cells show gene
complementation and therefore become $HGPRT^+$. The explanation
could be that the enzyme consists of several subunits each specified by
a different gene. If the parental cells differ with respect to which subunit is altered it is possible that the hybrid cell would be able to produce active enzyme. An alternative explanation is that enzyme deficiency may be caused not only by mutations in structural genes but also
by mutations in regulatory genes. Evidence in favor of this explanation

has been obtained by Watson *et al.* (1972) who found that *mouse* HGPRT activity was expressed in hybrids produced by fusing mouse A9 (HGPRT⁻) cells with human (HGPRT⁺) cells. Similarly Bakay *et al.* (1973) found *mouse* HGPRT activity to be restored in hybrids made by fusing mouse HGPRT⁻ cells + chick fibroblasts, while Croce *et al.* (1973b) found *rat* HGPRT activity in rat (HGPRT⁻) + human (HGPRT⁺) hybrids. Roy and Ruddle (1973) fused HGPRT⁻ mouse cells with HGPRT⁻ human cells (Lesch-Nyhan) and obtained hybrids by selection on HAT medium. Isoelectric focusing studies showed the hybrid cells to contain HGPRT enzyme with an isoelectric point intermediate between that of the mouse and the human enzyme suggesting that the enzyme activity was due to a heteropolymeric enzyme. These observations are interesting not only because they illustrate problems which may be encountered in using mutants and selective media to isolate hybrids but also because they show that cell hybridization can be used to dissect regulatory mechanisms controlling gene expression in eukaryotic cells.

3. Revertants

As already indicated, one important characteristic of mutant cells is that they can revert to a normal phenotype by undergoing back mutations. In experiments where hybrid cells are isolated such back mutations may present a problem since the revertants will grow on the selective medium and contaminate the hybrid cell population. It is important therefore to check cells which grow on the selective medium for karyotype and/or isozyme patterns so as to verify the hybrid nature of the cells. (For a discussion of these problems see Shin *et al.*, 1971, 1973.)

X

Chromosome Patterns in Hybrid Cells

A. CHROMOSOMAL CONSTITUTION OF HYBRID CELLS

Chromosome analysis played an important part in the discovery of hybrid cells. The presence in one cell of marker chromosomes characteristic of both of the two cell types originally mixed together was the strongest evidence that the newly emerging cells were in fact hybrids. Chromosome analyses have since then been an important method for the identification of hybrid cells.

The intraspecific mouse hybrid cells studied in the early work of Barski *et al.* (1961) and Scaletta and Ephrussi (1965) contained a total chromosome number which was very close to the sum expected if two cells, one of each parental type, had fused. Later studies have, however, shown that some hybrid lines may have more than two sets of chromosomes. Some of these hybrids may have arisen from fusions involving polyploid cells but there is also experimental evidence (Ricciuti and Ruddle, 1971) that two cells of one type and one cell of another type can fuse together to produce multiplying mononucleate hybrids. Another mechanism suggested for the formation of hybrids with a "2 + 1" chromosome complement is endoreduplication. If after a 1 + 1 fusion, only one of the two chromosome sets undergoes a replication cycle the first hybrid karyotype would be a 2 + 1 complement. Such 2 + 1 or even 3 + 1 hybrids have been observed by several groups of workers (Matsuya *et al.*, 1968; Nabholz *et al.*, 1969; Koyama *et al.*, 1970; Ricciuti and

162

Ruddle, 1971; Jami *et al.*, 1971; Marin and Pugliatti-Crippa, 1972; Labella *et al.*, 1973). Furthermore, it is worth noting that Nyormoi *et al.*
(1973b) obtained hybrid cell lines by fusing haploid cells (spermatids)
with diploid cells. These hybrids had a karyotype which approximated
that expected from a "1 + ½" fusion.

Although there are these interesting exceptions the majority of hybrids which have been reported appear to have originated from 1 + 1
fusions of diploid or near-diploid cells. A number of reports (e.g., Siniscalco *et al.*, 1969b; Kao *et al.*, 1969a,b; Silagi *et al.*, 1969; Hors-Cayala
et al., 1972; Glaser and O'Neill, 1972; Francke *et al.*, 1973; Bengtsson
et al., 1975), thus, confirm the early observations of Barski and
Ephrussi (Barski *et al.*, 1961; Sorieul and Ephrussi, 1961) that intraspecific hybrids show chromosome numbers which approximate or
are slightly less than the expected if one assumes that 1 + 1 fusion
has occurred. But these reports also show that the chromosome number may vary from one hybrid cell to another. Only rarely is the
mean value for the cell population an exact sum of the two parental
karyotypes.

It was not until interspecific hybrids were studied (Weiss and Green,
1967) that it became clear that hybrid cell lines may show extensive
chromosome elimination. This phenomenon, also referred to as *chromosome segregation* frequently involves the preferential elimination of chromosomes of one species while the chromosomes of the other species
are selectively retained. Thus, human chromosomes are selectively
eliminated in man + rodent hybrids. Weiss and Green (1967) were
the first to isolate hybrids which retained practically all the mouse chromosomes but had lost almost all of the human chromosomes. Chromosome segregation has now been observed in other types of interspecific
hybrids and also in intraspecific hybrids. From observations on interspecific hybrids certain rules of thumb have been established to indicate which set of chromosomes will be eliminated (segregated) in a
given combination of cells (Table X-1). Although these rules seem to
apply to most crosses involving established cell lines, later work has
shown that the pattern of chromosome elimination can in some cases be
reversed—a phenomenon known as *reverse chromosome segregation*.
However, little is known about the mechanism of chromosome loss and
the factors which determine the direction and rate of chromosome
losses. The fact that chromosome elimination can to some extent be controlled is interesting and useful from an experimental point of view and
has important implications in chromosome mapping with hybrid cells.

In this chapter we shall discuss the chromosomal evolution of inter-
and intraspecific hybrids and mechanisms which may be involved in

TABLE X-1

Preferential Loss of Chromosomes in the Interspecific Somatic Cell Hybrids

Hybrids	Specific origin of eliminated chromosomes	References
Man + mouse	Man	Weiss and Green, 1967; Migeon and Miller, 1968;
	Mouse[a]	Minna and Coon, 1974; Minna et al., 1974b
Man + Chinese hamster	Man	Westerveld et al., 1971; Kao and Puck, 1970
Man + Syrian hamster	Man	Grzeschik et al., 1972; Grzeschik, 1973b
Man + mosquito	Mosquito	Zepp et al., 1971
Mouse + monkey	Monkey	Cassingena et al., 1971
Mouse + mule	Mule	Deys, 1972
Mouse + rat	Rat	Weiss and Ephrussi, 1966a
	Mouse and rat	van der Noordaa et al., 1972
Mouse + Chinese hamster	Mouse and hamster	Scaletta et al., 1967; Graves, 1972b; Labella et al., 1973; Handmaker, 1971
Mouse + Syrian hamster	Mouse	Migeon, 1968
	Mouse and hamster	Wiblin and MacPherson, 1973
Mouse + chick	Chick	Schwartz et al., 1971
Chinese hamster + kangaroo-rat	Kangaroo-rat	Jakob and Ruiz, 1970
Chinese hamster + chick	Chick	Kao, 1973

[a] For a discussion of reverse chromosome segregation see Section C,1.

the chromosome elimination process. Since the new methods of chromosome staining and identification are of key importance in analyzing the complex karyotypes of hybrid cells and since the validity of gene mapping with hybrid cells depends heavily on these methods we shall start by briefly summarizing chromosome staining techniques.

B. IDENTIFICATION OF INDIVIDUAL CHROMOSOMES

1. Standard Methods

The standard method of chromosome staining in the early hybrid work was to stain hypotonically treated cells with orcein in alcohol acetic acid. With this method 7 different chromosome groups (A–G) could be distinguished in the normal human karyotype. It was possible to identify chromosomes 1, 2, 3, 9, 16, and Y by studying the morphol-

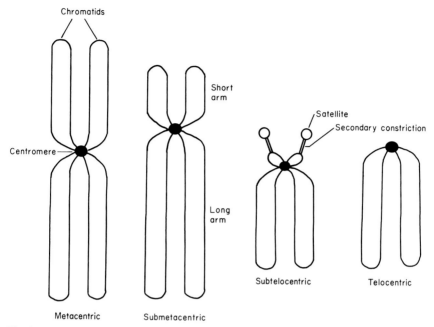

Fig. X-1. Classification of chromosomes on the basis of the position of the centromere.

ogy of the chromosomes, particularly the length of the chromosome arms, the position of the centromere, the localization of secondary constrictions, and the presence of satellites (Fig. X-1). These identifications were possible in normal diploid cells but with the more complex hybrid karyotypes, where the chromosomes of two different species often had to be distinguished from each other, and where chromosomal rearrangements were common, the possibilities of recognizing the individual chromosomes were much more limited. In early work in this field the chromosomes were, therefore, often only classified into groups depending on the positions of the centromeres (Levan *et al.*, 1964), i.e., into metacentric, submetacentric, acrocentric (subtelocentric) and telocentric chromosomes (see Fig. X-1).

2. Banding Techniques

A complete analysis of human and animal karyotypes and of hybrid karyotypes became possible only when the new chromosome banding techniques were introduced (Caspersson *et al.*, 1968, 1970, 1971; Sumner *et al.*, 1971; Arrighi and Hsu, 1971). When stained with quinacrine, animal and plant chromosomes show fluorescing bands with

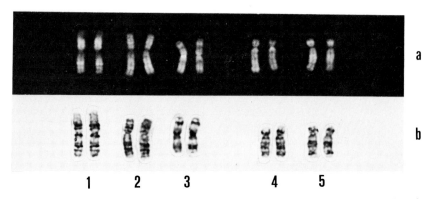

Fig. X-2. (a) Quinacrine stained human chromosomes number 1–5 photographed in the fluorescence microscope. (b) Giemsa banding ("G-banding") of human chromosomes number 1–5. (Courtesy Dr. L. Zech.)

darker bands between (Fig. X-2a). Each chromosome has its own characteristic Q-band pattern which makes it possible to identify each individual chromosome of man and also those of the common laboratory animals. The differential fluorescence pattern obtained after Q-staining has been attributed to differences in the extent of dye binding and/or to fluorescence enhancement and quenching phenomena. These phenomena in turn may be caused by differences in base composition, by local denaturation of the DNA double helix, or by differences in protein composition within different chromosome regions (for reviews see Caspersson and Zech, 1973; Hsu, 1973; Miller et al., 1973; Hecht et al., 1974). The standard method of Q-staining involves the staining of hypotonically treated metaphase cells which have been fixed in 3:1 methanol:acetic acid and then spread on glass slides.

In 1971, several groups of workers found that banding patterns could also be observed in Giemsa stained preparations if the chromosomes were subjected to various types of treatments before staining (Sumner et al., 1971; Drets and Shaw, 1971; Schnedl, 1971; Seabright, 1971). At present the most popular and widely used G-banding technique is the trypsin method of Seabright (1971) which gives good G-banding patterns with chromosomes from many different species including man (Fig. X-2b). Darkly staining G-bands are found in almost the same regions as the brightly fluorescing Q-bands but there are minor differences which are interesting and make it useful to combine both methods.

A modification of the Giemsa method, in which the cells are stained at an alkaline pH ("Giemsa-11 technique"), has been particularly useful

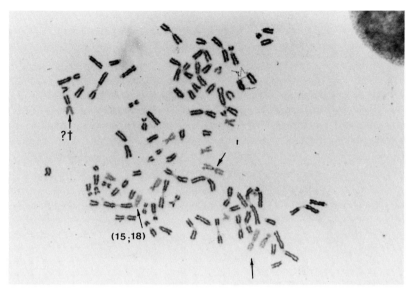

Fig. X-3. "Giemsa-11" staining of a mouse + human hybrid carrying 12 human chromosomes including a human translocation *t* (15; 18) chromosome and an unidentified translocation (? *t*) undetected by conventional analysis. Some of the human chromosomes are indicated by arrows (from Friend *et al.*, 1976).

in the analysis of mouse + human hybrids (Fig. X-3). Mouse chromosomes stain magenta with blue centromeres while human chromosomes stain blue with magenta centromeres (Bobrow *et al.*, 1972; Bobrow and Cross, 1974). This difference in staining pattern facilitates the identification of individual human or mouse chromosomes and can also be used to define human + mouse translocations (Friend *et al.*, 1976).

Methods which preferentially stain centromeric heterochromatin ("C-bands") (Pardue and Gall, 1970; Arrighi and Hsu, 1971; Hilwig and Gropp, 1972; Sumner, 1972) have also been useful in the analysis of hybrid karyotypes. All normal mouse chromosomes (except Y) have been found to contain large blocks of very strong centromeric heterochromatin bands whereas most human centromeric chromatin regions give considerably smaller C-bands. Among the human chromosomes 1, 9, and 16 contain greater amounts of centromeric heterochromatin than other chromosomes. The difference between mouse and human C-bands, however, is such that the C-banding can be used to distinguish most mouse and human chromosomes in hybrid karyotypes (Chen and Ruddle, 1971). Modified G- or C-banding methods or combination methods for the sequential staining of chromosomes may also be used for this purpose (Marshall, 1975; Kim and Grzeschik, 1974).

3. Nucleic Acid Annealing

Basically this technique involves denaturation ("melting") of double stranded DNA in cytological preparations by heating or using alkali to form single stranded regions. Single stranded radioactive DNA or RNA which is complementary (cDNA or cRNA) to specific genes is then allowed to anneal to the denatured chromosomal DNA. The exact localization of the labeled cDNA or cRNA is established by autoradiography. Using this technique it is possible to show that the large blocks of centromeric heterochromatin (C-bands) characteristic of mouse chromosomes contain highly repetitive DNA sequences. With this technique it is also possible to determine the species of unknown centromere regions in rearranged chromosomes present in interspecific hybrid cells. Because human cDNA does not cross react with mouse centromeric DNA and vice versa Boone *et al.* (1972) were able to identify a translocation between a mouse and a human chromosome which carried two centromere regions, one derived from a human chromosome and one from a mouse chromosome.

4. Nomenclature

The new banding techniques have made it possible not only to identify and number individual chromosomes but also to specify regions on the chromosome arms. The Paris Conference (1972) on "Standardization in Human Cytogenetics" has suggested a system of nomenclature which can be used to describe the normal human karyotype as well as karyotypes containing additional or rearranged chromosomes. Thus, the normal human male karyotype is described as "46, XY" and the female as "46, XX." Plus or minus signs before a chromosome designation are used to indicate additional or missing whole chromosomes. Thus, 45, XY -21 means a male karyotype missing chromosome 21.

Identification of individual chromosomes is based on the ratio of arm lengths (centromeric index), location of secondary constrictions, and the banding patterns obtained by the Q- and C-banding techniques. Each chromosome arm is considered to consist of continuous sequences of *regions* which are made up of fluorescent and nonfluorescent bands. The long arm of a chromosome is represented by the symbol q whereas the short arm is indicated by the symbol p. Starting from the centromere and proceeding distally towards the ends of the chromosomes successive regions of the chromosome arms are numbered consecutively q 1, 2, 3 etc. (long arm) and p 1, 2, 3, etc. (short

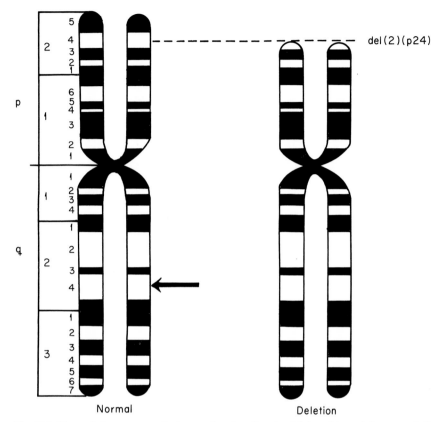

del(2)(p24)

Normal   Deletion

Fig. X-4. Normal chromosome 2 of man showing the short (p) and long (q) arms subdivided into regions and individual light and dark bands. Arrow indicates position of region 2 band 4 (2q 24). The left chromosome is an abnormal, human chromosome no. 2 which has undergone a deletion in the short arm, region 2 band 4.

arm). The midpoints of the *main* bands (not the margins) are used as "landmarks" to delimit the different regions (Fig. X-4). A chromosome region may in itself contain several smaller bands. Thus, 2q 24 indicates chromosome no. 2, long arm, region 2, band 4 (see arrow in Fig. X-4). The nonfluorescent bands after Q-staining (or weakly staining bands after G-staining) are sometimes referred to as "interbands" but should according to the Paris Conference (1972) also be classified as nonfluorescent or weakly staining bands.

It is also possible to describe rearranged chromosomes using the nomenclature of the Paris Conference (1972). In this case the first symbol is used to identify the type of rearrangement. Deletions for example are represented by *del,* whereas *t* indicates a translocation. The break points involved in the rearrangements are then described with the aid of the band designations indicated above. If there is a terminal deletion in the short arm of chromosome 2 with the break point in region 2 band 4 this is written as *del* (2) (p24). This chromosome aberration is illustrated in Fig. X-4). [For more detailed information about this nomenclature, the Paris Conference (1972) should be consulted.]

5. Heterogeneity Curves

Hybrid cell populations often show considerable variations in their chromosomal constitution from one cell to another. In some lines no two cells seem to have exactly the same chromosome complement. Great variation also can be found in cloned cell populations where all cells must be derived from a single fusion event. An example of the type

TABLE X-2

Cell by Cell Analysis of Human Chromosomes in a Hybrid Clone (Cl 6) of Hybrid Cells Obtained by Fusing Mouse A9 Cells with a Near-Diploid Human Lymphoblastoid Line[a]

Cell no.	Number of copies of each human chromosome[b]											
	2	10	11	12	13	17	19	20	22	X	Y	3p
1	-	1	1	1	2	-	-	-	-	1	-	-
2	-	1	-	1	1	-	1	-	1	1	-	-
3	-	1	-	1	2	1	1	-	-	-	-	-
4	-	-	1	1	2	1	-	-	-	1	-	-
5	-	-	1	1	2	1	-	-	-	1	-	-
6	-	1	-	-	2	1	-	1	-	1	-	-
7	-	1	1	1	1	-	1	-	1	1	-	-
8	1	-	-	1	1	1	-	-	1	-	-	-
9	-	-	-	-	2	1	-	-	1	1	-	-
10	-	-	2	-	2	-	-	-	-	1	-	-
11	-	1	1	1	1	-	-	-	-	1	-	-
12	-	-	-	-	2	1	1	-	-	1	1	1
13	-	1	-	1	1	1	-	-	-	1	-	-
14	-	-	-	-	1	2	-	-	1	1	-	-
15	-	-	1	-	2	1	-	-	-	1	-	-
16	-	1	-	-	2	-	-	-	-	1	-	-
17	-	-	1	-	1	1	-	-	-	-	-	-
18	-	-	-	-	2	1	-	-	-	-	-	-
19	-	-	-	-	2	-	-	-	-	-	-	-
20	-	-	-	-	1	-	-	-	-	-	-	-

[a] Allderdice *et al.*, 1973b.

[b] In addition to the human chromosomes listed here the hybrid cells contain a complete set of mouse chromosomes. One human chromosome (3p) underwent a morphological alteration.

of intercellular variation which can be encountered in a man + mouse hybrid (Allderdice *et al.*, 1973b) is given in Table X-2. The individual hybrid cells all contain a complete set of mouse chromosomes but vary greatly with respect to the number and type of human chromosomes. This heterogeneity can be described graphically (Figs. X-5 and X-6) by plotting heterogeneity curves (Allderdice *et al.*, 1973b). If one determines, for instance, the number of different human chromosomes present in the second cell analyzed but not in the first and present in the third but not in the first or the second, then one can plot the mean cumulative number of different human chromosomes against the number of cells analyzed. The shape of the curve depends to some extent on the order in which the cells are considered and therefore a mean curve is constructed (Fig. X-5) which is based on considering the cells in different orders. Figure X-6 shows that the heterogeneity curves for five different man + mouse hybrid lines may differ considerably. The slope of the curves gives a measure of the intercellular variation in human chromosome content. A cell population which shows little variation will give a flat curve. If, for example, one considers the unlikely event that all cells in a man + mouse hybrid contain the same 10 human chromosomes, a straight horizontal line will be obtained, a line which if extrapolated will cross the ordinate at a value of 10. If a total of 10 different chromosomes occur in the different hybrid cells but some cells have only one human chromosome and others have from 2 to 10, the

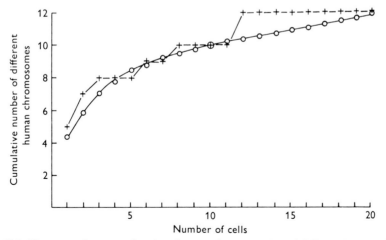

Fig. X-5. Heterogeneity curve showing the cumulative number of different human chromosomes in an A9/Daudi hybrid line; +, the cells considered in the order shown in Table X-2; ○, the cells considered in 20 different orders and mean values plotted (Allderdice *et al.*, 1973b).

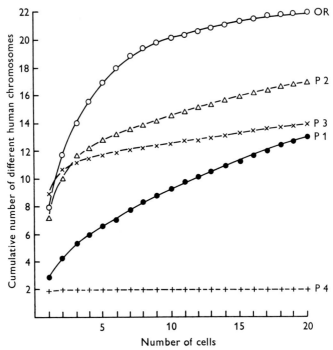

Fig. X-6. Heterogeneity curves for human chromosomes in 5 A9/Daudi hybrid cell lines (Allderdice *et al.*, 1973b).

curve will rise from a value of 2 to a value of 10 where the curve will level off.

The shape of the heterogeneity curves also gives an indication of how many cells should be analyzed to obtain a representative picture of the mean karyotype of the population. In Fig. X-6 the curves tend to level off after about 20 cells. In most work with hybrid cells the chromosome compositions of 20–30 cells have been analyzed.

C. INTERSPECIFIC HYBRIDS

1. Direction of Chromosome Losses

The discovery by Weiss and Green (1967) that mouse + man hybrids rapidly lose human chromosomes was soon confirmed by other similar reports (Migeon and Miller, 1968; Nabholz *et al.*, 1969; Ruddle *et al.*, 1970b). However, at least one exception was found. Jami *et al.* (1971) produced a series of man + mouse hybrids and obtained one line in which most of the human chromosomes were retained whereas the mouse component was drastically reduced. For other interspecific crosses, such as those between two different rodents, the direction of segregation varies. In rat + mouse hybrids the rat chromosomes are selectively eliminated whereas in hamster + mouse crosses both parental sets are reduced but with a preferential loss of mouse chromosomes.

In most interspecific crosses which had been examined for chromosome segregation patterns before 1971, the parental cells had been heteroploid cells of established lines. As hybrids involving diploid cells began to be studied it became apparent that the direction of chromosome loss depended not only on the species but also on the type of parental cells used. Among hybrids between diploid guinea pig peritoneal cells and human HeLa cells Colten and Parkman (1972) noted that both parental complements were reduced but with a preferential loss of guinea pig chromosomes. Reverse chromosome segregation was then observed by Minna and Coon (1974) and Minna et al. (1974b) in a large number of independent hybrids produced by fusing normal diploid mouse or rat cells freshly liberated from heart, liver, bone marrow, spleen, thymus, and the central nervous system with human cells of the aneuploid VA2 or D98 lines. In these studies the rodent chromosomes were preferentially lost in most hybrid lines. This segregation pattern is the opposite of that observed for interspecific crosses of cells from established human and rodent lines (Table X-1). In these crosses the rodent chromosomes are retained while the human chromosomes are eliminated.

Attempts have also been made to influence the direction of chromosome elimination by damaging the chromosomes of one of the parental cells before fusion. Pontecorvo (1971) exposed mouse or hamster cells to X or γ irradiation and then fused these cells with normal cells. In mouse + hamster fusions where the hamster parent had been irradiated, a preferential loss of hamster chromosomes was recorded, but if the mouse cells were irradiated the mouse complement suffered greater losses than the hamster component. With moderate doses of irradiation (600 R) the number of chromosomes lost was relatively small. When the cells were more heavily irradiated (1000 R) more extensive losses of the irradiated chromosomes occurred but this dose also induced more chromosome aberrations. As an alternative to irradiation Pontecorvo (1971) exposed cells to bromodeoxyuridine (BUdR) and light treatments. BUdR is incorporated into DNA and photosensitizes the DNA so that chromosome breaks occur when the cells are exposed to light. Although BUdR labeled hamster chromosomes were preferentially lost in a hamster + mouse hybrid this form of pretreatment obviously suffers from the same drawback as X irradiation. Although effective in directing chromosome losses, the chromosomes also undergo rearrangements which tend to make chromosome identification in the segregated hybrid cells very difficult. But as will be discussed in Chapter XIII, cytological chromosome identification is not always necessary for chromosome mapping experiments. When some gene loci are already known they can serve as landmarks to determine the location of new loci. This principle is used in a new method of chromosome mapping developed by Goss and Harris (1975) in which the chromosomes in one of the parental cells are extensively fragmented by X radiation.

Back selection with drugs has in some cases been successful in forcing the elimination of the complementing chromosomes of hybrids from HGPRT$^-$ + TK$^-$ fusions. In these cases the hybrids have first been isolated in HAT medium (see Chapter IX) and then been transferred to normal medium containing either thioguanine or BUdR. In the presence of thioguanine only hybrids that eliminate the X chromosome with the HGPRT gene will survive, whereas in the presence of BUdR the elimination of the thymidine kinase gene (in man on chromosome 17) is forced. Selection with TG or BUdR may not only select for cells which have lost the two chromosomes carrying the *HGPRT* or *TK* genes but may also select for cells which have undergone extensive losses of seemingly unrelated chromosomes (Marin and Littlefield, 1968; Marin, 1969; Marin and Pugliatti-Crippa, 1972). Wiblin and MacPherson (1973) in a study of hybrids between SV40 transformed Syrian hamster cells (BHK) and mouse 3T3 cells, combined BUdR selection against the hamster chromosomes with immunological selection against surface antigens of hamster cells by including cytotoxic antihamster antibodies in the medium. In this way elimination of hamster chromosomes was obtained in an interspecies combination which in other cases has shown elimination of mouse chromosomes. Later work (Collins *et al.* 1975) indicates that the chromosomes lost during selection with antisera are those carrying genes for species specific surface antigens.

From these examples it appears that chromosome segregation in populations of interspecific hybrids can to some extent be directed either by choosing the appropriate parental cells, by damaging the undesired chromosome complement before fusion, or by applying selective pressures against variants expressing the genes on the chromosome one is seeking to eliminate.

2. Timing of Losses

The rate at which chromosome losses occur may to some extent reflect the phylogenetic distance between the two parental cells in an interspecific fusion. With closely related species, such as two different rodents, (e.g., mouse + Chinese hamster) the rate is usually slow whereas with cells from different classes (man + chick) chromosome segregation is very rapid. Within the life history of a hybrid cell line, chromosome elimination may be rapid during an early phase. This phase is followed by a period during which the chromosome number is stabilized and chromosome segregation is slow (Weiss and Green, 1967; Nabholz *et al.*, 1969). In the man + mouse hybrid lines studied by Weiss and Green (1967) all but 2–15 human chromosomes were eliminated during the first 20 cell generations after fusion; but after this time the human chromosome count stabilized at 1–3 human chromosomes. There may, however, be great variations from one clone to another even within the same hybrid line (Ruddle *et al.*, 1970a). Thus, Miggiano *et al.* (1969) found one man + mouse hybrid clone which retained a full chromosome complement for 4 months while at the same time another clone lost 50% of its human chromosomes.

Although man + mouse hybrids have been extensively used for chromosome mapping it is possible that man + Chinese hamster hybrids may be even more favorable material for such studies. Kao and Puck (1970) and Puck (1974) found that hybrids containing a hamster complement plus only one or two human chromosomes can be obtained within 1–2 weeks. This rapid chromosome segregation is advantageous since it saves much time and minimizes the risk of chromosomal rearrangements. The fact that the Chinese hamster cell has only 22 chromosomes which can easily be distinguished from those of man is also an advantage.

3. Random or Nonrandom Chromosome Segregation

Chromosome segregation in interspecific hybrids is clearly a nonrandom process since chromosomes of one of the parental genomes are preferentially lost. It is also of interest and importance to know whether individual human chromosomes in a man + mouse hybrid are nonrandomly eliminated. Preliminary results obtained with the conventional methods of chromosome cytology suggested that the elimination of chromosomes from the segregating set was nonrandom (Yoshida and Ephrussi, 1967; Santachiara *et al.*, 1970; Ruddle *et al.*, 1970a). Since the new banding methods came into general use there have been several reports of departures from randomness in hybrids involving human cells. Douglas *et al.* (1973a) found that certain human autosomes and X were selectively retained during the early phase of rapid chromosome segregation in man + Chinese hamster hybrids. The presence of the X chromosome, however, was expected since the hybrid cells had to retain this chromosome in order to survive in the selective medium (HAT) used for the isolation of the hybrid cells.

Croce *et al.* (1973e), in some experiments where SV40 transformed human cells were fused with mouse TK⁻ cells, found that human chromosome 7 was consistently retained with chromosome 17 (which carries the complementing human *TK* gene). According to these investigators chromosome 7 carries the integrated SV40 genome (see Chapter XV). This chromosome may therefore confer a selective advantage in the form of a transformed phenotype and a fast growth rate. Selective elimination or retention of chromosomes has been observed in other cases (Norum and Migeon, 1974), but neither phenomenon has been studied extensively enough to be able to establish any rules by which one could make predictions about the segregation pattern likely to be found in a given fusion.

These observations, although difficult to explain, point to an interdependence between different chromosomes or groups of chromosomes. This aspect is discussed in more detail in Section E of this Chapter where the mechanism involved in chromosome elimination is considered.

D. INTRASPECIFIC HYBRIDS

Intraspecific hybrids in general show a greater chromosomal stability than do interspecific hybrids. Drastic reductions in chromosome numbers have only been observed in some instances where one or both of the parental cells was derived from an established cell line (Engel *et al.*, 1969a,b, 1971; Bregula *et al.*, 1971; Wiener *et al.*, 1971). Attempts have been made to provoke a more rapid chromosome segregation by drug treatments and low temperature (Littlefield, 1965; Marin, 1969) but have met with little success.

Contrary to the situation with interspecific hybrids there is little information as to whether chromosome losses in intraspecific hybrids specifically affect one or both of the parental sets. In order to analyze this problem it is necessary to use parental cells which differ with respect to chromosome markers or isozyme variants. Using this approach Bengtsson *et al.* (1975) obtained evidence suggesting that chromosome losses occurred from both parental genomes in a cross involving an established human cell line and normal human lymphocytes.

E. MECHANISM OF CHROMOSOME ELIMINATION

Clearly the loss of chromosomes observed in a hybrid cell population is an evolutionary process where new variant cells are created and selective forces operate. In order to reach an understanding of the mechanisms of chromosome elimination in a hybrid cell population it is necessary to consider (a) factors which cause genetic variation within the cell populations; (b) the nature of the selection pressures operating under a given set of culture conditions; and (c) the selective advantages and disadvantages of different chromosomal variants.

1. Sources of Genetic Variation

In those cases where the parental cells are heteroploid and vary in chromosome constitution a considerable amount of genetic variation will exist within the uncloned hybrid cell populations. Further variation is caused by phenomena which take place shortly after fusion. Many heterokaryons formed after fusion of asynchronously multiplying cells will be heterophasic, i.e., contain nuclei representing different stages of the cell cycle. As already discussed in Chapter VI this will result in premature chromosome condensation and other forms of mitotic abnormalities which may increase the chromosomal variation among surviving mononucleate hybrids. Although these hybrids, in order to survive, must show some degree of coordination with respect to the replication of the two sets of parental chromosomes (Marin and Manduca, 1972), the individual chromosomes retain their characteristics with respect to the timing of their replication during the cell cycle (Sonnenschein, 1970; Marin and Coletta, 1974; Lin and Davidson, 1975).

Furthermore there have been several suggestions that the generation times of the parental cells are of importance for the rate of chromosome elimination in a hybrid cell population. Thus, in several cases the chromosomes of the slow-growing parent have been selectively lost (Kao and Puck, 1970; Marin and Pugliatti-Crippa, 1972). Segregants which eliminate chromosomes from the slow-growing parent grow progressively faster. Matsuya *et al.* (1968), examining different hybrid sublines from the fusion of a slow-growing human cell line with a fast-growing mouse cell line, found that there was a correlation between increased growth rate and elimination of chromosomes from the slow-growing human parent. Another illustration of the role of generation times in chromosome elimination is that in interspecific hybrids the direction of chromosome elimination is reversed if one of the parent cells is a normal diploid cell from a nondividing tissue while the other one is derived from an established cell line (Minna and Coon, 1974).

Another reason why chromosomes are lost may be a *faulty interaction between chromosomes and the spindle fibers.* In polykaryons an unusual situation arises because there are probably more than two centrioles and a much larger number of chromatids to be separated than in a normal cell. If under these conditions the spindle fibers do not interact normally with all the chromosomes an unequal distribution of chromosomes to daughter cells could result. Multipolar spindles, metaphase delay, and other mitotic abnormalities which could be due to faulty interactions between centrioles, spindle threads, and kinetochores have in fact been observed in polykaryons (Heneen et al., 1970; Oftebro and Wolf, 1967). Although there is no evidence for it, species incompatibility between microtubules and centrioles of one species and chromosomes of another species could also be a factor in chromosome segregation in interspecies hybrids.

2. Selection of Chromosomal Variants

In a heterogeneous mixture of hybrids the fastest growing cell enjoys a clear advantage and will overgrow all other cell types. However, the particular cell type with this advantage will probably vary depending on the culture conditions used. It is not possible at present to predict which human chromosomes, if any, would confer a selective advantage on a man + mouse hybrid growing in vitro except in those cases where the human chromosome complements for a genetic defect in a mouse cell and this complementation is therefore required for survival (see Chapter IX and XIII). It is clear though that in many cases the culture conditions select heavily in favor of the chromosomes of one species. This is the case for interspecific hybrids from phylogenetically unrelated species such as insect + man (Table X-1). Since insect cells require very special media, the standard culture conditions designed primarily to support growth of mammalian cells probably favor cells which retain human chromosomes and eliminate insect chromosomes. Furthermore, cells of different species vary with respect to their optimal temperature for growth. Toad cells grow best at 28°C, mammalian cells at 37°C, and bird cells at 40°C. If a man + chick hybrid is cultured at 37°C, theoretically this may select against the chick component and for the human component.

Furthermore it should be realized that in the procedures used to isolate hybrid cells there are several factors operating which may tend to generate and select for broken and rearranged chromosomes. First, if Sendai virus is not completely inactivated the virus may be carried by the cells for a long period of time without manifesting itself in the form of a massive infection (see Chapter IV). There are indications both from Sendai and from other, related viruses, that this may generate chromosome aberrations. Second, the PCC phenomenon (Chapter VI) may be a factor causing chromosome breaks and translocations (Matsui, 1973). Third, the HAT medium itself may select for chromosome abnormalities such as broken chromosomes carrying only the minimum genetic material required for survival. In the case of chick erythrocyte + mouse HGPRT⁻ cells the chick fragment retained is so small that it is not visible in the light microscope. It manifests itself only in that it specifies the synthesis of chick HGPRT in a cell which is otherwise 100% mouse (Harris and Cook, 1969; Cook, 1970; Schwartz *et al.*, 1971; Klinger and Shin, 1974).

A particular type of chromosome rearrangement which can apparently occur in hybrid cells is a translocation of a chromosome piece from one genome to a chromosome of another genome (Boone *et al.*, 1972; Douglas *et al.*, 1973a; Friend *et al.*, 1976). If identified these translocations may be helpful in gene mapping with hybrid cells. It is perhaps more likely, however, that most of these aberrations pass unnoticed and cause discrepancies and upset experiments designed to establish gene maps and linkage relationships.

XI

Phenotypic Expression in Hybrid Cells

A. PARENTAL CELLS AND MARKERS FOR PHENOTYPIC STUDIES

1. Inter- and Intraspecific Fusions

Both interspecific and intraspecific hybrids have been used to study gene regulation and cell differentiation. Interspecific crosses have the advantage that chromosome identification is facilitated and since homologous proteins from two different species usually differ slightly in amino acid composition and electrophoretic mobility they can be distinguished in such hybrids. For example, many human and mouse enzymes can be separated by electrophoresis and so it is possible to determine whether genes in one or in both of the parental genomes of a mouse + human hybrid are expressed. A major disadvantage of some interspecific hybrids (e.g., man + mouse) is the rapid elimination of chromosomes (see Chapter X). It is often difficult or impossible to decide whether the loss of a given property such as the expression of an enzyme, is due to a normal regulatory event, to the elimination of the chromosome carrying the structural gene, or to loss of chromosomes with regulatory genes. Another problem is the uncertainty of whether or not regulatory molecules specified by one species will interact in a normal way with the genome of the other species. Studies with heterokaryons (Chapter VII) indicate that the signals governing the overall activity of cell nuclei cannot be species specific, but it is nonetheless impossible to know if this is true also for the point regulation of all individual genes.

180

Some of the problems with interspecific hybrids can be avoided by using intraspecific hybrids. This type of hybrid offers greater chromosomal stability (see Chapter X) and eliminates the possibility that one genome does not recognize the regulatory signals of the other. On the other hand, chromosomal analysis and the identification of the gene products specified by each of the two genomes is very difficult in intraspecific hybrids. By using parental cells which have different marker chromosomes it is possible to distinguish hybrid cells from polyploid cells of either parental type. If one of the parental cells is derived from an established cell line and the other cell is a normal diploid, it may even be possible to identify a series of rearranged chromosomes that come from the established cell line. To demonstrate that proteins such as enzymes are specified by one genome of an intraspecific fusion, the two parental cells must differ in their isozyme types. Fortunately cells from different inbred mouse strains frequently do differ in one or several isozymes (for example see Ruddle and Nichols, 1971). For example, the A and C3H strains have different isozymes of glucose phosphate isomerase (GPI) and therefore in A + C3H hybrids both forms of the enzymes can be detected by electrophoresis. Alternatively one of the parental cells may be genetically defective for a certain enzyme which the other parent can produce. This aspect will be discussed in more detail under the heading of gene complementation (Chapter XIII).*

2. Phenotype and Developmental Potentiality of Parental Cells

Four main types of parental cells have been used to generate mononucleate hybrid cells:

(a) *Totipotent* or *multipotent* cells like fertilized eggs, early cleavage stages, and teratoma cells (an unusual type of tumor cell arising in the gonads)

(b) Normal diploid cell strains from embryos at late stages of development or from adult tissues

(c) Established cell lines from differentiated tumors

(d) Undifferentiated heteroploid cell lines which have been maintained in culture for a very long time and no longer exhibit any differentiated markers

These types of cells show important differences in their developmental capacity, chromosome number, growth properties *in vitro,* and ability to give rise to tumors *in vivo.* Cells within the first group (a) can give rise to some or all of the differentiated cell types found in the adult

---

* For a list of cell hybrids produced up until 1971–1972 see Sell and Krooth (1972).

organism while cells belonging to the other three groups are restricted in their capacity to differentiate.

Mouse eggs and blastomeres from early embryonic stages have been successfully fused with somatic cells (see Chapter VII) but the resulting heterokaryons do not seem to give rise to synkaryons capable of long term multiplication or differentiation (Graham 1969, 1974). Mouse teratoma cells have been more useful since they grow well and also differentiate in cultures (Lehman et al., 1974; Martin and Evans, 1975). If injected back into animals, teratoma cells give rise to tumors containing highly differentiated tissues (Kahan and Ephrussi, 1970; Rosenthal et al., 1970). The biological properties of these cells will be discussed further in conjunction with teratoma hybrids.

Normal diploid cells from tissues of later embryonic stages and from adult tissues differ from the multipotent cells of the first group in that they are programmed (determined) to undergo specific forms of histiotypic differentiation (erythropoiesis, myogenesis, chondrogenesis, etc.). In order to obtain cells for fusion experiments, primary cultures may be initiated from explants taken from a variety of different tissues and organs. By subculturing primary cultures it is then possible to obtain diploid cell strains which may be cloned to ensure that the cultures only contain one type of cell. One important characteristic of diploid cell strains is that their capacity for growth is limited. After a certain number of cell divisions the cells will become senescent and die.

Established cell lines capable of indefinite multiplication may be obtained by treating cell strains with chemical carcinogens, X irradiation, and other transforming agents (see Chapter XIV). Cell lines established from differentiated tumors often grow well in vitro while still retaining the ability to differentiate. They differ from normal cells not only in the loss of senescence but also in that they have an abnormal karyotype and abnormal growth regulation. Hybrid cells have been obtained by fusing cell lines established from hepatomas (liver), melanomas (pigment cells), gliomas (glia), neuroblastomas (brain), plasmocytomas (plasma cells), and lymphoblastoid tumors (lymphoid cells).

Undifferentiated cell lines with a heteroploid karyotype have been used as the undifferentiated partner in many fusions with differentiated cells. HeLa cells of human origin and L cells established from a mouse source are typical examples of such "undifferentiated" cells. L cells are sometimes described as fibroblasts but this description is misleading. This cell line was established more than 30 years ago and its ability to form hyaluronic acid and collagen, two macromolecules characteristic of the fibroblast phenotype, has not been convincingly documented in recent years. In this presentation these cells are referred to by their laboratory designation and the term fibroblast is reserved for those cells which produce collagen and hyaluronic acid. Most of the undifferentiated cell lines have been in culture for a very long time and thus been strongly selected to grow on glass or plastic in the presence of calf serum, pH indicators, and other substances which are not present in the normal tissue environment. In consequence these cells have escaped many of the factors which control cell growth *in vivo* and have become well adapted to laboratory conditions. L cells will nevertheless give rise to tumors at a low frequency when injected back into mice. As with the differentiated tumor cell lines, the karyotypes of the undifferentiated laboratory cell lines contain supernumerary chromosomes and show chromosome rearrangements (Chapter X). Many mouse cell lines are subtetraploid and contain chromosomes which have been so extensively modified that it is difficult or impossible to trace the origin of a given chromosome in the normal karyotype. The fact that these cells, after a long time in culture, have lost whatever differentiated markers they originally possessed cannot be taken as evidence that the differentiated state is reversible. The *dedifferentiation* characteristic of these cells is more likely the result of a long period of selection for variants which are well adapted to tissue culture. During this process there has been an accumulation of gene mutations which may have favored rapid growth *in vitro* while impairing functions not immediately required for survival or growth.

3. Types of Markers

Hybrid cells have been examined for (a) general characteristics such as morphology, growth rate, contact inhibition, and senescence; (b) complex physiological or immunological properties, e.g., electrophysiological, nutritional or antigenic characteristics; (c) specific gene products, e.g., enzymes, hormones, immunoglobulins, etc.; and (d) sensitivity to specific drugs.

Any of these characteristics can be used as a marker if the two parental cells differ with respect to it. Obviously then a distinction has to be made between properties common to all cells of a given organism (*constitutive markers*) and properties which are expressed only by certain determined cells as they enter the differentiated state (*facultative markers*). Enzymes which are part of vital metabolic pathways, for example, the citric acid cycle and oxidative phosphorylation, or which are important to the basic transcription and replication mechanisms, are constitutive markers. These types of ubiquitous proteins are essential to the basic "housekeeping" functions of all growing cells and are necessary for their survival. Another term which has been used to describe

these proteins has been *household proteins* (Ephrussi, 1972). Facultative markers, on the other hand, would be more or less synonymous
with *luxury molecules* (Ephrussi, 1972) since they represent properties
which are only expressed by cells representing certain epigenotypes.
Facultative markers are not essential to the survival of the single cell but
may be of vital importance to the organism as a whole. Hemoglobin,
myosin, melanin, the brain specific protein S-100, organ-specific surface antigens, electrophysiological processes peculiar to nerve cells, formation of axons, and many other properties characteristic of tissue-
specific (histiotypic) differentiation would qualify as facultative
markers or luxury items. With many other markers it becomes debat-
able whether they should be classified as constitutive or facultative
markers. Myosin and actin are found in great quantities in muscle and
could therefore be considered as facultative markers characteristic of
myogenic differentiation. But these proteins also occur in small
amounts in other cell types making this classification doubtful. Some
enzymes, although of the constitutive type, occur in different forms
(isozymes) in different tissues, or occur in different forms at different
stages of embryological development. It should be remembered therefore that the distinction between the two types of markers in many
cases is very hazy and that facultative and constitutive may be the ex-
tremes of a continuum. A further complication is that the degree to
which a marker is expressed may vary depending on culture conditions. This form of variation is sometimes referred to as phenotypic
modulation.

Resistance to drugs, in most cases, is an artificially induced marker,
expressed only by mutant cells isolated after treatment with mutagens
and selection with appropriate drugs. Although artificial this type of
marker has been extremely useful not only in the isolation of hybrid
cells but also in the genetic dissection of metabolic pathways and membrane functions.

B. GENE EXPRESSION PATTERNS

The nomenclature used to describe the properties of hybrid cells has
to a great extent been borrowed from the field of genetics. Thus, hybrids which express a marker typical of one of the two parental cells
have been said to show *dominance*. Examples of dominant phenotypes
have been observed in crosses of drug resistant and sensitive cells in
which the hybrids have shown a drug resistance similar to that of the
resistant parent. The term *codominance* has been used to describe the
simultaneous expression of homologous genes of both parental cells.

Cells	Phenotypic markers	Expression
Parental cells I	A B C D	
Parental cells II	a b c d x y z	
Hybrid cells		
clone 1	A a B b C c D d x y z	Coexpression of all markers
clone 2	A a B b C c D d	Extinction of x y z
clone 3	A a B b C c D d X x Y y Z z	Cross-activation of X Y Z
Segregant		
hybrid clone 2	A a B b C c D d x y	Reexpression of x and y Extinction of z

Fig. XI-1. Schematic illustration of gene expression patterns in hybrids.

Human + mouse hybrids which produce both the human and the mouse form of the same enzyme can be quoted as examples of this gene expression pattern. Many hybrids between differentiated and undifferentiated cells have shown an undifferentiated phenotype. The loss of markers typical of histiotypic differentiation has then been interpreted as *recessiveness*.

In many ways this terminology is unfortunate since the analysis of hybrid cell phenotypes is different from the analysis of progenies arising from sexual crosses. For this reason many workers use a descriptive "nongenetic" terminology. Hybrids exhibiting a parental marker are simply said to show *expression* while failure to express the marker is referred to as *extinction*. *Coexpression* is used to describe the simultaneous expression of homologous genes of both parental genomes. The difference between expression and extinction is not always a question of all or none. Varying degrees of intermediate expression have also been observed.

Extinction of a marker may be due to regulatory phenomena, to loss of the chromosome carrying the structural gene for the marker, or to disturbances caused by the loss of regulatory genes. Usually it is very difficult to distinguish between these alternatives. In some cases, however, the extinguished marker has been reexpressed after prolonged culture of the hybrids. *Reexpression* (or reappearance) makes it possible to conclude that the initial extinction was due to regulatory phenomena rather than loss of structural genes. If two different markers are always extinguished together and reexpressed together they are said to show *coordinate expression*. In a few cases hybrids have expressed new properties which have not been present in the parental cells. We will refer to this as *activation*.

In the following pages we will review work in which cell hybrids

TABLE XI-1

Examples of Coexpression of Constitutive Markers

Parental cells	Markers	References
Interspecific crosses		
Rat + mouse	Lactate dehydrogenase	Weiss and Ephrussi, 1966b; Sonnenschein et al., 1971
Rat + mouse	β-Glucuronidase	Weiss and Ephrussi, 1966b
Syrian hamster + mouse	Lactate dehydrogenase	Davidson et al., 1966, 1968
Syrian hamster + mouse	Malate dehydrogenase	Davidson et al., 1966, 1968
Man + mouse	28 S rRNA	Stanners et al., 1971
	Glucose-6-phosphate dehydrogenase, malate dehydrogenase, lactate dehydrogenase, isocitrate dehydrogenase	Boone and Ruddle, 1969a
Man + Chinese hamster	Surface antigens	Weiss and Green, 1967; Kano et al., 1969
	Isozymes	Westerveld, 1971
	Surface antigens	Westerveld, 1971
Intraspecific crosses		
Mouse	H-2 antigens	Gershon and Sachs, 1963; Spencer et al., 1964; Silagi, 1967; Coffino et al., 1971; Mohit and Fan, 1971; Gordon et al., 1971
Human	HL-A antigens	Kennett et al., 1975
Mouse	β-Glucuronidase isozymes	Ganshow, 1966
Mouse	Phosphoglucomutase	Minna et al., 1972
Mouse	Glucosephosphate isomerase	Minna et al., 1972; Gordon et al., 1971
Human	Glucose-6-phosphate dehydrogenase	Silagi et al., 1969

have been used as an experimental tool in studies of gene regulation and differentiation in eukaryotic cells. The discussion will be focused on certain types of hybrids which have been studied in greater detail or which have provided important new information. For practical purposes most of the references have been tabulated to provide a better overview of the work done so far. Table XI-1 lists fusions in which constitutive markers are coexpressed in the hybrids. Table XI-2 gives examples of hybrids showing extinction of facultative (differentiated) markers, whereas hybrids that retain facultative markers are listed in Table XI-3. Examples of hybrids showing activation of new phenotypes are given in Table XI-4. The terminology will be the "nongenetic" descriptive one (summarized in Fig. XI-1), but in a few cases where we discuss the conclusions of other authors we will use their own terminology even though it belongs to the "genetic" category.

C. EXPRESSION OF CONSTITUTIVE MARKERS

1. Autosomal Markers

Coexpression of constitutive markers has been demonstrated for a very large number of inter- and intraspecific hybrids (Table XI-1). Among these markers are enzymes of intermediary metabolism [e.g., lactate dehydrogenase (LDH), and β-glucuronidase (β-Glu), and surface antigens (e.g., human HL-A and mouse H-2 antigens)]. In addition to the parental enzyme forms, new heteropolymeric enzymes (see Chapter III) have been found (Table XI-5). Levels of expression which are intermediate between those in the parental cell lines have been reported for constitutive markers such as folate reductase (Littlefield, 1969), enzymes involved in pyrimidine synthesis (Silagi *et al.*, 1969), and surface antigens (Weiss and Green, 1967). A relative dominance of the enzymes of one species has sometimes been observed in interspecific hybrids. Thus, in human + mouse hybrids a stronger expression of constitutive mouse enzymes was noted (Boone and Ruddle, 1969a). The explanation of this phenomenon may be that more mouse genes are present or that the human genes are repressed by some regulatory mechanism.

2. X-Linked Markers

Constitutive markers specified by genes on X chromosomes are of particular interest since in normal female cells one X chromosome is inactivated at an early stage of embryogenesis. This inactivation randomly affects either the paternal or the maternal X chromosome. In cer-

TABLE XI-2

Extinction of Differentiated Markers in Hybrid Cells

Marker	Differentiated parental cell	Undifferentiated parental cell	Remarks	References
Melanin	Melanoma (Syrian hamster)	L cell (mouse)		Davidson et al., 1966, 1968
	Melanoma (mouse)	L cell (mouse)		Silagi, 1967
	Melanoma (Syrian hamster)	L cell (mouse)	Gene dosage effect	Fougère et al., 1972
	Melanoma (Syrian hamster)	L cell (mouse)	Gene dosage effect	Davidson, 1972
	Melanoma (Syrian hamster)	L cell (mouse)		Davidson and Yamamoto, 1968
Dopa oxidase	Hepatoma (rat)	3T3 (mouse)		Schneider and Weiss, 1971
TAT, high baseline activity	Hepatoma (rat)	SV40 transformed cells (human)		Schneider and Weiss, 1971
TAT inducibility	Hepatoma (rat)	L (mouse)		Schneider and Weiss, 1971
	Hepatoma (rat)	3T3 (mouse)		Schneider and Weiss, 1971
	Hepatoma (rat)	Epithelial (rat)	Reappearance	Weiss and Chaplain, 1971
	Hepatoma (rat)	3T3 (mouse)		Benedict et al., 1972
	Hepatoma (rat)	WI-38 (human)	Reappearance linked to loss of human X	Croce et al., 1973a, 1974d,f
Albumin	Hepatoma (rat)	3T3 (mouse)	Gene dosage effect and crossactivation	Peterson and Weiss, 1972;
Aldolase B	Hepatoma (rat)	3T3 (mouse)		Bertolotti and Weiss, 1972a
	Hepatoma (rat)	L (mouse)		Bertolotti and Weiss, 1972a
	Hepatoma (rat)	BRL-1 (rat)	Reappearance	Bertolotti and Weiss, 1972b,c, 1974
	Hepatoma (rat)	DON (hamster)	Reappearance	Weiss et al., 1975

Alcohol dehydrogenase (ADH)	Hepatoma (rat)	3T3 (mouse)		Bertolotti and Weiss, 1972b
Alanine aminotransferase	Hepatoma (rat)	BRL-1 (rat)		Bertolotti and Weiss, 1972b
	Hepatoma (rat)	BRL-1 (rat)		Sparkes and Weiss, 1973
Growth hormone	Pituitary cells (rat)	L cell (mouse)		Sonnenschein et al., 1968, 1971
Protein S 100	Diploid glia cell (rat)	L cell (mouse)		Benda and Davidson, 1971
	"near tetraploid" glia cell	L cell (mouse)		Benda and Davidson, 1971
GPDH inducibility	Glia cell (rat)	L cell (mouse)		Davidson and Benda, 1970
Immunoglobulins	Plasmocytoma (mouse)	3T3 (mouse)		Periman, 1970
	Myeloma (mouse)	L cell (mouse)	Incomplete extinction	Coffino et al., 1971
	Lymphoblast (human)	Fibroblast (mouse)		Klein and Wiener, 1971
	Thymocyte (human)		First + then −	Parkman et al., 1971
	Myeloma (mouse)	Lymphoma (mouse)	Light chain synthesis not extinguished	Mohit and Fan, 1971
Esterase 2	Kidney cell (mouse)	L cell (mouse)		Klebe et al., 1970
	Kidney cell (mouse)	WI-38 (human)		Klebe et al., 1970
Macrophage specific surface receptors	Macrophage (mouse)	L cell (mouse)	Reappearance	Gordon et al., 1971
Steroid sulfatase	Neuroblastoma (mouse)	L cell (mouse)		McMorris et al., 1974
Sensitivity to 6-hydroxydopamine	Neuroblastoma (mouse)	Fibroblast (human)		Cronemeyer et al., 1974

TABLE XI-3

Retention of Facultative Markers in Hybrid Cells

Marker	Parental cells		References	Remarks
	Expressor parent	Nonexpressor or other expressor parent		
Hyaluronic acid	Mouse 3T6 fibroblasts	Mouse NCTC 2555 + 2472 cells	Green *et al.*, 1966	Intermediate expression
Collagen	Mouse 3T6 fibroblasts	Mouse NCTC 2555 + 2472 cells	Green *et al.*, 1966	Intermediate expression
GPDH baseline activity	Rat glia cell	Mouse L cell	Davidson and Benda, 1970	Intermediate expression
Hemoglobin, DMSO induced	Mouse erythroleukemia (induced)	Mouse erythroleukemia (non-induced)	Paul and Hickey, 1974	Intermediate expression
Membrane ATPase	Mouse macrophage	Mouse L cell	Gordon *et al.*, 1971	Intermediate expression
Excitable membrane	Mouse neuroblastoma	Mouse L cell	Minna *et al.*, 1971	
Acetylcholinesterase	Mouse neuroblastoma	Mouse L cell	Minna *et al.*, 1972	
	Human neuroblastoma	Mouse L cell	McMorris and Ruddle, 1974	
Neuron specific protein 14-3-2	Mouse neuroblastoma	Mouse L cell	McMorris *et al.*, 1974	
Electrical excitability	Mouse neuroblastoma	Human fibroblast	Peacock *et al.*, 1973	
Sensitive to acetylcholine	Mouse neuroblastoma	Mouse L cell	Peacock *et al.*, 1973	
Neuronal morphology	Mouse neuroblastoma	Mouse L cell	McMorris and Ruddle, 1974; Minna *et al.*, 1971, 1972	
Electrical coupling	Human Lesch–Nyhan cell	Mouse L cell (ClID)	Azarnia *et al.*, 1974	Extinction upon loss of human chromosomes

Trait	Parent 1	Parent 2	Reference	Remark
Light chain of Ig	Mouse myeloma	Mouse lymphoma	Mohit and Fan, 1971	
Light chain of Ig	Human lymphoblastoid cell	Mouse 3T3	Orkin et al., 1973	Coexpression
Immunoglobulins	Human lymphocytic line	Human lymphocytic line	Bloom and Nakamura, 1974	
Immunoglobulins	Mouse myeloma	Mouse myeloma	Cotton and Milstein, 1973	Cross-activation
Albumin	Rat hepatoma	Mouse 3T3 fibroblast	Peterson and Weiss, 1972	
Transferrin	Mouse hepatoma	Mouse and rat fibroblasts	Szpirer and Szpirer, 1975	
Complement (C3)	Mouse hepatoma	Mouse and rat fibroblasts	Szpirer and Szpirer, 1975	
Catalase	Rat hepatoma	Mouse L cell	Levisohn and Thompson, 1973	
Complement factors	Rat hepatoma	Mouse L cell	Levisohn and Thompson, 1973	
Arylhydrocarbon hydroxylase	Mouse 3T3 fibroblast	Rat hepatoma	Benedict et al., 1972	
Killed by dexamethasone	Mouse myeloma (sensitive)	Mouse lymphoma (resistant)	Gehring et al., 1972	Hybrid killed
Desmosterol conversion to cholesterol	Human fibroblast line	Mouse L cell	Croce et al., 1974a	Gene complementation phenomenon ?

TABLE XI-4

Appearance of New Activities Not Present in Parental Cells

Product or property	Parental cells	References
Cross-activation		
Mouse serum albumin	Rat hepatoma + mouse 3T3 cells	Peterson and Weiss, 1972
	Rat hepatoma + mouse lymphoblasts	Malawista and Weiss, 1974
Human immunoglobulins	Mouse myeloma + human lymphocytes	Schwaber and Cohen, 1973
		Schwaber, 1975
Human serum albumin	Mouse hepatoma + human leukocytes	Darlington et al., 1974a,b
		Bernhard et al., 1973
Human complement factor (C4)	Guinea pig macrophages (C4⁻) + human HeLa cells	Colten and Parkman, 1972
Possible cross-activation		
Complement factor (C5)	Mouse spleen cells (C5⁻) + chick erythrocytes	Levy and Ladda, 1971
Other new activities		
Hyaluronic acid	Mouse cell line + Chinese hamster cell line	Koyama and Ono, 1970
Esterases	Chinese hamster ovary cells + human fibroblasts	Kao and Puck, 1972a
Radioresistance	Rat pituitary cells + mouse L cells	Little et al., 1972
Choline acetyltransferase	Mouse neuroblastoma + human fibroblasts	McMorris and Ruddle, 1974
	Mouse neuroblastoma + rat glioma	Amano et al., 1974; Klee and Nirenberg 1974
Morphine receptors	Mouse neuroblastoma + rat glioma	Klee and Nirenberg, 1974

TABLE XI-5

Some Examples of Heteropolymeric Enzymes in Hybrid Cells[a]

Enzyme	Hybrid	References
Lactate dehydrogenase (LDH) (tetrameric)	Rat + mouse	Weiss and Ephrussi, 1966b
	Syrian hamster + mouse	Davidson et al., 1966
	Human + mouse	Ruddle and Nichols, 1971
Malate dehydrogenase (MDH) (dimeric) (NAD dependent)	Syrian hamster + mouse	Davidson et al., 1966
	Human + mouse	Ruddle and Nichols, 1971
6-Phosphogluconate dehydrogenase (PGD)	Syrian hamster + mouse	Migeon, 1968
Malate dehydrogenase (MDH) (NADP dependent)	Human + mouse	Ruddle and Nichols, 1971
Glucose-6-phosphate dehydrogenase (G6PD)	Human + mouse	Ruddle and Nichols, 1971
	Human + human	Glaser et al., 1973b; Migeon et al., 1974
Isocitrate dehydrogenase (IDH)	Human + mouse	Ruddle and Nichols, 1971
Glutamic-oxaloacetic transaminase (GOT)	Human + mouse	Ruddle and Nichols, 1971
Glucosephosphate isomerase (GPI)	Human + mouse	Ruddle and Nichols, 1971
Peptidase A (Pep-A)	Human + mouse	Ruddle and Nichols, 1971
Indophenol oxidase (IPO)	Human + Chinese hamster	Westerveld et al., 1971a

[a] Modified from Davidson, 1972.

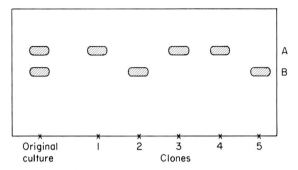

Fig. XI-2. Schematic illustration of the electrophoretic separation of the two forms (A and B) of glucose-6-phosphate dehydrogenase (G6PD) from a culture of female cells heterozygous for G6PD. The original cell population established from tissue explants is heterogeneous and consists of a mixture of cells. Some have an inactive parental X chromosome while other cells have an inactive maternal X chromosome. The cloned cells only have one type of enzyme.

tain genetic diseases more than two X chromosomes may be present but all except one are inactivated and remain condensed during interphase as sex chromatin bodies. Exactly when inactivation of X chromosomes takes place in man is not known but sex chromatin bodies are present in 12–18 day embryos. Once inactivation has occurred in a given cell the same X chromosome remains stably inactivated in all daughter cells arising from this cell. The tissues of adult females are mosaics of two different cell types; those with an inactive paternal and those with an inactive maternal X chromosome. These two cell types occur intermingled with each other but each cell type can be isolated in a pure form by cloning cultured cells. In those cases where a woman is heterozygous for genes specifying X-linked isoenzymes it is possible to analyse which X chromosome is active by enzyme electrophoresis (Fig. XI-2).

Hybrid cells have been found to contain more than one active X chromosome by the demonstration of coexpression of X-linked enzyme variants of both parental cells (Silagi *et al.*, 1969; Siniscalco *et al.*, 1969a; Migeon, 1972; Migeon *et al.*, 1974; Croce and Bakay, 1974). Migeon *et al.* (1974) fused cells from a male Lesch–Nyhan patient (HGPRT[−]) with leukocytes from a normal male (HGPRT[+]). The two cell types produced different forms of the X-linked enzyme, glucose-6-phosphate dehydrogenase (G6PD). The hybrid cells were found to contain both parental enzyme forms and in addition a heteropolymeric enzyme. These results indicate that inactivation of supernumerary X chromosomes is restricted to a specific stage of embryogenesis and does not occur in hybrid cells. Whether X chromosomes that are inactive in the parental cells can undergo activation in hybrid cells is uncertain. Kahan and DeMars

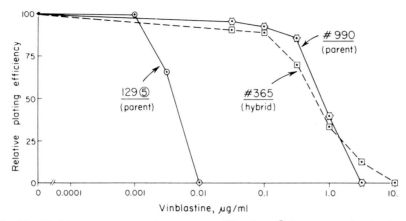

Fig. XI-3. Vinblastine resistance of a sensitive cell line [129 ⑤], a resistant line (990), and a hybrid line (365) obtained by crossing the sensitive and the resistant line. The hybrid appears to be as resistant to vinblastine as the resistant parent. Resistance to the drug was measured by scoring the plating efficiency in media containing different concentrations of the drug (M. Harris, 1973).

(1975) obtained evidence for infrequent derepression of the *HGPRT* locus by creating situations where the hybrid cells had to activate an *HGPRT*$^+$ gene on an inactive X chromosome to survive selection on HAT medium.

D. DRUG RESISTANCE MARKERS

Hybrids between drug resistant and sensitive cells exhibit varying degrees of sensitivity. Fusion of vinblastine resistant cells with sensitive cells (M. Harris, 1973) gives hybrids that are fully resistant to the drug (Fig. XI-3). A similar dominance of the resistant phenotype has been observed for actinomycin D (Sobel *et al.*, 1971) and ouabain (Baker *et al.*, 1974).

Other forms of drug resistance, however, behave differently (Fig. XI-4). Crosses where one parent is sensitive and the other resistant to cytosine arabinoside result in hybrids which are sensitive to the respective drugs (M. Harris, 1973).

Most likely, the reason why resistance to one drug is expressed in hybrid cells while resistance to another is extinguished has to do with the molecular mechanisms causing resistance. Vinblastine, colchicine, and actinomycin D resistance appear to be due to reduced permeability of the cell surface (Siminovitch, 1974), while resistance to purine and pyrimidine analogs seems to be caused by mutational loss of specific intracellular enzymes. Thus, in most cases resistance to bromodeoxy-

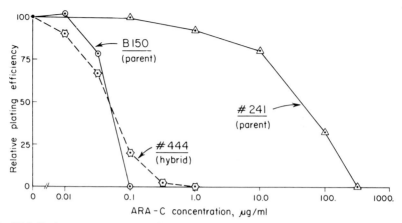

Fig. XI-4. Resistance to cytosine arabinoside of a sensitive cell line (B150), a resistant line (241) and a hybrid line (444). The hybrid resembles the sensitive parental cell line in its sensitivity to the drug (M. Harris, 1973).

uridine (BUdR) is associated with a loss of thymidine kinase (TK) (see Chapter IX). In fusions of sensitive + resistant cells the sensitive parent will contribute genes which make TK synthesis possible. These hybrids are drug-sensitive since BUdR will be phosphorylated and incorporated into DNA, thus killing the hybrid cell.

E. FACULTATIVE MARKERS

1. Melanoma Hybrids

The earliest studies of histiotypic differentiation in hybrid cells were performed with pigmented Syrian hamster melanoma cells (Davidson *et al.*, 1966, 1968). As with the normal pigment cells of the skin, the melanocytes, the melanoma cells house in their cytoplasm a large number of granules which contain the black pigment melanin. Differentiated melanoma colonies appear black and can easily be detected. In order to express pigmentation the cell must produce the enzyme dopa oxidase (tyrosinase), which catalyzes two steps in the conversion of tyrosine to melanin.

Davidson *et al.* (1966) mixed pigmented Syrian hamster melanoma cells (HGPRT⁻) with mouse L cells (TK⁻) and were able to isolate spontaneously arising hybrid cells by selection on HAT medium. A large number of hybrid clones were isolated and all were found to be unpigmented. Karyological analysis demonstrated that the melanoma hybrids contained an average of 101 chromosomes which was very close to the sum expected if one melanoma cell (average 51 chromosomes) had fused with one L cell (52 chromosomes). Some hybrid clones had, however, already lost chromosomes by the time their karyotype could be analyzed. But since *all* hybrid clones were unpigmented and because they were unpigmented from the time of their detection (approximately 5 cell cycles after fusion) it seemed very unlikely that the extinction of pigmentation resulted from the loss of specific chromosomes carrying structural genes involved in pigment formation.

The possibility that the extinction of pigmentation was caused by a nonspecific repression of a large number of hamster genes was eliminated when Davidson *et al.* (1968) demonstrated that the unpigmented melanoma hybrids produced both hamster and mouse forms of MDH, LDH-1, and LDH-5 in addition to the hamster TK that enabled the hybrids to survive on HAT medium. Thus, it appeared that constitutive markers were coexpressed by the hamster and mouse genomes, while facultative markers, like dopa oxidase and melanin production were selectively extinguished. While the evidence presented thus far clearly suggests that the extinction of pigmentation is due to some regulatory process, it remains unclear at what level this process takes place. Experiments with mixed cell extracts of melanoma cells and unpigmented cells (Davidson and Yamamoto, 1968) suggest that the loss of dopa oxidase activity in the hybrids is not due to inhibition or destabilization of the enzyme and makes it likely that the regulatory process takes place at the transcription or translation level.

Silagi (1967) examined a series of intraspecific melanoma hybrid clones that were obtained in a fusion of mouse melanoma and mouse L cells. These hybrids were also found to be unpigmented. It appears, therefore, that extinction of pigmentation is not merely a characteristic of interspecific hybrids but also occurs in intraspecific hybrids. This observation rules out the possibility that extinction of pigmentation is caused by species incompatibility at the intracellular level.

In more recent experiments Fougère *et al.* (1972) and Davidson (1972) have fused near-tetraploid (2 s) melanoma cells with near-diploid (1 s) L cell mutants. Some of the hybrids obtained were pigmented and produced dopa oxidase whereas *all* hybrids containing one (1 s) melanoma genome and one (1 s) L cell genome were unpigmented. These results show that the *gene dosage* (i.e., the ratio between the number of chromosomes from the differentiated parent relative to the number from the undifferentiated parent) is important in the expression of pigmentation.

2. Hepatoma Hybrids

Hepatoma cells are differentiated tumor cells derived from liver. Rat and mouse hepatoma cell lines have been established *in vitro* where they grow well while still retaining facultative markers characteristic of hepatic differentiation. The hepatoma phenotype includes high activities of tyrosine aminotransferase (TAT), alanine aminotransferase (AAT), arylhydrocarbon hydroxylase (AHH), aldolase, and alcohol dehydrogenase (ADH), tryptophan pyrrolase, and a capacity to synthesize and secrete serum albumin. With both TAT and AAT there are two types of enzyme activities, a high baseline activity and an additional activity which can be induced by glucocorticoid steroid hormones (induced activity). The two types of activity are believed to be regulated by two separate genes (Tomkins *et al.*, 1969).

TAT activity has been studied in both heterokaryons (Thompson and Gelehrter, 1971) and synkaryotic hybrid cells (Schneider and Weiss, 1971; Weiss and Chaplain, 1971; Benedict *et al.*, 1972; Croce *et al.*, 1973a). Since the heterokaryon work is discussed in Chapter VII we shall concentrate here on hepatoma hybrids.

Schneider and Weiss (1971) produced the first hepatoma hybrids by fusing rat hepatoma cells with mouse 3T3 cells (TK$^-$) in the presence of Sendai virus. The cells were grown on HAT medium where only hepatoma cells and hybrid cells can grow. The hybrids were then separated from the hepatoma parent by cloning. Chromosomal analysis performed 20–30 cell generations after fusion showed preferential loss of rat chromosomes. The hybrids had a baseline activity of TAT which was twice that of the 3T3 parent but which only represented 6% of the activity of the hepatoma parent. Heat inactivation curves revealed that the enzyme responsible for the residual baseline activity of the hybrids was that of the rat type. None of the hybrid clones were found to be inducible with steroids. These data suggested that, analogous to the situation in melanoma hybrids, the differentiated functions of the hepatoma parent were suppressed in the hybrid cells. However, in view of the chromosomal losses it was impossible to exclude the possibility that the lack of TAT inducibility was due to a loss of rat chromosomes carrying either structural genes for TAT or genes concerned with regulatory functions.

But extinction of TAT inducibility was also observed (Weiss and Chaplain, 1971; Deschatrette and Weiss, 1975) in intraspecific rat hybrids. These hybrids were chromosomally stable and showed virtually no chromosome elimination during an observation period of several months. These studies are interesting because, first, they strongly support the idea that the extinction phenomenon is due to a regulatory phenomenon and not to loss of structural genes. Second they show that a hepatoma cell which has lost markers characteristic of hepatic differentiation is capable of extinguishing these markers from a hepatoma cell which retains these properties. It appears that further studies of this type of regulatory interactions can provide important information about factors which control liver cell differentiation.

In order to select for possible reexpression of hepatoma properties Weiss and Chaplain (1971) examined a large number of hybrid colonies selecting only those which showed a colony morphology resembling that of the hepatoma parent. One of the clones thus obtained was of particular interest. This clone and all of its subclones showed a low baseline activity of TAT but were induced by steroids to produce significantly higher TAT activity. It appeared that baseline activity of TAT and TAT inducibility were *independently reexpressed* and that reexpression was the result of chromosomal loss. These data thus support the hypothesis put forward by Tomkins *et al.* (1969) that the two types of TAT activity are controlled by two different genes.

The possibility that a specific chromosome is involved in the extinction of TAT in hepatoma hybrids has been examined in a more direct way by Croce *et al.* (1973a) using rat hepatoma cells (HGPRT⁻) and cells from a human diploid strain (WI 38) as parental cells. Hybrid clones were isolated by selecting against the hepatoma parent on HAT medium and then separating the hybrid cells from the human parental cell by cloning. Both baseline and inducible TAT activity were found to be extinguished in the same way as in a number of clones studied by Weiss and coworkers. Some clones did, however, show baseline activity of TAT at a level which was similar to that of the rat hepatoma parent. Neither these cells nor those which showed complete extinction of baseline activity could, however, be induced by steroids to produce more TAT.

Croce *et al.* (1973a) then proceeded to show that the extinction in hybrids could be linked to the presence of a human X chromosome. Since all the hybrids had been obtained after selection on HAT medium where the cells depended on the X-linked salvage enzyme HGPRT for their survival, one could be reasonably sure that all cells contained an active X chromosome. This assumption was confirmed by chromosome cytology and by assaying for another X-linked enzyme (G6PD). The hybrids showing extinction of TAT inducibility were backselected on a normal medium containing the drug 8-azaguanine. Only cells which lacked HGPRT (having lost their active X chromosomes) survived this treatment. In this way a series of clones were obtained which had all lost their active X-chromosome but which still retained a wide variety of human autosomal chromosomes. *All* of these clones showed reexpression of TAT inducibility. This was consistent with the hypothesis that the X chromosome carries a regulatory locus which suppresses one of the differentiated traits of the hepatoma cell, namely the inducibility of TAT synthesis.

Alanine amino transferase (AAT) is another enzyme which shows high activity in liver and hepatoma cells. In addition to a high baseline activity there is also steroid inducible AAT activity. Sparkes and Weiss (1973) analyzed different subclones of an intraspecific rat hepatoma hybrid (rat hepatoma + rat epithelial cell). The original hybrid clones showed extinction of both baseline and inducible AAT activity. Reexpression of both types of enzyme activity did, however, occur in the subclones. The same subclones which were analyzed for AAT were also analyzed for TAT. No correlation was found between reexpression of AAT and reexpression of TAT. The main points brought out by the analysis of AAT reexpression were that the baseline and inducible forms of the enzyme are expressed independently of each other and independently of TAT reexpression.

Extinction and reexpression has also been reported for aldolase B and for alcohol dehydrogenase (Bertolotti and Weiss, 1972a,b,c, 1974). The hepatoma cells normally produce three different forms of aldolase (A, B, and C forms), but only the B form is characteristic of hepatic differentiation. The A and C forms of the enzyme can be considered as constitutive markers and the B form as a facultative marker. It is interesting therefore that only the B form is extinguished in the hybrids while the A and C forms persist. Reexpression of the B form of activity occurred in subclones after loss of chromosomes. In some cases the reexpression of enzyme activity was paralleled by a change in cellular morphology. Thus, one hybrid line which initially resembled the epithelial parent gave rise to hepatoma-like subclones (Fig. XI-5).

Production of serum albumin is another trait typical of hepatic differentiation which is retained by hepatoma cells. The albumin is secreted into the tissue culture medium where, after dialysis and lyophilization, it can be detected by immunological methods. Peterson and Weiss (1972) have examined a series of rat hepatoma + mouse 3T3 fibroblast hybrids for both rat and mouse serum albumin. Eight different hybrid clones containing one chromosomal set (1 s) from the rat hepatoma parent and one (1 s) from the 3T3 parent produced rat albumin but not mouse albumin. Since the rat albumin production was less than that of the hepatoma parent, the result is best described as an *intermediate expression* or *partial extinction* of a facultative marker. The hybrid clones were derived from a hybrid cell population which had previously lost two other facultative markers of hepatic differentiation: TAT (Schneider and Weiss, 1971) and aldolase B (Bertolotti and Weiss, 1972a). Production of rat albumin in these hybrids, therefore, is a trait which can be

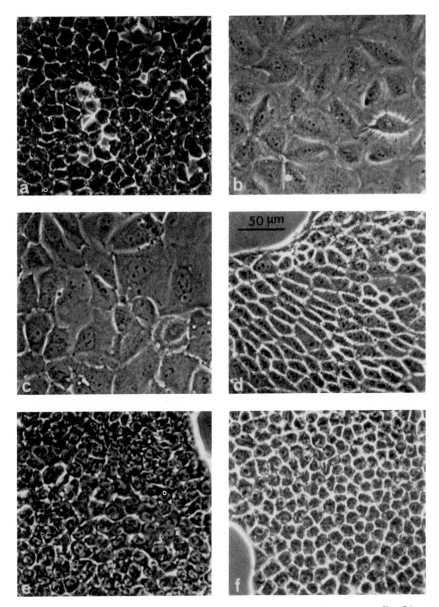

Fig. XI-5. Phase contrast photomicrographs of (a) Fu5-5 mouse hepatoma cells, (b) rat epithelial cells, and (c) hybrid clone containing complete chromosome sets of both parents. Hepatoma morphology extinguished. (d) Segregant where aldolase B activity is extinguished and (e) and (f) segregants reexpressing aldolase B. These cells resemble the hepatoma parent in having a denser and highly granular cytoplasm. From Bertolotti and Weiss, *Differentiation* **2,** 5–17 (1974).

TABLE XI-6

Serum Albumin Production by Rat Hepatoma + 3T3 Hybrids[a]

Cells	Number of chromosomes	Rat albumin	Mouse albumin
Parental			
3T3 (mouse fibroblast)	76	−	−
Fu5 (rat hepatoma)	52	+	−
2 s Fu5 (rat hepatoma)	99	+	−
Hybrid			
3T3 + Fu5			
Clone 1–8	107	+	−
3T3 + 2 s Fu5			
Clone VIII	205	+	+
III	146	−	+
Vb	150	−	+
IV	131	−	−
Va	128	−	−

[a] Peterson and Weiss, 1972.

expressed even if these enzyme markers are extinguished, i.e., *independent or noncoordinate* expression. Another example of this type of regulation is the observation that hybrids between mouse hepatoma cells and mouse or rat fibroblasts continue to produce transferrin and the third factor of complement at the same time that albumin and α-fetoprotein synthesis is extinguished (Szpirer and Szpirer, 1975).

Another series of hepatoma hybrids was produced by fusing hypertetraploid (>2 s) rat hepatoma cells with mouse 3T3 (1 s) fibroblasts. Most of the hybrid clones in this series had 15–30% fewer chromosomes than the expected chromosomal sum of the two parental cells. Of five clones isolated, one produced both rat and *mouse* albumin, two produced only mouse albumin and two did not produce either rat or mouse albumin (Table XI-6). The extinction of rat albumin synthesis in four out of five clones of the 2 s hepatoma + 1 s fibroblast hybrids was somewhat surprising since 8 of the 1 s hepatoma + 1 s fibroblast hybrids did show detectable rat albumin synthesis. Even more remarkable was the synthesis of *mouse* serum albumin in three out of five clones in the 2 s hepatoma + 1 s fibroblast hybrids. These observations were verified in another type of hybrid which involved fusion of rat hepatoma cells with mouse leukemic lymphoblasts (Malawista and Weiss, 1974). In this case activation of mouse albumin synthesis was observed in a large number of independent clones. Similar observations have been made by Dar-

lington *et al.* (1974a,b) and Bernhard *et al.* (1973) in fusions of mouse hepatoma cells with human leukocytes. These hybrids were of a different
type from those previously studied since the hepatoma chromosomes
were retained in this case while the chromosomes of the nonhepatoma
parent (human leukocyte chromosomes) were eliminated. Although the
hybrids examined contained very few human chromosomes some
clones did nevertheless synthesize both human and mouse albumin.
These results indicate that the hepatoma parent in a hybrid can *repro-
gram* or *cross-activate* the genome of the other partner in a fusion so that
homologous genes are expressed.

Other liver specific functions which have been analyzed in hepatoma
hybrids include catalase and complement factors (Levisohn and
Thompson, 1973). These markers were found to be expressed at a level
which was intermediate between those of the hepatoma cell and the
nonexpressing parental cell. Rintoul *et al.* (1973b), on the other hand,
recorded yet other examples of the extinction of liver specific functions.
Intraspecific mouse hybrids obtained by fusing fetal liver cells and fibroblasts showed extinction of glycogen accumulation and inducibility
of tryptophan pyrrolase by steroid hormones.

3. Lymphoid Cell Hybrids

Plasmocytoma (myeloma) lymphoblastoid cells, and normal thymo-
cytes from mouse and man have been used in fusions designed to analyze the regulation of antibody (immunoglobulin, Ig) production. Extinction of Ig synthesis has been observed in many of these crosses
(Periman, 1970; Coffino *et al.*, 1971; Klein and Wiener, 1971; Parkman *et
al.*, 1971; Mohit and Fan, 1971) but notable exceptions have occurred. In
the intraspecific mouse plasmocytoma + L cell fusions reported by
Periman (1970) the extinction may not have been complete. The mouse
myeloma + mouse lymphoma hybrids studied by Mohit and Fan (1971)
and Mohit (1971) showed extinction of heavy chain synthesis but continued production of kappa light chains. Similar results have also been
obtained by Orkin *et al.* (1973) who found production of light chains in
hybrids formed by fusing HGPRT⁻ human lymphoblastoid cells with
mouse 3T3 cells (TK⁻). These results, thus, demonstrate that although
some hybrids exhibit varying degrees of extinction others continue to
produce immunoglobulins. In later work there have also been reports
that intraspecific hybrids obtained by crossing human lymphocytic
lines continue to produce immunoglobulins characteristic of both
parental cells (Bloom and Nakamura, 1974). A similar case of coexpression has been observed in fusions of immunoglobulin producing
mouse myeloma cells (Cotton and Milstein, 1973).

To complete the list of different gene expression patterns in hybrids of immunoglobulin producing cells Schwaber and Cohen (1973, 1974) and Schwaber (1975) have recorded activation of Ig synthesis. Human peripheral blood lymphocytes, not synthesizing immunoglobulins were fused with mouse myeloma cells producing immunoglobulin. The hybrids produced heavy human γ- and α-chains and human light chains as well as mouse heavy α-chains and mouse light chains. In addition some "hybrid" antibody molecules containing both human and mouse heavy chains were observed. At present it is rather difficult to integrate these pieces of information into an overall regulatory scheme since they range from coexpression via intermediate expression to cross activation and extinction. Further information about chromosome patterns and further knowledge about immunoglobulin synthesis and secretion appear essential. In most studies immunoglobulins are detected in the culture medium. Secretion of Ig into the medium may, however, be a complex phenomenon which depends not only on the synthesis of the immunoglobulin chains but also on the presence of the appropriate intracellular organization required for secretion. It is possible therefore that before Ig can appear in the medium several different properties characteristic of specialized cells must be coordinately expressed.

4. Glia Cell Hybrids

When glioma cells are cultured *in vitro*, they may express at least three facultative markers typical of glia cell differentiation: synthesis of a brain specific acidic protein referred to as S-100; high baseline activity of glycerolphosphate dehydrogenase (GPDH), and hydrocortisone induced GPDH activity. The synthesis of S-100 was found by Benda and Davidson (1971) to be extinguished both in hybrids which contained two glia cell genomes and one fibroblast genome (2 + 1 hybrids) and in hybrids containing one glia and one fibroblast genome (1 + 1 hybrids). The baseline activity of GPDH in the hybrids was between that of the two parental cells in both types of hybrids. Inducibility of GPDH was completely extinguished in 1 + 1 hybrids but 2 + 1 hybrids showed some inducibility (Davidson and Benda, 1970). This disparity is probably the effect of gene dosage. The observation that complete extinction of S-100 is not necessarily associated with extinction of GPDH activity indicates that these two facultative markers are independently regulated.

5. Fibroblast Hybrids

In one of the first studies of differentiation in cell hybrids Green *et al.* (1966) fused 3T6 mouse fibroblasts with two different established mouse cell lines. The 3T6 cells expressed two characteristic markers of the fibroblast phenotype since they produced collagen and hyaluronic acid. The established mouse cell lines used as the other parental cell produced very little of these macromolecules. Hybrids formed spontaneously in cell mixtures and gradually overgrew the parental cells. Like many other intraspecific hybrids these contained a chromosome set which came close to being the sum of the two parental sets. Both hyaluronic acid and collagen were produced by the hybrids at a level which was intermediate in relation to the parental cell types.

6. Teratoma Hybrids

An interesting form of hybrid where one of the parental cells was a multipotent teratoma cell was produced by Finch and Ephrussi (1967) and by Jami et al. (1973). The teratoma cells used in these experiments did not show any distinct differentiation in vitro. When injected subcutaneously into mice they did, however, produce tumors containing a spectrum of differentiated tissues, e.g., yolk sac epithelium, neuroepithelium, mesenchyme, cartilage, bone, ciliated epithelium, etc. Hybrids made in vitro by fusing teratoma cells with L cells failed to show differentiation both in vitro and in vivo (Fig. XI-6). When injected into mice the hybrids gave rise to malignant, sarcoma-like tumors showing no tissue-specific differentiation. This indicates that the ability of the teratoma partner to give rise to highly specialized cell types is extinguished in the hybrids. Interpretation of these results is difficult, however, since the tumors could arise from a minority of fast growing, undifferentiated segregants from the initial hybrid population.

7. Neuroblastoma Hybrids

Neuroblastoma cells grown in culture exhibit a series of highly specialized properties which may be considered as facultative markers characteristic of neuronal differentiation. Among the traits which make up the neuroblastoma phenotype are: electrically excitable membranes; high acetylcholine esterase activity; acetylcholine sensitivity; a neuron-specific protein 14-3-2; steroid sulfatase activity; and a characteristic morphology (Fig. XI-7) with neurites containing neurofilaments. The electrical response is of two different types described as A and B response. These responses differ from that of L cells and other non-neuronal cells (C-type response).

Most studies on neuroblastoma hybrids have used the mouse C1300 neuroblastoma as one of the parental cells. The original cell line established by Augusti-Tocco and Sato (1969) has been found to be heterogeneous in several respects. Single cell clones can be divided into cholinergic, adrenergic, inactive and axon-negative cell types (Amano et al.,

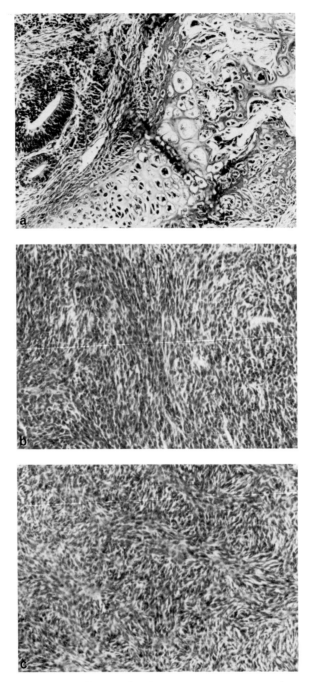

Fig. XI-6. Mouse teratoma cells inocculated into appropriate mice produce tumors which contain a variety of differentiated tissues. Mouse L cells and teratoma + L cell hybrids produce undifferentiated, fibrosarcoma-like tumors. (a) Section through mouse teratoma

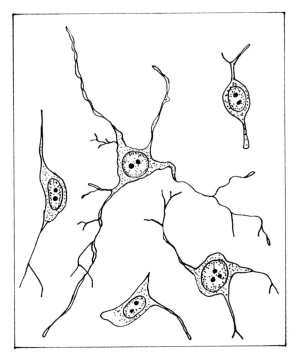

Fig. XI-7. Schematic drawing illustrating the morphology of mouse neuroblastoma cells cultured *in vitro*. The number of neurites per cell, the length and degree of branching of neurites, and the frequency of cells with neurites can be used to quantify the expression of the neuronal phenotype.

1972). It is not known whether this heterogeneity is a reflection of the great variation in chromosome constitution among subclones or if neuroblastoma cells are primitive and possible multipotent cells from the neural crest which can differentiate into a spectrum of neuronal phenotypes.

Mouse neuroblastoma cells have been hybridized with mouse L cells (Minna *et al.*, 1971, 1972; McMorris and Ruddle, 1974; McMorris *et al.*, 1974), rat glia cells (Amano *et al.*, 1974; McMorris *et al.*, 1974; Klee and Nirenberg, 1974), mouse sympathetic ganglion cells (Chalazonitis *et al.*, 1975; Greene *et al.*, 1975), human fibroblasts (McMorris and Ruddle, 1974; McMorris *et al.*, 1974), and human neuroblastoma cells (McMorris and Ruddle, 1974). A variety of neuronal phenotypes have been found among these hybrids. Contrary to the situation observed in many other

showing neuroblastic tissue with neural tube (N) formation and well developed cartilage nodules (C) surrounded by mesenchyme. (× 160.) (b) Teratoma + L cell hybrid tumor also showing fibrosarcoma-like histopathology. (Jami *et al.*, 1973) (× 115). (c) Fibrosarcoma like morphology of L cell tumor (× 115.)

fusions of differentiated + undifferentiated cells (e.g., in fusions with melanoma and hepatoma cells) the hybrids obtained by fusing neuro-blastoma cells with undifferentiated (nonneuronal) cells continue to ex-press many differentiated markers; these include neuronal morphol-ogy (Minna et al., 1971, 1972; McMorris and Ruddle, 1974; Daniels and Hamprecht, 1974), acetylcholine sensitivity (Peacock et al., 1973), ace-tylcholinesterase activity (Minna et al., 1971, 1972; McMorris and Ruddle, 1974), neuronspecific 14-3-2 protein (McMorris et al., 1974), and action potentials (Minna et al., 1971, 1972; Peacock et al., 1973). However, extinction of individual neuronal markers has also been ob-served. Thus, McMorris et al. (1974) found extinction of steroid sulfa-tase activity in some hybrids while Cronemeyer et al. (1974) found ex-tinction and later, after chromosome losses, reexpression of sensitivity to 6-hydroxydopamine. In one clone derived from a mouse neuroblas-toma + human fibroblast hybrid McMorris and Ruddle (1974) found choline acetyltransferase activity which was a hundred-fold greater than the baseline activities of the two parental cells. The new appear-ance of this marker, which is a characteristic of cholinergic neurons, may have been a result of altered chromosome constitution since this clone had lost all but 9 of the human chromosomes. It is interesting, however, that Amano et al. (1974) observed a similar activation of choline acetyl-transferase in hybrids between mouse neuroblastoma and rat glioma cells (a cross between two rodent species), in which chromosome se-gregation can be expected to be insignificant. In order to understand the regulatory interactions which result in the synthesis of a new en-zyme not present in the parental cells further studies are required. It would, for instance, be of interest to know if the enzyme is specified by the mouse genome or the genome of the other parental cell.

Another example of activation of new properties in hybrid cells was observed by Klee and Nirenberg (1974). Hybrids between mouse neuroblastoma cells and rat glioma cells resembled the neuroblastoma parent in some respects but they also expressed a number of neuronal properties not found in the parental cells; for instance, morphine re-ceptors, choline acetyltransferase, intracellular acetylcholine, and clear and dense core vesicles.

There is a considerable range in the quantitative expression of neuronal markers in the neuroblastoma hybrids examined by Minna et al. (1971, 1972) and McMorris and Ruddle (1974). After scoring clones for acetylcholinesterase activity, neurite formation, and electrical activ-ity Minna et al. (1972) found at least seven phenotypic classes (Table XI-7). Some properties appeared to be expressed independently of other properties. For instance, high levels of acetylcholinesterase activity oc-

TABLE XI-7

Phenotypic Classes Detected in Mouse Neuroblastoma + Mouse L Cell Hybrid Clones[a]

Class of hybrid clones	AChE level	Neurite formation	Silver staining neurofilaments	Electrical excitability B+	Electrical excitability A+
A	+	+	+	+	+
B	+	+	+	+	−
C	+	+	−		
D	+	−	−	+	−
E	−	+	−	+	−
F	−	+	−	−	−
G	−	−	−		

[a] Modified from Minna et al., 1972.

curred independently of neurite formation and neuron-specific electrical response ("A-type response"). Other markers, however, appeared to be expressed coordinately. Thus, some clones that produced neurofilaments also expressed acetyl cholinesterase and formed neurites. On the basis of these observations Minna et al. (1972) suggested that the spectrum of neuronal phenotypes observed among different neuroblastoma hybrid clones reflected different steps in neuronal differentiation.

The experimental results of McMorris and Ruddle (1974) are similar to those of Minna et al. (1972) in that there is continued expression of a large number of neuronal markers. McMorris and Ruddle (1974), however, interpret their results in a slightly different way. They emphasize the quantitative variation in the expression of neuronal phenotypes. Clones which show a strong expression of one neuronal marker are also likely to show a strong expression of other neuronal markers. This indicates that the neuronal phenotypes are expressed as a block (i.e., coordinately) and that their level is regulated by a single integrating control system. Further work is clearly needed to establish whether or not there exists such a system and also to identify if any particular group of markers is expressed independently of others.

A somewhat different type of neuroblastoma hybrids is comprised of those that have been generated by fusing neuroblastoma cells with sympathetic ganglion cells (Greene et al., 1974, 1975; Chalazonitis et al., 1975). In this case the neuroblastoma parent induces the nuclei of nonmultiplying ganglion cells to synthesize DNA and to replicate as one component of synkaryotic hybrid cells. This type of hybrid has been found to exhibit a variety of neuronal markers and to do so at a level of expression which is considerably above the level represented by the neuroblastoma parent. Furthermore, these hybrids resemble their ganglion parent in the morphology of their processes, tyrosine hydroxylase activity, ability to synthesize dopa and dopamine, electrical activity, and other neuronal markers.

Compared with other hybrids made by fusing differentiated and undifferentiated cells, neuroblastoma hybrids retain a greater number of facultative markers than usual (Table XI-3). There may be several explanations for this. The relative dominance of the neuronal phenotype could be a characteristic of this particular form of differentiation. Because the neuroblastoma cell line was derived from early embryonic cells it is possible that it retains a certain degree of multipotency which makes it possible for the hybrids to express many different neuronal phenotypes. Another explanation could be that the neuroblastoma cells are heteroploid and have a near-tetraploid chromosome constitution. Fusions with this cell type should therefore be compared with 2 + 1 fusions with other cells rather than with 1 + 1 fusions. As already discussed in Section E, 1, melanoma hybrids containing two genomes from the pigmented and one from the nonpigmented parent often express pigmentation while 1 + 1 fusions rarely do so. The double input of chromosomes from the differentiated parent may be one reason why neuroblastoma retain differentiated markers; and the variation of expression of neuronal phenotypes in clones of neuroblastoma cells and their hybrids could relate to the heteroploidy of the neuroblastoma lines.

8. Erythroleukemic Hybrids

In vitro cultures of Friend erythroleukemia, a virus-induced form of leukemia in mice, show a low level of constitutive hemoglobin synthesis. This level can be amplified by including dimethyl sulfoxide (DMSO) in the culture medium. Under these conditions a large proportion of the cells are induced to undergo erythroid differentiation. Five days after adding DMSO as many as 60% of the cells become positive in the benzidine reaction for hemoglobin (Friend *et al.*, 1971). Also, those cells that do not accumulate hemoglobin in quantities sufficient for cytochemical detection have a potential to undergo differentiation since more than 94% of single cell clones undergo erythroid differentiation when induced with DMSO (Singer *et al.*, 1974). Although the mechanisms by which DMSO induces Friend erythroleukemia cells to differentiate are completely unknown, the experimental advantages offered by this cell system make it useful for cell fusion studies.

Hybrids have been obtained by fusing DMSO inducible sublines of Friend erythroleukemia with mouse L cell mutants (Orkin *et al.*, 1975; Deisseroth *et al.*, 1975a,b), mouse erythroblasts (Deisseroth *et al.*, 1975a,b), and noninducible sublines of Friend erythroleukemia (Paul and Hickey, 1974). Extinction of inducible hemoglobin synthesis was observed in fusions with nonerythroid cells whereas some hemoglobin or globin mRNA synthesis could be detected in hybrids where the other parent was an erythroid cell or a noninducible erythroleukemic cell (Orkin *et al.*, 1975; Paul and Hickey, 1974; Deisseroth *et al.*, 1975a,b).

The observation that the inducibility of hemoglobin synthesis is extinguished when erythroleukemia cells are fused with nonerythroid cells is in agreement with the results obtained with heterokaryons (Chapter VII). In heterokaryons prepared by fusing immature chick erythrocytes with undifferentiated mouse and hamster cells (Harris, 1970a; Davis and Harris, 1975) or avian fibroblasts (Alter and Ingram, 1975) hemoglobin synthesis is extinguished.

F. CELL HYBRIDIZATION AND THE ANALYSIS OF CELL DIFFERENTIATION

A number of different gene expression patterns have been observed in the cell fusion studies reviewed in the preceding sections of this chapter: *coexpression* of constitutive and facultative markers (Table XI-1), *dominance* and *recessiveness* of drug resistance markers, *extinction* (Table XI-2) or *retention* (Table XI-3) of facultative markers, and, in some cases, *activation* of new properties not expressed by the parental cells (Table XI-4). Extinction has sometimes been followed by reexpression of the extinguished markers as the hybrids underwent chromosome segregation. Although these observations are important for our understanding of how gene expression and cell differentiation are controlled in eukaryotic cells they do not provide a clear picture of how gene activity is regulated. Still it seems useful to emphasize some of the principles which have been illustrated by the results obtained in cell hybrid studies:

(a) The fact that extinguished markers can reappear (e.g., liver specific enzymes in the hepatoma hybrids, studied by Weiss and coworkers) after a considerable number of cell cycles indicates that the epigenotype, that is the programming for a certain type of cell differentiation, is quite stable and can be retained for long periods of time in the absence of its expression.

(b) There are no clear-cut examples of a complete reprogramming of a genome from one type of differentiation to another but there are several observations (Table XI-4) which suggest that genes for individual facultative markers may be activated. The synthesis of *human* as well as mouse serum albumin in mouse hepatoma + human leukocyte hybrids is one example of this phenomenon.

(c) Gene dosage effects have been demonstrated in hepatoma and melanoma hybrids since differentiated markers are extinguished in 1 + 1 hybrids but expressed in 2 + 1 hybrids where the input of chromosomes from the differentiated parental cell has been doubled.

(d) There are many examples of independent expression of one out of several markers characteristic of a specific form of cell differentiation. Extinction and reexpression of individual, liver specific enzymes can occur independently of the expression of other facultative markers. At the same time it is evident from studies on neuroblastoma hybrids that groups of markers may be coordinately expressed.

In spite of all the uncertainties and difficulties of interpretation which are associated with cell hybridization studies there is little doubt that this technique is a very valuable addition to the methodology of somatic cell genetics. Further analysis of the hybrid systems discussed on the preceding pages will certainly throw light on the mechanisms regulating gene expression and cell differentiation in higher cells. Several of the systems appear very promising and a considerable amount of information has already been accumulated.

XII

Cell Organelles in Hybrid Cells

A. NUCLEI

Studies on interspecific heterokaryons show that there is a fairly rapid exchange of macromolecules between the nuclear and cytoplasmic compartments. This phenomenon can be studied by immune fluorescence techniques using species-specific antibodies against nuclear components. In hybrid myotubes formed by spontaneous fusion of chick and rat myoblasts chick and rat nuclear antigens are first restricted to the respective types of nuclei, but with increasing time after fusion, chick nuclear antigens appear in the rat nuclei and vice versa (Carlsson *et al.,* 1974b) (Fig. XII-1). Similarly, chick erythrocyte nuclei undergoing reactivation in heterokaryons made with human HeLa cells accumulate large quantities of human nuclear proteins and antigens (see Chapter VII). In this system the nucleocytoplasmic exchange of macromolecules seems to involve mainly protein. No backward migration of RNA from the cytoplasm to the nucleus could be detected (Goto and Ringertz, 1974) although it has been shown to occur in protozoa (Goldstein, 1974).

There is little information about the composition of nuclei in synkaryons. On the basis of the results obtained in heterokaryons it may be assumed that their nuclei are molecular mosaics containing proteins characteristic of both parental cells. Direct evidence for this is, however, difficult to get. Jost *et al.* (1975) studied the spectrum of DNA binding proteins in human and murine cells by acrylamide gel electro-

213

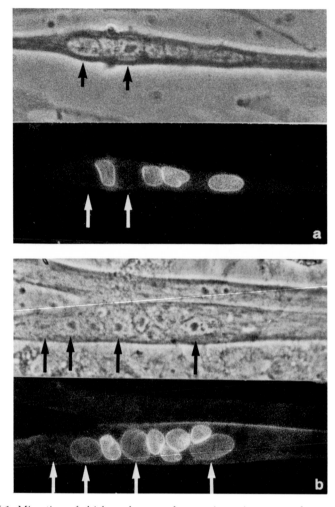

Fig. XII-1. Migration of chick nuclear envelope antigens into rat nuclear envelopes in chick myoblast + rat myoblast hybrid myotubes. (a) Three days after fusion only chick nuclei contain chick nuclear antigens. Rat nuclei (arrows) are negative. (b) Six days after fusion began, rat nuclei (arrows) have taken up chick nuclear antigens (from Carlsson *et al.*, 1974b).

phoresis. Few if any species specific differences were found and it was not possible to identify individual human or murine proteins in extracts of nuclei isolated from hybrid cells. The fact that in a number of species both histone and nonhistone nuclear proteins are very similar, if not identical, (Elgin and Bonner, 1970; Ohlson and Busch, 1974; Jost *et al.*, 1975) is perhaps the reason why interspecific hybrids can grow so well and express genes typical of both species.

B. RIBOSOMES

Ribosomal 28 S RNA from a variety of mammalian species differ from each other in their mobility during electrophoresis on polyacrylamide gels. Furthermore, since the 28 S rRNA determines the ability of ribosomes to form dimers at specific concentrations of Mg^{2+} and salt, the extent of dimerization also varies from species to species. Because of these differences it is possible to examine interspecific hybrid cells for the synthesis of 28 S rRNA characteristic of the parental cells (Table XII-1). These studies show that hybrids between two rodent species synthesize mitochondrial 28 S rRNA characteristic of both species (Stanners *et al.*, 1971; Eliceiri, 1972, 1973b,c; Sayavedra and Eliceiri, 1975). In man + mouse hybrids, however, two groups have reported that only mouse 28 S rRNA is synthesized (Eliceiri and Green, 1969; Bramwell and Handmaker, 1971). This observation caused a great deal of interest since F_1 hybrid animals from intrageneric interspecific matings show selective inactivation of nucleoli of one of the parental species. Similarly, plant hybrids sometimes show suppression of the nucleolus organizing capacity of chromosomes of one parental species (Honjo and Reeder, 1973).

The lack of human 28 S rRNA synthesis in man + mouse hybrid cells could be explained as an inactivation phenomenon but equally well as an elimination of human chromosomes carrying nucleolar organizing regions and rRNA genes. In a more recent study Marshall *et al.* (1975) analyzed the ability of man + mouse heterokaryons to synthesize ribosomal RNA. In order to carry out this analysis, HGPRT⁻ mouse cells that grew attached to a surface were fused with human HGPRT⁺ cells that were unable to attach. After fusion, the nonattached human cells were rinsed away so that the monolayer only consisted of HGPRT⁻ cells and HGPRT⁺ heterokaryons. ³H-Hypoxanthine was then used to selectively label RNA synthesized in heterokaryons. Polyacrylamide-gel electrophoresis of RNA extracted from heterokaryon cultures showed that both human and mouse 28 S rRNA were synthesized (Fig. XII-2). Thus, in a situation where all human and mouse chromosomes could be expected to be present, the hybrid cells produced both types of RNA. Although the results with synkaryons have not been fully explained, it seems most likely that the absence of human 28 S rRNA reported for some of the human + mouse hybrids (Eliceiri and Green, 1969; Bramwell and Handmaker, 1971) is due to elimination of chromosomes and not to inactivation phenomena.

Attempts to analyze whether ribosomal proteins of both parental cell types are synthesized in interspecific hybrids have been hampered by the fact that these proteins too are very similar in different mammalian

TABLE XII-1

Synthesis of Ribosomal Components in Hybrid Cells

28 S rRNA	Species of 28 S RNA in hybrids	References
Heterokaryons		
Ehrlich cells (mouse) + Lesch-Nyhan cells (HGPRT⁻, human)	Mouse + human	Marshall et al. (1975)
BHK₃ (Syrian hamster HGPRT⁻) + Daudi (human)	Syrian hamster + human	Marshall et al. (1975)
Synkaryons		
WI-38 (human) + L cell (cl 1D) mouse	Only mouse[a]	Eliceiri and Green (1969)
SV40 transformed fibroblasts (human) + diploid fibroblast (mouse)	Only mouse	Eliceiri and Green (1969)
SV40 transformed fibroblasts (human) + 3T3 (mouse)	Only mouse	Eliceiri and Green (1969)
D98/AH-2 (human) + 3T3 (mouse)	Only mouse	Eliceiri and Green (1969)
Lesch-Nyhan diploid fibroblasts (human) + Ehrlich cells (mouse)	Only mouse	Bramwell and Handmaker (1971)
T6a (Syrian hamster) + 3T3 (mouse)	Hamster + mouse	Stanners et al. (1971); Eliceiri (1972, 1973b); Sayavedra and Eliceiri, 1975
ts BHK[b] (Syrian hamster) + 3T3 and 3T6 (mouse)	Hamster + mouse[c]	Toniolo and Basilico (1974)
Leukemia cells (rat) + A9HT (mouse)	Rat + mouse	Kuter and Rodgers (1975)

Ribosomal proteins	Species of ribosomal protein	References
Synkaryons		
Leukemia cells (rat) + A9HT (mouse)	Rat + mouse	Kuter and Rodgers (1975)

[a] The hybrid contained 8–40 human chromosomes.

[b] Temperature sensitive (ts) mutants of BHK cells unable to grow at 39°C because of block in 32 S → 28 S rRNA processing.

[c] ts Defect complemented in hybrid.

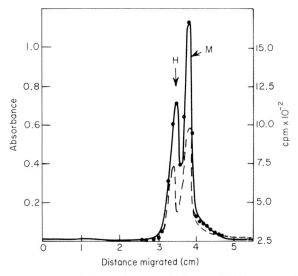

Fig. XII-2. Polyacrylamide gel electrophoresis of ribosomal RNA extracted from hetero-karyons formed between Lesch–Nyhan fibroblasts (human) and Ehrlich ascites tumor cells (mouse) and labeled with ^{3}H-hypoxanthine. ---, absorbance at 265 nm; ●—●, ^{3}H cpm × 10^{-2}. Peak marked "H" represents human rRNA, "M" is mouse rRNA (from Marshall *et al.*, 1975).

species. Kuter and Rogers (1975) compared the two dimensional electrophoretic patterns of rat and mouse ribosomal proteins and only found one protein where the homologous rat and mouse variants differed in mobility. Rat + mouse hybrid cells were shown to produce both the rat and the mouse variant of this particular protein.

C. MITOCHONDRIA

Mitochondria are equipped with their own genome in the form of circular DNA molecules. In mammalian cells the mitochondrial DNA has a molecular weight of approximately 10^7 daltons and a circumference of approximately 5 μm. This genome is believed to specify the large and the small RNA components of mitochondrial ribosomes, 11–20 different tRNA species and a small number of proteins. The proteins synthesized on mitochondrial ribosomes probably represent less than 5% of all the proteins present in these cell organelles. Most proteins (>95%) are specified by nuclear genes and are synthesized on cytoplasmic ribosomes (Fig. XII-3). Attempts have been made to identify which proteins are made in mitochondria and which are taken up from the cytoplasm (for reviews see Sager, 1972; Kroon and Succone, 1973; Schatz and Mason, 1974). The results obtained can be summarized as follows: *Neurospora*

MITOCHONDRION

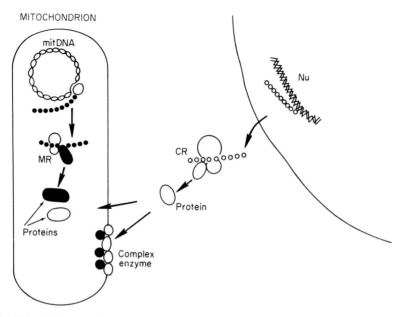

Fig. XII-3. Schematic illustration of the interrelationship between cytoplasmic and mitochondrial protein synthesis. Most mitochondrial proteins are specified by the nuclear genome (Nu) and are synthesized on cytoplasmic ribosomes (CR). A few polypeptide chains, tRNA, and mitochondrial rRNA are coded for by the mitochondrial genome. Some mitochondrially specified polypeptides (black) combine with polypeptides specified by the nuclear genome (white) and form complex enzyme molecules. MR, mitochondrial ribosomes; mit-DNA, circular mitochondrial DNA.

and yeast mitochondria specify and synthesize polypeptide chains which become part of cytochrome oxidase, oligomycin-sensitive ATPase and cytochrome b. The situation with respect to cytochrome oxidase is particularly interesting since this enzyme is made up of seven polypeptide subunits, three of which are specified by the mitochondrial genome while four are specified by the nuclear genome. Although little is known about animal mitochondria, there too one may expect a complex interaction and cooperation between two genetic systems, one nuclear and one mitochondrial. In hybrid cells the situation becomes even more complex since there are then two nuclear and two mitochondrial genetic systems. Furthermore, both systems change during chromosome segregation and the multiplication of the mitochondrial populations. The complexity can to some extent be reduced if instead of fusing two cells together one fuses an anucleate cell with an intact cell or a minicell with an anucleate cell (Chapter VIII).

The possibilites of using cell fusion techniques in mitochondrial

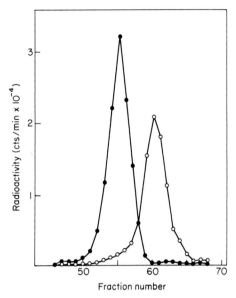

Fig. XII-4. Density gradient centrifugation on CsCl separates mouse mit-DNA (O—O) and human mit-DNA (●—●). (From Coon *et al.*, 1973.)

genetics depends to a great extent on being able to identify mitochondrial DNA and proteins with one of the parental cells and on the availability of mitochondrial mutants. Mitochondrial DNA of some species, for instance, mouse and man, can be separated by CsCl density gradient centrifugation (Fig. XII-4) and can be measured separately by molecular hybridization methods which detect species specific nucleotide sequences (Coon *et al.*, 1973; Horak *et al.*, 1974).

1. Mitochondrial DNA (mit-DNA)

The first studies of mitochondrial DNA (mit-DNA) in hybrid cells were carried out by Clayton *et al.* (1971) and Attardi and Attardi (1972). Both groups examined a series of human + mouse hybrids which contained almost complete sets of mouse chromosomes and a considerable number of human chromosomes (Table XII-2). In spite of this only mouse mit-DNA was present in the hybrids.

Later work by Coon and co-workers (1973; Horak *et al.*, 1974) has shown that both human and rodent mit-DNA can be retained over many cell generations, but that the proportion of human–rodent mit-DNA changes as the hybrid cells undergo chromosome segregation (Fig. XII-5). Among the hybrid clones examined by Coon *et al.* (1973), many showed reverse chromosome segregation, that is, extensive

TABLE XII-2

Mitochondrial DNA in Hybrid Cells (Synkaryons)

Parental cells	Species of mit-DNA in hybrid cells	Remark	References
Normal lymphocyte (human) + 3T3 (mouse)	Only mouse	8–11 Human chromosomes	Clayton *et al.* (1971)
HeLa cells (human) + 3T3 (mouse)	Only mouse	17 Human chromosomes	Clayton *et al.* (1971)
Daudi lymphoma (human) + A9 (mouse)	Only mouse	—	Clayton *et al.* (1971)
Normal lymphocyte (human) + L (mouse)	Only mouse	7–16 Human chromosomes	Clayton *et al.* (1971)
VA 2-B (human) + ClID (mouse)	Only mouse	6–24 Human chromosomes	Attardi and Attardi (1972)
Diploid FH10 (human) + 3T3 (mouse)	Only mouse	11–31 (?) Human chromosomes	Attardi and Attardi (1972)
T6a (Syrian hamster) + 3T3 (mouse)	Hamster + mouse	—	Eliceiri (1973a)
VA2 and D98/AH-2 (human) + normal diploid (mouse)	Human + mouse	Varying ratios human/rodent chromosomes; recombinant mit-DNA	Coon *et al.* (1973)
VA2 and D98/AH-2 (human) + normal diploid (rat)	Human + rat	Varying ratios human/rodent chromosomes; recombinant mit-DNA	Horak *et al.* (1974)

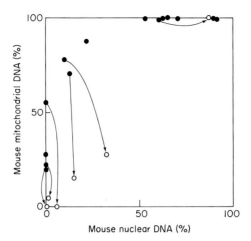

Fig. XII-5. DNA composition of mouse + human hybrid cell strains. Solid circles represent populations when first tested about 40–60 doublings after fusion. Strains which at that time contained no detectable mouse nuclear or mitochondrial DNA were omitted from the figure. Arrows lead from the position at the first testing to the position at the second testing of the same strain about 25 doublings later. (From Coon *et al.*, 1973.)

losses of rodent chromosomes, while others showed the expected loss of human chromosomes (see Chapter X). A rough correlation was found between the loss of mit-DNA from one species and loss of chromosomes from the same species. In a series of 23 mouse + human hybrid clones, nine had lost all human chromosomes and all human mit-DNA when first tested, seven were still hybrid with respect to their nuclear DNA and chromosomes but contained only mouse mit-DNA, while another seven were hybrid both with respect to their nuclear and mitochondrial DNA. Thus, in these results too, there is a marked tendency for the mitochondrial population to segregate away from equal proportions. Furthermore, there is some correlation between the segregation of the nuclear and mitochondrial genomes in the sense that hybrids which have lost virtually all chromosomes of one species tend also to have lost mit-DNA of that species. It is also apparent from these studies that segregation of mit-DNA is more rapid than segregation of the nuclear genomes. So far there is very little information about what causes mitochondrial segregation and to what extent the elimination of mit-DNA depends on losses of *specific* chromosomes. It appears, however, that cell hybridization could be an effective way to acquire more understanding of mitochondrial dependency on nuclear functions.

It would be interesting to know more about how mitochondria interact with each other in eukaryotic cells. In yeast there is evidence that mit-DNA molecules undergo molecular recombination resulting in

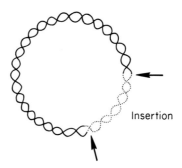

Insertion

Fig. XII-6. Schematic illustration of a recombinant mitochondrial genome in which mouse (dotted double helix) DNA has been inserted into a human mit-DNA molecule (solid) double helix.

phenotypically recombinant cells (Michaelis *et al.*, 1973). The only information about mitochondrial recombination in eukaryotic cells comes from the work of Horak *et al.* (1974) in which a series of human + rodent hybrid lines were examined for recombinant mit-DNA molecules. Among hybrids which contained both types of parental mit-DNA such molecules were frequent (Fig. XII-6). Further studies of this type of genetic recombination would be greatly facilitated if mammalian cell lines carrying specific mitochondrial gene mutations could be isolated.

In protozoa, yeast, and *Neurospora* there is evidence that under some conditions acridine dyes (e.g., acriflavine and ethidium bromide) act as mitochondrial mutagens with little effect on nuclear genes. Furthermore resistance to drugs that selectively inhibit mitochondrial protein synthesis (chloramphenicol, erythromycin, spiramycin, and mikamycin) appears to be due to mitochondrial gene mutations. (Beale, 1969; Adoutte and Beisson, 1970; Beale *et al.*, 1972; Linnane *et al.*, 1968; Sager, 1972).

Much less is known about mitochondrial gene mutations in mammalian cells. However, Eisenstadt and co-workers (Spolsky and Eisenstadt, 1972; Kislev *et al.*, 1973; Bunn *et al.*, 1974) have described chloramphenicol resistant mutants of human HeLa and mouse A9 cells, and Wallace and Freeman (1975) have isolated mouse cells resistant to Tevenel, a sulfamoyl analog of chloramphenicol. Bunn *et al.* (1974) also demonstrated that chloramphenicol resistance could be transferred to sensitive cells by Sendai virus-induced fusion of anucleate resistant cells with sensitive intact cells. These findings indicate that drug selection can be used to isolated mammalian mitochondrial mutants. Such mutants in combination with virus-induced fusion of cell fragments offers new approaches to the analysis of mitochondrial genetics of animal cells.

2. Mitochondrial Enzymes

Chromosome segregation in interspecific hybrids has made it possible to assign genes for mitochondrial enzymes to specific chromosomes and to establish with certainty which enzymes are specified by the nuclear genome. Using man + mouse hybrids, van Heyningen *et al.* (1973) have shown that two mitochondrial enzymes, citrate synthetase (CS), and malate dehydrogenase (MDH-2) are under nuclear control. Some of the hybrids examined had lost all human mit-DNA (Clayton *et al.*, 1971) while still retaining some human chromosomes. Of three hybrid lines of this type, two produced human CS and MDH-2 while one line had lost the human variants of these enzymes. This indicates that the presence of human mit-DNA is not required for the expression of some human mitochondrial markers and that mitochondria in interspecific hybrids become molecular mosaics containing proteins of both species. A further indication of this was the finding of heteropolymeric enzyme forms in some hybrids but not in mixtures of human and mouse cell extracts.

Kit and Leung (1974) used human + mouse hybrids to study the genetic control of mitochondrial and cytosol thymidine kinase (TK). The hybrids were obtained by fusing mouse (cytosol TK$^-$, mitochondria TK$^+$) cells with human (cytosol TK$^+$, mitochondria TK$^+$) cells and then selecting under conditions where the hybrids had to retain human cytosol TK in order to survive. Extracts of the hybrids were then subjected to polyacrylamide gel electrophoresis and tested for TK activity. This method makes it possible not only to distinguish between cytosol and mitochondrial TK but also to discriminate between the respective human and mouse enzymes. The results showed that mouse mitochondrial and human cytosol enzymes were retained at the same time as the human mitochondrial TK activity was lost. Since the hybrid cells had lost most of the human chromosomes this finding indicates that human mitochondrial TK is specified by a different chromosome from that coding for the human cytosol enzyme.

XIII

Gene Mapping and Gene
Complementation Analysis

The formal genetics of mammalian cells has, until recently, been dependent on breeding experiments with whole organisms whose long generation times and small numbers of offspring in higher mammals have limited the speed of progress. Somatic cell hybridization, sometimes described from a narrow perspective as "an alternative to sex," makes it possible to obtain hybrids between cells from the same or different species very quickly. Cell hybrids also show segregation of genes, albeit of a different type from that observed in the progeny of sexual crosses. The selective elimination of chromosomes of one species in interspecific hybrids is particularly important since detailed comparisons of the phenotypes and chromosome patterns make it possible to establish gene linkage groups, to assign genes to specific chromosomes, and to map the location of genes on individual chromosomes. Moreover, both inter- and intraspecific hybrids can be used for complementation studies in which it is possible to establish whether genetic defects in two mutant cell types are of the same or different genetic loci.

A. LOCATING GENES ON HUMAN CHROMOSOMES

The analysis of the human genome with the aid of man + mouse hybrids is based on the fact that human chromosomes are generally selectively eliminated (see Chapter X), thus, resulting in hybrid cells that contain a more or less intact mouse genome plus various parts of the human genome. Which group of human chromosomes is retained varies from clone to clone. The ideal hybrids for assigning genes to chromosomes are those which only retain a single human chromosome. Unfortunately, the isolation of such hybrid clones is time consuming and may require some luck. Instead, gene assignments are often based on detailed analysis of a number of human + mouse hybrid clones which each retain a different combination of human chromosomes. The cells are analyzed for the presence of specific human chromosomes, using the new methods of chromosome identification discussed in Chapter X, and for the presence of human proteins. Because many homologous human and mouse enzymes differ in electrophoretic mobility, it is possible to establish whether or not there is coexpression of both genomes or whether or not a human enzyme activity has been lost because of loss of a specific human chromosome.

Hybrid cells suitable for gene mapping procedures have been obtained by two different methods. One is based on random segregation of human chromosomes in a fairly large number of independent human + mouse hybrid lines. By analyzing the expression of a variety of human markers in many different clones it is possible to establish that some markers always occur together and that they are always lost as a group (i.e., whether they are linked). In some cases the retention or loss of a marker can be correlated with a specific chromosome. The other method is based on the use of mouse cells deficient in some phenotypic character and culture conditions which select against the deficient cells. Thus, while unfused mouse cells die, any mouse + man hybrid cells that retain a human chromosome carrying genetic information that can overcome the defect survive. The human component complements the mouse genetic defect. HGPRT deficiency is an example of a selective marker which can be used to fix the retention of a specific chromosome, the X chromosome. In a mouse HGPRT$^-$ + human HGPRT$^+$ hybrid growing on HAT medium (see Chapter IX) all human chromosomes may be eliminated except an active human X chromosome carrying an intact *HGPRT* gene.

1. Linkage Analysis in Randomly Segregating Hybrid Clones

The use of nonselective markers in establishing linkage groups and in assigning genes to specific human chromosomes is based on the assumption that the elimination of human chromosomes in man + rodent hybrids is a random process and that the absence of a marker is due to loss of the corresponding chromosome.

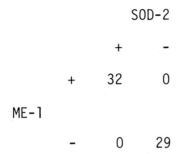

Fig. XIII-1. Synteny of human cytoplasmic malic enzyme (ME-1) and (mitochondrial superoxide dismutase (SOD-2) in mouse + human hybrid clones. Figures indicate number of clones with a certain phenotype (from Tischfield *et al.*, 1974, see New Haven Conference, 1974).

One begins by isolating a large number of different hybrid clones. It is important that these are primary clones (each clone arising from a single fusion event). After these clones have been propagated for some time so as to allow chromosome elimination and selection for variants (segregants) with a reduced human chromosome complement, the hybrid cells are analyzed for phenotypic markers and for the presence of individual human chromosomes.

In order to assign gene loci to different chromosomes, phenotypic and chromosomal data from a large number of clones are analyzed by making comparisons of the occurrence of pairs of markers (illustrated in Fig. XIII-1) and for a chromosome and an enzyme (illustrated in Fig. XIII-2). If two human enzymes always occur together and are always lost together, it is likely that their structural genes belong to the same linkage group and map on the same chromosome.

Markers which are assigned to the same chromosome are said to be *syntenic* (Renwick, 1971a,b), and the analysis just described is some-

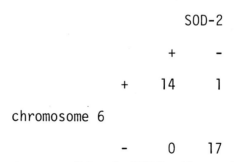

Fig. XIII-2. Enzyme–chromosome linkage for SOD-2 and human chromosome 6 in human + mouse hybrids. One discordant clone contained chromosome 6 in 3% of the cells but did not produce sufficient enzyme to be scored as positive. (From Tischfield *et al.*, 1974, see New Haven Conference, 1974.)

times referred to as synteny testing. The opposite to synteny is asynteny meaning that two genes are on different chromosomes.

The first step in assigning genes to a specific chromosome has usually been the establishment of enzyme–enzyme linkage groups. The next step has been to assign linkage groups or individual markers to a specific chromosome. If an enzyme marker and a specific chromosome show concordant segregation [the presence and absence of the enzyme correlate with the presence and absence of a certain chromosome (Fig. XIII-2)], then an enzyme–chromosome linkage is established. Thus, the structural gene for that enzyme is assigned to that chromosome. The validity of the assignment is strengthened if it can be shown that there is a gene dosage effect, that is there is a stronger expression of the enzyme if several copies of the specific chromosome are present (Green *et al.*, 1971b).

The number of clones which needs to be analyzed in assignment tests depends on the number of chromosomes which have been retained from the segregating genome. If for instance, only one human chromosome is retained in man + mouse hybrids, three to four clones could be enough for the assignment of a large number of genes to one specific chromosome. Hybrid clones which retain several human chromosomes make it possible to assign genes to several different chromosomes (for example see Ruddle *et al.*, 1971). Provided such clones have been selected so that each contains a unique subset of chromosomes, eight to ten clones may be sufficient for unequivocal assignments to all of the 24 different human chromosomes (Ruddle and Kucherlapati, 1974). The principle underlying the use of such *hybrid clone panels* was first suggested by Creagan and Ruddle (1975) and is illustrated schematically in Fig. XIII-3. A panel of three clones, each containing four different

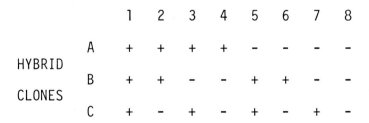

Fig. XIII-3. Principle for the use of a hybrid clone panel of three clones in making gene assignments to eight different chromosomes. If a given enzyme is found in clone A and B but not in clone C the corresponding gene can be assigned to chromosome 2. (Modified from Ruddle and Creagan, 1975.)

human chromosomes is sufficient to define eight different chromosomes from a segregating genome. Theoretically, five clones would be the minimum number of clones required to define the haploid number of human chromosomes. In practice, however, a somewhat greater number of clones may be required since it is unlikely that anyone would be able to isolate a set of five clones retaining a perfect combination of human chromosomes, which would make it possible to assign genes to all 24 different human chromosomes (22 pairs of autosomes and the X and Y chromosomes).

2. Fixed Retention of One Chromosome by the Use of Selective Markers

Obviously a marker is selective only if a system of culture exists that can isolate hybrids retaining the marker while other phenotypic characters of the segregated genome are lost. So far there are at least six different selective markers and corresponding selective media which can be used for the isolation of man + mouse or man + Chinese hamster hybrids retaining specific human chromosomes (Table XIII-1). Four of the selective systems use enzyme deficient cells ($HGPRT^-$, TK^-, $APRT^-$, and dCD^-) whereas two use auxotrophic mutants. These selective methods make it possible to fix the retention of the human X chromosome or of autosomes 12, 16, and 17. One of the systems (Ade^-) appears to retain a human chromosome which is either number 4 or number 5 while another selective system (dCD^- cells and HAM medium) has not yet been exploited for the localization of the complementing human gene.

When the appropriate hybrids have been obtained on selective medium they are analyzed for chromosomes and isozyme markers. Synteny tests and gene assignment tests are then carried out as described for randomly segregating hybrids. The great advantage in using selective markers is that experiments can be designed to retain specific chromosomes. This reduces the number of clones which have to be analyzed. Furthermore, it is sometimes possible to check gene assignments by performing back-selection experiments. A man + mouse hybrid made by crossing $HGPRT^+$ human cells with $HGPRT^-$ mouse cells must retain the human X chromosome carrying an active $HGPRT^+$ gene in order to survive on HAT medium. But when such a hybrid cell population is transferred to normal medium and treated with 8-thioguanine the majority of the cells die. Only a small minority of segregants survive. The absence of human HGPRT and human X chromosomes in hybrids surviving back-selection confirms the assignment of the $HGPRT$ gene to the X chromosome. Back-selection can also be used to obtain

TABLE XIII-1

Selective Retention of Human-Chromosomes in Man + Rodent Hybrids

Genetic defect in rodent cell	Rodent parent	Human complementing gene	Human chromosome retained in hybrid	Selective medium	Drug for back selection on permissive medium	References
HGPRT⁻	Mouse	HGPRT	X	HAT	Thioguanine	Littlefield, 1964a; Nabholz et al., 1969
TK⁻	Mouse	TK	17	HAT	BUdR	Miller et al., 1971
APRT⁻	Mouse	APRT	16	Alanosine + adenine	Fluoroadenine	Kusano et al., 1971; Tischfield and Ruddle, 1974
dCD⁻	Mouse	dCD	?	HAM	BCdR	Chan et al., 1975
Gly⁻A[a]	Chinese hamster	Serine hydroxymethylase	12	MM(−glycine)[b]	—	Kao et al., 1969a,b; Jones et al., 1972
Ade⁻B[a]	Chinese hamster	Unknown	4 or 5	MM(−adenine)[c]	—	Kao and Puck, 1972b

[a] Auxotrophic mutants.
[b] MM(−glycine), minimal medium lacking glycine.
[c] MM(−adenine), minimal medium lacking adenine.

hybrid segregants which have eliminated other specific chromosomes. A number of different gene loci can be used as "targets" in back selection experiments, for instance, those that specify enzymes involved in drug metabolism or surface receptors that determine the sensitivity of the cells to viruses or toxins. For example, fluoroadenine selects for cells that have lost the chromosome carrying the adenophosphoribosyltransferase (APRT) gene (Kusano *et al.*, 1971) subsequently identified as chromosome 16 (Tischfield and Ruddle, 1974), while poliovirus kills hybrids containing chromosome 19 specifying the poliovirus receptor on the cell surface (Miller *et al.*, 1974). Methods for obtaining hybrids which have eliminated other chromosomes will probably soon be available. (For a review see Ruddle and Creagan, 1975.)

3. Discrepant Clones

In synteny testing it frequently is the case that the majority of hybrid clones confirms a certain assignment, but that at the same time there are a few discrepant or discordant clones. Discrepant clones may be of two types: the gene product (as, for example, an enzyme) may be present while the chromosome is not, *or* the chromosome may be present while the enzyme is not. The first type of discrepant clones can be due to chromosomal rearrangements which complicate the identification of the chromosome. A translocation may, for instance, divide a human chromosome into two parts and, thus, break up a linkage group into two subgroups that in the hybrids may segregate independently. Sometimes the chromosome rearrangements involve fragments which are too small to be detectable in the light microscope. Although this often complicates assignment tests, sometimes it may be of great help in mapping the location and order of genes. This aspect will be discussed in more detail in a later section of this chapter.

The second type of discrepant clones raises several different possibilities. For one, the chromosome may have undergone a submicroscopic deletion causing a loss of the structural gene for the enzyme being selected for or a mutation which causes the enzyme molecule to lose its catalytic activity. Another possibility is that the expression of the enzyme is subject to regulation. Failure to detect the expression of the enzyme marker could also be due to purely technical problems in the enzyme assay.

B. REGIONAL MAPPING OF CHROMOSOMES

Chromosome mapping consists of establishing the linear order in which genes occur on individual chromosomes, measuring distances between genes, and relating these data to the structure of the chromo-

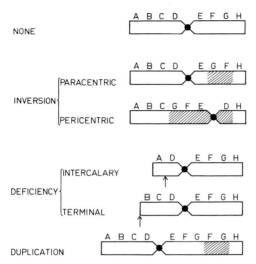

Fig. XIII-4. Main types of intrachromosomal aberrations.

somes. Hybrid cells can be used for chromosome mapping in several different ways. One method uses parental cells with rearranged or fragmented chromosomes (Ricciuti and Ruddle, 1973a,b; Grzeschik *et al.*, 1972; Grzeschik, 1973a; Shows and Brown, 1974). Another method is based on the induction of massive chromosome fragmentation by treating one of the parental cells (M. Harris, 1972; Goss and Harris, 1975) or the hybrids (Burgerhout *et al.*, 1973) with chemical or physical mutagens and, thus, causing chromosome breaks.

In regard to the first method, a number of human chromosomal abnormalities are known (Figs. XIII-4 and XIII-5; for review see Bartalos and Baramki, 1967), some of which are associated with diseases while others are not. The most useful of these abnormalities for mapping are those that involve translocations to chromosomes carrying known selective loci (e.g., to chromosomes X, 16, or 17; see Table XIII-1). When human cells with such chromosomal rearrangements are fused with appropriate mouse mutants it is possible to select for man + mouse hybrids that retain only the translocated chromosome. One can then determine which additional human markers have cosegregated with the selective markers, and by identifying the translocated piece, additional markers can be assigned to one of the normal chromosomes. In order to use this technique for establishing detailed maps of the human chromosomes one needs not only a battery of mutant mouse cells carrying genetic defects that can be complemented by each human chromosome but also a library of human cells with different types of well defined breaks or translocations at different points along each of the chromo-

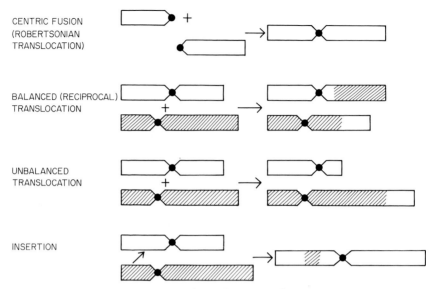

Fig. XIII-5. Main type of interchromosomal rearrangements.

somes.* Alternatively one may make use of spontaneous chromosome rearrangements that occur after fusion during the growth of hybrid cells (Boone *et al.,* 1972; McDougall *et al.,* 1973; Elsevier *et al.,* 1974; Douglas *et al.,* 1973a,b).

The second method (Fig. XIII-6), sometimes referred to as the "disruptive strategy" of chromosome mapping (Siniscalco, 1974), has been developed by Goss and Harris (1975). The main advantages of this method are that it permits high resolution mapping, is independent of karyological data, makes it possible to measure distances between genes, and can be used to determine the order in which genes occur along the chromosome. The limitations of this method are that it depends on selective systems which can fix the retention of specific genes and that the location of some gene loci must be known in advance. These loci serve as "landmarks" and make it possible to correlate the mapping data with the light microscopical appearance of the chromosomes.

Goss and Harris (1975) have demonstrated the usefulness of the disruptive method in model experiments designed to map four genes which have previously been assigned to the X chromosome. X-Irra-

* A repository of mutant human cells with chromosome aberrations and/or biochemical variants as well as normal human cells has been organized at the Institute for Medical Research, Copewood Street, Camden, New Jersey 08103. For listings of cells available see Rotterdam Conference, 1975.

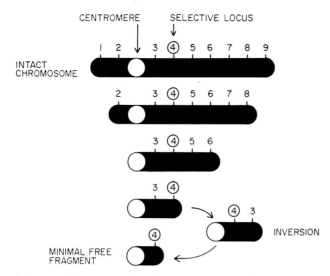

CENTROMERE   SELECTIVE LOCUS

Fig. XIII-6. Schematic illustration of the disruptive strategy of chromosome mapping as developed by Goss and Harris (1975). The chromosomes of the segregating genome are fragmented by X irradiation before fusion. In order to survive on HAT-selective medium the cells must retain a fragment carrying the selective *HGPRT* locus on the X chromosome. The frequency at which other loci cosegregate with the selective locus reflects the distance between these loci and the selective locus. Gene locus 3, situated between the selective locus 4 and the centromere, will be retained by practically all chromosome fragments whereas locus 9 will easily be lost. The latter locus is furthest away from the selective locus which means that the probability is high that locus 9 is separated from the selective locus 4 by a chromosome break.

diated human cells were fused with HGPRT⁻ Chinese hamster cells. The resulting hybrid cells were isolated by selection on HAT medium. Under these conditions the human *HGPRT* gene, which is known to map on the X-chromosome, had to be retained. The hybrids were found to contain human HGPRT enzyme but had lost one or more of three other (nonselective) X-chromosome markers: PGK, α-galactosidase, and G6PD. The extent to which these other X-chromosome markers were lost was found to depend on the dose of irradiation. Hybrids made with unirradiated human cells usually retained all four markers indicating that intact human X chromosomes were present. As the dose of irradiation was increased, fewer and fewer of the nonselective markers remained in the hybrids. By analyzing in what combinations and at what rate these markers were lost, Goss and Harris were able to measure distances between the corresponding genes and also to determine in which order they occurred. The mathematical basis for this analysis is the assumption that the greater the distance separating two genes on a chromosome the higher the probability that chromosomal breaking agents will separate them, and, similarly, the smaller the distance the lower the probability. The results showed for instance, that *PGK* segregated from *HGPRT* more frequently than either *α-gal* or *G6PD*, indicating that *PGK* is further removed from *HGPRT* than are the other two genes. Furthermore, segregation of *G6PD* from *HGPRT* was found to be independent of the segregation of *PGK* from *HGPRT*. This made it possible to conclude that *G6PD* and *PGK* must be on opposite sides of *HGPRT*. With the use of this type of analysis it was deduced that the gene order was *PGK–α-gal–HGPRT–G6PD*. This elegant analysis confirms data obtained by the use of spontaneous translocations (Ricciuti and Ruddle, 1973b; see also New Haven Conference, 1974; Rotterdam Conference, 1975).

C. GENE COMPLEMENTATION ANALYSIS

Several of the methods used in the isolation of mononucleate hybrid cells (Chapter IX) illustrate how gene complementation between genetically defective cells can be exploited for a practical experimental purpose, namely to isolate hybrid cells. But as already discussed for heterokaryons (Chapter VII), gene complementation tests in hybrid cells are also a method that permits a detailed analysis of the organization of the genome. In principle it is possible to distinguish between two forms: intergenic and intragenic (interallelic) gene complementation.

Intergenic complementation is the interaction between two different sets of genes within a hybrid cell which permits it to function under conditions where the parental cells would not. This form of complementation is illustrated by the survival on HAT medium of a hybrid between an $HGPRT^-$, TK^+ and a $HGPRT^+$, TK^- cell (see Chapter IX). Other examples of intergenic complementation in mononucleate hybrid cells have been reported by Silagi *et al.* (1969) who fused two enzyme deficient human cell lines. One (AUC) was derived from a patient suffering from orotic aciduria and was deficient in two enzymes necessary for uridylic acid synthesis: orotidine-5'-monophosphate (OMP) decarboxylase and OMP-pyrophosphorylase. The other (D98/AH) was a drug-resistant $HGPRT^-$ cell, probably of HeLa cell origin. The two parental cell lines also differed with respect to the type of glucose-6-phosphate dehydrogenase that they produced. The cloned hybrid cells showed genetic traits typical of both parents (Table XIII-2). OMP-decarboxylase, OMP-pyrophosphorylase, and HGPRT were produced at a level which was intermediate between those of the two parental lines. The hybrids also had both electrophoretic variants of G6PD. This

TABLE XIII-2

Characteristics of Parental and Hybrid Lines[a]

	Line		
Characteristics	AUC-parent cell	Hybrid	D98/AH-parent cell
Cellular morphology	Fibroblastlike	Epithelioid	Epithelioid
Colonial morphology	Cometlike	Diffuse	Compact
Total chromosome number	46	97 (modal)	62 (modal)
Growth in HAT medium	+	+	−
HGPRT activity	+	+	−
OMP-decarboxylase activity	−	+	+
OMP-pyrophosphorylase activity	−	+	+
Glucose-6-phosphate dehydrogenase electrophoretic phenotype	B	AB	A

[a] Silagi *et al.*, 1969.

showed that two X chromosomes were present and that both *G6PD* loci were expressed in the hybrids. Supernumerary X chromosomes, therefore, do not seem to be inactivated in hybrid cells as they are in early embryonic cells (see also Chapter XI).

Siniscalco *et al.* (1969a) fused two male fibroblast strains carrying different X-linked gene deficiencies. One cell strain obtained from a Lesch–Nyhan patient was HGPRT⁻ while the other came from a patient suffering from a congenital form of anemia known to be associated with a deficiency in glucose-6-phosphate dehydrogenase (G6PD⁻). Although no hybrid cells could be isolated, cultures of the fusion mixture contained cells which incorporated ³H-hypoxanthine and were positive for G6PD in a cytochemical test. This result suggests that these cells were hybrids that, because of intergenic complementation, were able to produce both HGPRT and G6PD.

Fusion of auxotrophic cell lines has been found to be a powerful method in the genetic dissection of nutritional requirements and metabolic pathways. This type of work was pioneered by Puck and his collaborators (Kao *et al.*, 1969a,b) who fused Chinese hamster cells requiring glycine (Gly⁻), proline (Pro⁻), adenine (Ade⁻), serine (Ser⁻), inositol (Ino⁻), or a combination of adenine and thymidine (AT⁻) or glycine, adenine, and thymidine (GAT⁻). Among the glycine-requiring cell lines four complementation groups (A–D) were identified by pairwise fusion of 13 different Gly⁻ lines. The biochemical defects corresponding to these four complementation groups have not yet been identified, but in one case (Gly A⁻) the cells have been found to be deficient in serine hydroxymethylase, an enzyme which converts serine to glycine.

Using a similar approach, Patterson et al. (1974) analyzed a family of 30 adenine-requiring mutants. The complementation behavior of six different Ade⁻ mutants (Ade⁻A − Ade⁻F), the double mutant AT, and the triple mutant GAT is illustrated in Table XIII-3. These data when combined with biochemical analysis of the cells using radioactively labeled intermediates (Patterson et al., 1974; Patterson, 1975) have made it possible to identify tentatively the biochemical defects in the different types of mutants. Clearly most of these Ade⁻ mutants differ with respect to the enzymes affected by the mutations, suggesting that different genes are involved. The complementation then is most appropriately classified as intergenic. In some cases, however, mutants with the same enzyme defect complement each other suggesting intragenic complementation.

A case which may represent intragenic complementation in human cell hybrids was recorded by Nadler et al. (1970) who fused cells from seven different patients suffering from galactosemia. This disease is associated with a deficiency in the enzyme galactose-1-phosphate uridyltransferase. Cells from one of the patients complemented the genetic defects of six of the other patients. As a result the hybrid cells produced a hybrid enzyme which was active but different from the normal ("wild-type") enzyme in its specific activity and kinetic properties. The authors interpreted this finding as a case of interallelic complementation. Similar conclusions have also been made by Sekiguchi and co-workers (1975a; Sekiguchi and Sekiguchi, 1973) who found complementation between different HGPRT⁻ cell lines (see Chapter X). It must be pointed out, however, that the molecular background for this form of complementation is not at all clear. One possibility is that the enzyme studied consists of two subunits, both of which are required for catalytic activity. If one of the parental cells is defective with respect to one subunit and the other parental cell with respect to the other the hybrid cell will be able to produce active "heteropolymeric" enzyme (Fig. XIII-7). Another possibility is somatic recombination. If two parental cells are defective with respect to the same gene but differ with respect to the precise location of the nucleotide changes within the gene, part of one gene could be recombined with part of the other gene to give one normal gene. However, there is very little experimental evidence to support this hypothesis. In order to more certainly demonstrate intragenic (interallelic) complementation one must isolate the comple-

TABLE XIII-3

Summary of the Complementation Behavior Found in Analysis of All of the Adenine-Requiring Mutants of the CHO-K1 Cell So Far Obtained[a]

	Ade A	Ade B	Ade C	Ade D	Ade E	Ade F	AT	GAT	Presumed or established site of defect[b]
Ade A	−								PRPP-amidotransferase or GAR-synthetase
Ade B	+	−							FGAM-synthetase
Ade C	+	+	−						PRPP-amidotransferase or GAR-synthetase
Ade D	+	+	+	−					AIR-carboxylase or SAICAR-synthetase
Ade E	+	+	+	+	−				GAR-formyltransferase
Ade F	+	+	+	+	+	−			AICAR-formyltransferase
AT	+	+	+	+	+	+	−		Tetrahydrofolate metabolism
GAT	+	+	+	+	+	+	+	−	Tetrahydrofolate metabolism

[a] Patterson et al., 1974; Patterson, 1975.
[b] Enzyme abbreviations are explained in the list of abbreviations.

INTERGENIC COMPLEMENTATION

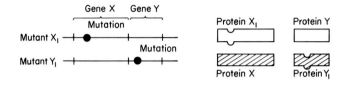

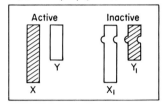

Proteins in X₁+ Y₁ hybrid cells

INTRAGENIC COMPLEMENTATION

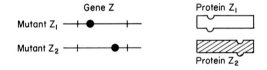

Proteins in Z₁ + Z₂ hybrid cells

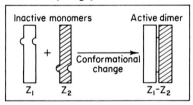

Fig. XIII-7. Schematic illustration of intergenic and intragenic (interallelic) complementation in hybrid cells. Intergenic complementation is illustrated for genes X and Y which specify two different monomeric proteins; protein X and protein Y. Mutant X_1 has a mutation in gene X and specifies an abnormal protein X_1. Mutant Y_1 in a similar way, codes for abnormal protein Y_1. In hybrids between mutant X_1 and mutant Y_1 enough functional protein X and Y will be produced to enable the cell to survive. It is assumed in this example that both mutants are homozygous and have the same mutations on both chromosomes.

Intragenic (interallelic) complementation is illustrated for gene Z which specifies a dimeric enzyme consisting of two identical subunits. Mutants Z_1 and Z_2 have undergone mentation product and demonstrate that it carries markers native to the mutant genes (Ruddle and Creagan, 1975).

D. HUMAN GENE MAPS

Table XIII-4 is a 1975 summary of genes assigned to specific human chromosomes by cell hybridization. Some of the assignments are provisional and the dating of the table is important because additions and corrections are sure to follow. The table does not include gene assignments which have been made exclusively on the basis of family studies, nucleic acid annealing, virus induced chromosome changes, or by other methods. For a more complete listing of human gene assignments the review by Ruddle and Creagan (1975) should be consulted. Close to 100 genes have been assigned to the 22 different autosomal chromosomes in man. Cell hybridization has established or confirmed 56 of these assignments. Nineteen genes have been assigned to chromosome 1 and a considerable amount of information has also accumulated concerning the regional localization of individual genes on this chromosome (Fig. XIII-8). More than 90 different traits have been assigned to the X chromosome in family studies [for a review see McKusick's "Mendelian Inheritance in Man" (McKusick, 1971)]. Five of these assignments have been confirmed in cell hybrid studies. In the case of the X chromosome, cell hybridization has played a relatively small role in the assignment of new genes (only one has been added). It has been an important method, however, in the regional mapping of this chromosome. The fact that the X chromosome carries a selective locus (*HGPRT*) which makes it possible to use the "disruptive strategy" of Goss and Harris (1975), together with the wealth of information which can be derived from family studies and clinical studies of sex-linked diseases, makes it likely that the human X chromosome will soon be the best mapped of all human chromosomes. The current situation with respect to the mapping of the human X chromosome is summarized in Fig. XIII-9. The information available is still fragmentary in the sense that the number of genes mapped is too small to see any correlations between the organization of the genome and more complex cell functions and their regulation. However, some of the information is already of medical interest since it establishes the location of gene loci that are defective in some forms of human genetic disease.

---

mutations at different points in gene Z. The monomeric and dimeric forms of the proteins are inactive in the mutants but when mutant Z_1 is fused with Z_2, the hybrid cells will contain both Z_1 and Z_2 monomers. When these combine with each other to form Z_1-Z_2 dimers the polypeptide chains undergo conformational changes which restore function.

TABLE XIII-4

Human Gene Assignments Based on Cell Hybrid Studies[a,b]

Chromosome	Gene symbol	Marker	Status
1	AK-2	Adenylate kinase-2	C
	FH	Fumarate hydratase	C
	GuK_1	Guanylate kinase	C
	Pep-C	Peptidase C	C
	PGD	Phosphogluconate dehydrogenase	C
	PGM_1	Phosphoglucomutase 1	C
	PPH	Phosphopyruvate hydratase (enolase)	C
	UGPP	Uridyldiphosphate glucose pyrophosphorylase	P
	UK	Uridine kinase	P
	UMPK	Uridine monophosphate kinase	C
2	Gal^+-Act	Galactose enzyme activator	P
	Gal-1-PT	Galactose-1-phosphate uridyltransferase (see also chromosome 3)	I
	IDH-1	Isocitrate dehydrogenase (cytoplasmic form)	C
	If_1	Interferon-1	P
	MDH-1	Malate dehydrogenase	P
3	ACO	Aconitase	I
	Gal-1-PT	Galactose-1-phosphate uridyltransferase (see also chromosome 2)	I
4	PGM_2	Phosphoglucomutase 2	P
4 or 5	Ade^+B	Human gene complementing Chinese hamster adenine B auxotroph (formylglycinamide, ribotide aminotransferase)	P
	E	Gene activating Chinese hamster esterase	P
5	DTS	Diphteria toxin sensitivity	P
	Hex B	Hexosaminidase-B	C
	If_2	Interferon-2	P
6	LA	HL-A antigen	C
	P	P blood group	C
	ME-1	Malic enzyme (cytoplasmic)	C
	PGM_3	Phosphoglucomutase-3	C
	SOD-2	Superoxide dismutase-2 (mitochondrial)	C
7	MDH-2	Malate dehydrogenase (mitochondrial)	P
	SV40-T	SV40 virus integration site (see Chapter XIV) (T-antigen gene)	P
8	GR	Glutathione reductase	P
9	AK-1	Adenylate kinase (red cell form)	P
	β-Glu	β-Glucuronidase	P
10	GOT-1	Glutamate oxaloacetate transaminase (soluble)	C
	HK-1	Hexokinase-1	C
11	ACP_2	Acid phosphatase-2 (lysosomal)	P
	Es-A_4	Esterase-A_4	C
	LDH-A	Lactate dehydrogenase-A	C
	AL	Species antigen 1 (lethal antigen)	C

TABLE XIII-4 Continued

Chromosome	Gene symbol	Marker	Status
12	*CS*	Citrate synthase (mitochondrial)	P
	LDH-B	Lactate dehydrogenase-B	C
	Pep-B	Peptidase B	C
	TPI	Triosephosphate isomerase	C
	Gly^+A	Serine hydroxymethyl-transferase ($GlyA^+$-complementing)	P
13	*Es-D*	Esterase-D	C
14	*PNP*	Purine nucleoside phosphorylase	C
15	B_2m	β_2-Microglobulin	P
	Hex-A	Hexosaminidase-A	P
	Hex-C	Hexosaminidase-C	P
	MPI	Mannosephosphate isomerase	P
	PK-3	Pyruvate kinase-3	P
16	*APRT*	Adenine phosphoribosyltransferase	C
17	*GaK*	Galactokinase	C
	TK	Thymidine kinase	C
18	*Pep-A*	Peptidase A	C
19	*GPI*	Glucosephosphate isomerase	C
	PVS	Poliovirus sensitivity	P
20	*ADA*	Adenosine deaminase	P
	DCE	Demesterol: cholesterol conversion enzyme	P
21	*IFS*	Interferon sensitivity	C
	SOD-1	Superoxide dismutase-1 (cytoplasmic)	C
22		No assignment with hybrid cells	
X	α-*GalA*	α-Galactosidase-A	C
	G6PD	Glucose-6-phosphate dehydrogenase	C
	PGK	Phosphoglycerate kinase	C
	HGPRT	Hypoxanthine guanine phosphoribosyl-transferase	C
	TATr	Tyrosine aminotransferase regulator	P
Y		No assignment with hybrid cells	

[a] Modified from Ruddle and Creagan (1975) and the New Haven (1974) and Rotterdam (1975) conferences on Human Gene Mapping.

[b] This table only lists results obtained by or confirmed by cell hybrid studies. For detailed references see Ruddle and Creagan (1975). The status of the gene assignments are classified into "confirmed" (C), "provisional" (P) or "inconsistent" (I).

E. GENE MAPS OF OTHER SPECIES

Genetic studies on isozyme variants of G6PD have shown that this enzyme is X linked not only in man but also in horses and donkeys (Trujillo *et al.*, 1965 Mathai *et al.*, 1966), European hares (Ohno *et al.*, 1965), field voles (Cook, 1975), and marsupials (Richardson *et al.*, 1971).

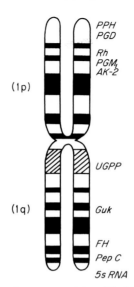

Fig. XIII-8. Gene map of human chromosome 1 (modified from Hamerton and Cook 1975; for abbreviations see Abbreviation list).

Cell hybrid studies have shown that in mice the X chromosome carries not only *G6PD* but also *HGPRT*, *α-gal*, *PGK*, and other loci known to be X-linked in man (Ruddle and Creagan, 1975). Together these data indicate a homology among X chromosomes with respect to their genetic information, thus, supporting Ohno's hypothesis (Ohno, 1967) about the evolutionary stability of the X chromosome.

About 300 linkage groups and synteny relationships have been established in mice (Green, 1966). A comparison of these data with the preliminary gene map of man reveals that there are few similarities with respect to autosomal genes. Loci which are syntenic in mice are asyntenic in man and vice versa. This is hardly surprising in view of the

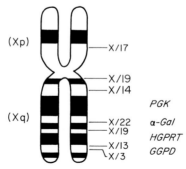

Fig. XIII-9. Gene map of the human X chromosome (modified from Pearson *et al.*, 1975).

phylogenetic distance between these two species and the fact that human and mouse chromosomes differ in number, morphology, and banding patterns. More interesting from the point of human karyotype evolution is the situation in our primate cousins, the hominoid apes. The chromosome morphology and banding patterns in primates are similar to that of man and a homology appears to exist also with respect to the genetic information content of the chromosomes. For example, cell hybrid studies have shown that the genes for galactokinase and thymidine kinase (chromosome 17 in man) are present on an African Green Monkey chromosome which resembles human chromosome 17 morphologically (Croce *et al.*, 1974e). Two enzymes, LDH-B and Pep-B, specified by genes on chromosome 12 in man are syntenic in chimpanzee + mouse hybrids with a chimpanzee chromosome which resembles human chromosome 12 (Ruddle and Creagan, 1975). Since monkey + mouse hybrids selectively eliminate monkey chromosomes (see Chapter X) cell hybridization may be used for mapping of monkey chromosomes. Such studies may be very useful in the mapping of the human genome. One example of a situation where monkey chromosomes may contribute information relevant to the human gene map is human chromosome 2. Studies of banding patterns suggest that the corresponding gene material in the chimpanzee is organized into two separate chromosomes, one corresponding to the long arm and the other corresponding to the short arm of human chromosome 2 (Turleau *et al.*, 1972). In hybrids the two chimpanzee chromosomes can be expected to undergo independent segregation and therefore to provide information about which genes to expect on the long and short arms, respectively, of human chromosome 2 (Ruddle and Creagan, 1975). It should be emphasized, however, that further studies are required before it will be clear how far the homologies between human and other primate chromosomes extend.

F. WHY GENE MAPPING?

Cell hybrids clearly represent a powerful tool for constructing detailed gene maps of the chromosomes of man and of other species. It should be pointed out, however, that it is not the only method used in gene mapping. A number of gene assignments have been made by family studies, nucleic acid annealing with cDNA or cRNA, by studies of virus induced changes on specific chromosomes, etc. These techniques and the cell hybridization method complement each other in several ways, and a large number of laboratories are now using these methods to study the organization of eukaryote genomes. In view of the effort and money which goes into this it may be appropriate to ask why genes should be mapped.

First, it is clear that gene maps may improve our understanding of genetic disease in man. Hybrids made by crossing cells from patients can be used in complementation analysis to establish if two patients with similar symptoms carry the same or different gene defects.

Second, genetic maps will be of importance in the study of many cellular processes, in particular regulatory phenomena. The knowledge of whether the synthesis of a specific protein is controlled by one or several genes and whether control genes and structural genes are located on the same or different chromosomes may be fundamental to our understanding of gene regulation.

Third, as gene maps of many different species become available new perspectives on the evolution of the human gene material and of animal species will be gained.

XIV

Analysis of Malignancy
by Cell Hybridization

A. GROWTH OF TUMOR CELLS IN VIVO AND IN VITRO

1. Malignancy and Transformation

Malignancy is a clinical term which is used to describe the property of tumor cells which enables them to grow progressively in and ultimately to kill their hosts. This concept is equally applicable to a discussion of the malignancy of cultured hybrid cells, since they behave in the characteristic clinical fashion when introduced into laboratory animals. It should be pointed out, however, that malignancy is also used in a wider sense to describe the histological appearance of tumors or their ability to metastasize.

Malignant tumors may arise *in vivo* from exposure to oncogenic agents such as chemical carcinogens, radiation, or tumor viruses, but in many cases the inducing agent is unknown ("spontaneous tumors"). Similarly, tumor cells may also arise *in vitro* when normal cells are cultured for long periods of time or when they are treated with oncogenic agents *in vitro* (Vogt and Dulbecco, 1962; Shein and Enders, 1962; Borek and Sachs, 1966; Todaro and Green, 1964; Stoker, 1963; Dulbecco, 1969). Under these conditions normal cells undergo *transformation,* a process which is believed to be in many ways similar to the formation of tumor cells *in vivo.* One property of transformed cells is their ability to give rise to malignant tumors when inoculated into histocompatible labora-

245

tory animals. This experimental method is also used to test the malignancy of cell hybrids obtained by fusing transformed cells with normal cells. But before discussing this, it is useful to consider a brief summary of the properties of normal and transformed cells growing *in vitro*, and of malignant cells growing *in vivo*. [For a more penetrating discussion of these fields see Clarkson and Baserga (1974) and Tooze (1973).]

2. Normal and Transformed Cells in Vitro

Cell cultures are usually initiated by inoculating suspensions of trypsinized or mechanically dispersed cells into tissue culture dishes made of glass or plastic. Under these conditions the cells will settle, attach, and flatten out on the bottom surface of the culture vessel. Attachment and flattening are important because normal cells (with the exception of blood cells) require a solid substratum in order to multiply ("anchorage dependence"; Stoker et al., 1968). In contrast, transformed cells are capable of multiplying in semisolid media such as soft agar (MacPherson and Montagnier, 1964) or in suspension cultures. Once cultures of normal or transformed cells have started to grow, several differences may be noted between them (Table XIV-1). One difference is that transformed colonies often consist of several layers of cells piled on top of one another, whereas normal colonies usually consist of only a single layer of cells (Fig. XIV-1).

There are also differences in the geometry of the individual cells within normal and transformed colonies. Normal cells often grow in regular arrangements and are well attached to their substratum (except during mitosis). Transformed cells are loosely attached, rounded, and randomly oriented (Fig. XIV-2).

The most clearly defined differences between normal and transformed cells concern cell motility and the regulation of cell multiplication *in vitro*. Normal cells migrate over their substratum and multiply until they form a confluent monolayer. At this stage the cells are immobilized. This phenomenon is a characteristic of normal cells and is known as contact inhibition of cell movement (Abercrombie and Heaysman, 1954; Abercrombie and Ambrose, 1958, 1962). Soon after cells have become contact inhibited, cell multiplication also stops. This phenomenon is now often referred to as density dependent inhibition of cell growth (Stoker and Rubin, 1967) or topoinhibition (Dulbecco, 1970a,b). While there is a link between topoinhibition and contact inhibition of cell movement, the two phenomena can to some extent be dissociated by frequently renewing the culture medium. Under such conditions cells may continue to synthesize DNA and to divide even

TABLE XIV-1

Comparison between Transformed and Normal Cells

Property	Normal cells (diploid)	Transformed cells (heteroploid)	References
Growth *in vitro*			
Terminal cell density	Low (monolayer)	High (multilayering)	Stoker and Rubin (1967)
Serum dependence	High	Low	Dulbecco (1970b); Holley (1971, 1974)
Life span *in vitro* (senescence)	Limited number of cell cycles	Infinite growth	Hayflick and Moorhead (1961); Hayflick (1965)
Attachment	Anchorage dependent	Growth in suspension or soft agar	MacPherson and Montagnier (1964); Stoker *et al.* (1968)
Morphology	Flat	Rounded form; defective; formation of lamellar cytoplasm	Domnina *et al.* (1972); Cherny *et al.* (1975)
Motility	Contact inhibited	Loss of contact inhibition	Abercrombie and Ambrose (1958)
	Actin filaments	Disorganized actin filaments	Weber *et al.* (1975); Wickus *et al.* (1975)
Permeability		Enhanced uptake of sugars and amino acids	Hatanaka *et al.* (1969, 1970); Foster and Pardee (1969); Venuta and Rubin (1973)
Agglutination with lectins	Low agglutinability of interphase cells	Increased agglutinability of interphase cells; increased mobility of lectin binding sites	Burger (1969, 1973); Sachs *et al.* (1974)
Chemical characteristics of cell surface	Lectin binding sites masked by cell surface protein	Loss of surface protein and unmasking of lectin receptors; glycoprotein changes	Burger (1969); Gahmberg and Hakomori (1973)
	Low proteolytic activity	Increased proteolytic activity (plasminogen activator)	Ossowski *et al.* (1973); Hynes (1974); Pollack *et al.* (1974)

247

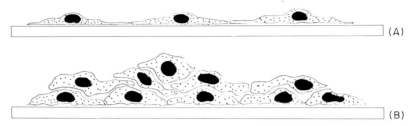

Fig. XIV-1. Schematic illustration of the growth pattern of normal (A) and transformed (B) cells on a solid substratum. Normal cells become contact inhibited and stop multiplying when a confluent monolayer has formed while transformed cells continue to multiply and therefore give rise to multilayering of cells.

though confluency has been achieved (Rubin, 1966; Todaro *et al.*, 1966).

In contrast, transformed cells lack both inhibitions. Even after confluency they continue to multiply and move over each other so that they pile up into multilayered masses. Eventually, however, cell multiplication also stops in transformed cell cultures. But because of the lack of contact inhibition, transformed cells reach much higher terminal cell densities (saturation densities) than do normal cells.

The rate of growth and the magnitude of terminal cell densities of both normal and transformed cells depend very much on how often the medium is renewed and on its serum concentration. In order to multiply rapidly, normal cells require a much greater concentration of serum factors than do transformed cells (Holley and Kiernan, 1968; Holley, 1971, 1974; Benjamin, 1974).

There are also genetic differences between normal and transformed cells. While normal cells retain a strictly diploid chromosome number, transformed cells are usually heteroploid and show extensive chromosome aberrations. In virus-transformed cells, new genetic information in the form of viral genes can often be detected directly by nucleic acid hybridization or indirectly by the presence of nuclear or cell surface antigens characteristic of the tumor virus and presumably specified by the viral genome (see Chapter XV).

Another important difference between normal and transformed cells concerns the life span of cells during serial passages in culture. Normal cells show a finite life span. They become senescent and stop multiplying after a certain number of cell generations which for human fibroblasts is usually about 50–70 cell generations (Hayflick and Moorhead, 1961; Hayflick, 1965; for a review see Orgel, 1973). On the other hand, transformed mammalian cells seem to have an infinite life span for they continue to multiply indefinitely, passage after passage.

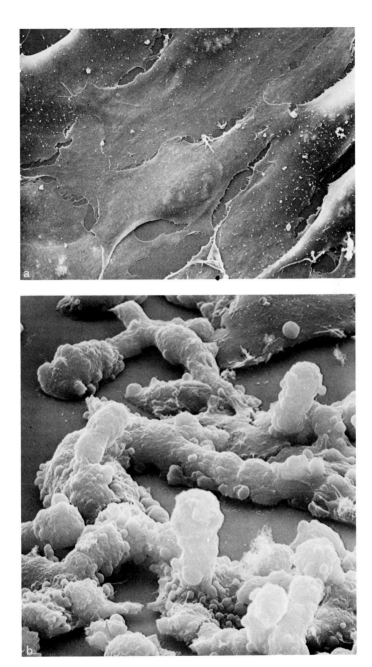

Fig. XIV-2. Electron stereoscan micrographs showing (a) normal and (b) transformed cells growing on a solid substratum. The normal cells are flat and well attached while the transformed cells are rounded and show a tendency to pile up and form several layers of cells (from Goldman *et al.*, 1975).

3. Variants of Transformed Cells

While transformation is thus associated with many different phenotypic changes, the expression of these new properties is far from being
an all-or-none phenomenon; they may vary quantitatively. Furthermore, not all the alterations listed above occur invariably in every transformed cell line. The fact that different lines of transformed cells vary in
their expression of the transformed phenotype is due to their genetic
heterogeneity and the fact that different culture conditions select for
different variants. The most striking illustration of this is the isolation
of untransformed revertant cells from populations of transformed cells
(Pollack *et al.*, 1968). It appears that transformed-cell populations contain small numbers of cells which are sensitive to contact inhibition. If
such populations are treated with BUdR or ^{3}H-thymidine at high cell
densities, that is conditions under which transformed cells replicate
while normal cells do not, the transformed cells are selectively killed
(see Chapter X). The surviving cells suppress some or most of the properties of transformation and therefore have regained many of the characteristics of normal cells (for a review see Vogel and Pollack, 1974). It is
important to remember, however, that there is no evidence that such
revertants return to a diploid chromosome number or regain senescence. So far, no one has been able to isolate a completely normal "revertant" cell.

B. EXPRESSION OF THE TRANSFORMED PHENOTYPE
IN HYBRID CELLS GROWING IN VITRO

1. Patterns of Expression

The transformed phenotype thus consists of a number of properties:
cell surface changes, loss of contact inhibition, cell multilayering, proliferation to high terminal cell densities, loss of senescence, capacity to
grow in soft agar, and tumorigenesis *in vivo*. What then are the properties of cell hybrids obtained by fusing transformed cells with untransformed* cells? Taken together, fusion experiments of this sort have
yielded three different types of results. Some hybrids have expressed a
transformed phenotype, another set of hybrids has been untransformed with respect to some characteristics, and a third group has
shown intermediate properties.

---

* The term "untransformed" will be used for normal diploid cells as well as for cell lines
(for definitions see Glossary) which show density dependent inhibition of cell proliferation.

2. Hybrids Expressing a Transformed Phenotype

Hybrids with a transformed phenotype show absence of senescence (Scaletta and Ephrussi, 1965; Weiss *et al.*, 1968a; Davidson and Ephrussi, 1970; Goldstein and Lin, 1972a), lack of contact inhibition and/or growth to high terminal cell densities (van der Noordaa *et al.*, 1972; Wiblin and MacPherson, 1973; Croce and Koprowski, 1974b,c, 1975), agglutination with concanavalin A, growth in soft agar, and a low serum requirement (van der Nordaa *et al.*, 1972; Wiblin and Mac-Pherson, 1973; Basilico and Wang, 1971; Jha and Ozer, 1976).

The experiments of Croce and Koprowski (1974a,b,c, 1975; Croce *et al.*, 1973c,e, 1974c,e, 1975a,b,c) are of particular interest because they show that in cells transformed with a DNA tumor virus, the transformed phenotype (a) behaves as a dominant genetic trait and (b) is controlled by viral genes integrated on a specific chromosome. The hybrid cells used in these experiments were prepared by fusing human SV40 virus transformed fibroblasts from a Lesch–Nyhan patient (LN-SV40) with three types of cells: mouse Cl1D cells (Croce *et al.*, 1973c,d; Croce and Koprowski, 1974a), WI-38 human diploid fibroblasts (Croce and Koprowski, 1974b), and normal mouse macrophages (Croce and Koprowski, 1974c, 1975; Croce *et al.*, 1975b). In all these combinations the hybrid cells showed a transformed phenotype since they grew to high saturation densities, piled up into several cell layers, grew in soft agar, and could be cultured for long periods of time. The hybrids obtained by crossing human SV40 transformed cells with normal mouse macrophages, in addition to being transformed *in vitro*, gave rise to tumors in immunologically defective mice ("nude mice") (Croce *et al.*, 1975b,c). Analysis of a number of different clones showed that both the transformed phenotype *in vitro* and the oncogenic phenotype *in vivo* were strongly correlated with the retention of human chromosome 7 and with the presence of SV40 specified T antigen (Croce and Koprowski, 1975; Croce *et al.*, 1975a,b,c) (Table XIV-2; Fig. XIV-3). The fact that some of the transformed macrophage hybrids contained a normal diploid mouse chromosome set plus only one human chromosome, number 7, is particularly interesting since it means that the transformed phenotype is controlled by viral genes integrated on human chromosome 7. It also means that the transformed phenotype, in this case, behaves as a dominant trait.

3. Hybrids Expressing an Untransformed Phenotype

Hybrids expressing an untransformed phenotype are contact inhibited and grow to lower saturation densities than the transformed parent, as reported by Weiss (1970) in crosses of SV40 virus-

TABLE XIV-2

Transformed Phenotype in Hybrids between Mouse and SV40-Transformed Human Cells[a]

Cells	Saturation density[b] (cells/cm²)	Colony formation (soft agar)
Human LN fibroblasts	0.4×10^5	−
Human LN-SV (SV40 transformed)	2.0×10^5	+
Mouse peritoneal macrophages	ND[c]	−
Mouse fibroblasts	0.6×10^5	−
LN-SV + mouse macrophage hybrids		
Clone 1	2.0×10^5	+
Clone 2	2.1×10^5	+
Clone 6	2.3×10^5	+
Clone 13	ND[d]	+
Clone 17	1.9×10^5	ND

[a] From Croce et al., 1975a.

[b] Parental diploid human and mouse fibroblasts and hybrid cells were counted 10 days after seeding the same number (2×10^6) of cells in 75-cm² plastic Falcon flasks.

[c] The saturation density of mouse peritoneal macrophages was not determined because the macrophages are nondividing cells.

[d] ND, not done.

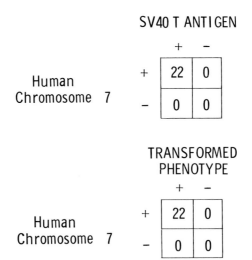

Fig. XIV-3. Positive correlation between the expression of the SV40 T antigen, the expression of the transformed phenotype and human chromosome 7 in a series of somatic cell hybrid clones derived from the fusion of mouse peritoneal macrophages and SV40 transformed human cells. +, presence of a given property in the 22 mouse + human hybrids examined;−, absence of a given property (from Croce et al., 1975a).

transformed human cells with contact inhibited mouse 3T3 cells. Similar observations were again made by Weiss (1974) on hybrids between rat hepatoma cells (Fu5-5) and diploid epithelial rat cells (BRL-1). Although the rat hepatoma cells lack contact inhibition the hybrid cells were highly contact inhibited, resembling the epithelial parent. In a different sort of cross Levisohn and Thompson (1973) fused mouse L cells with rat HTC hepatoma cells. Both these cell types lack contact inhibition and show multilayering, but the hybrids made from these cells showed contact inhibition.

The finding that two different transformed cell lines, when fused with each other, sometimes produce hybrids with an untransformed phenotype could be due to gene complementation. However, neither the hybrids nor their parents have been studied sufficiently well to permit any such conclusions. It should also be noted that all the "untransformed" hybrids studied so far appear to be capable of long term growth *in vitro*, that is, it must be assumed that they lack senescence. Thus, even if these hybrids grow to low saturation densities and appear to be contact inhibited, doubt remains about their true phenotype.

4. Hybrids Showing Intermediate Phenotypes

Van der Noordaa *et al.* (1972) reported about hybrids which expressed some transformed and some untransformed characteristics. Similar observations were made on segregants that arose from hybrids which initially showed a transformed phenotype (Marin and Littlefield, 1968; Wiblin and MacPherson, 1973). Since the revertant hybrid cells provide some insights into the chromosomal control of the transformed phenotype we shall discuss them in more detail.

5. Hybrid Cells Reverting from a Transformed to an Untransformed Phenotype

Marin and Littlefield (1968) and Marin (1971) obtained revertants from hybrids of two mutant sublines of Syrian hamster fibroblasts. One of the sublines lacked HGPRT (HGPRT⁻) while the other was deficient in thymidine kinase (TK⁻). The hybrids exhibited mutual gene complementation and were HGPRT⁺ and TK⁺ and therefore sensitive to both 6-thioguanine and 5-bromodeoxyuridine (see Chapter IX). After transformation with polyoma virus, the hybrids were cultured on normal medium containing 6-thioguanine in order to select variants which had lost chromosomes. Although this form of back selection demands that survivors have eliminated X chromosomes carrying active *HGPRT*⁺ loci (see Chapter XIII), additional chromosomes are lost at the same time as the X chromosome(s). Thus, among the sublines which survived were some which had lost a number of chromosomes. Some of the cell lines were transformed and grew in soft agar while others had reverted to a more or less normal morphology and gave a low plating efficiency in soft agar. One subline showed intermediate properties. The revertant cells showed chromosome losses, but the relationship between chromosome elimination and the phenotype of the different sublines could not be assessed.

Wiblin and MacPherson (1973) analyzed hybrids between SV40-transformed Syrian hamster cells (HGPRT⁻) and contact-inhibited 3T3 mouse cells (TK⁻). The hybrids showed a transformed phenotype. Selection for hybrid segregants with a reduced hamster chromosome complement (the "transformed" genome) was carried out by treating the cells with BUdR and rabbit antisera to hamster surface antigens. These treatments resulted in a series of sublines, some of which had eliminated all hamster chromosomes. Most segregant sublines retained properties typical of transformed cells even when the hamster chromosome complement was greatly reduced. Others exhibited phenotypes intermediate between transformed and untransformed. Two sublines which had lost all hamster chromosomes showed an untransformed phenotype. These results suggest the possibility of using revertant hybrid cells as an experimental system for defining genes controlling the transformed phenotype.

6. Interpretation of Transformation in Cell Hybrids

Early studies of hybrids made with transformed cells suggested that the transformed phenotype was dominant. This interpretation also receives strong support from more recent studies by Croce and Koprowski (1974a) which have been discussed in a preceding section of this Chapter. But, and as already mentioned, there are also studies which suggest that not only transformed + untransformed cells but even transformed + transformed cells give hybrids with an untransformed phenotype. Before attempting an interpretation of these results at least three complicating factors should be stressed:

(a) Many of the early studies were carried out with mouse cells that showed a high spontaneous transformation rate (Todaro and Green, 1963). It is possible, therefore, that some hybrids with a transformed phenotype arose by spontaneous postfusion transformation or that an undetected minority of transformed cells was present in what was taken to be the untransformed parental cell population. A few transformed hybrid cells would rapidly overgrow a majority of slow-growing, un-transformed hybrids and shortly after fusion give the impression that the total population of hybrids was transformed. This explanation, however, is not sufficient to explain all the data.

(b) In most of the studies published so far the phenotypes were classified as transformed or untransformed on purely morphological grounds. Morphology alone is a very ambiguous criterion of transformation and one which is likely to be very much affected by subjective errors. The more penetrating studies have used quantitative measures of transformation (e.g., growth of soft agar, saturation densities, serum dependence). Even in these cases it is uncertain however, whether "untransformed" represents a condition which could be described as "normal." An important characteristic of normal cells is senescence. This phenomenon has not yet been detected in "untransformed" hybrids.

(c) The chromosome analyses reported for hybrid cells and their segregants do not exclude the possibility that chromosome losses influence the phenotype expressed. If the hybrids have lost chromosomes from one or both of the parental genomes it becomes very difficult to decide whether the absence of a marker is due to recessive expression or to a loss of genetic information.

These technical difficulties make it difficult to interpret many of the earlier studies of transformation in hybrid cells. The recent studies of Croce and Koprowski (1974c) in which SV40 transformed human cells were fused with mouse macrophages are, however, quite convincing since the transformed phenotype *in vitro* is well documented and because the hybrids have also been shown to give tumors in nude mice. In this case the hybrids have also been well characterized with respect to their karyotype. The presence of a normal and complete set of mouse chromosomes and human chromosome 7 from the SV40-transformed parent is strong evidence that the transformed and oncogenic phenotype is dominant and controlled by the only remaining human chromosome. This interpretation, however, differs from that of Sachs and co-workers (Rabinowitz and Sachs, 1970; Hitotsumachi *et al.*, 1971; Sachs, 1974) and Harris and Klein (1969) (to be discussed below). Sachs and co-workers studied nonhybrid cells reverting from a transformed to a nontransformed state and found chromosomal changes which indicated that this change was associated with changes in the balance between groups of chromosomes carrying suppressor genes and expressor genes for transformation. Harris and Klein's groups, on the other hand, studied malignancy *in vivo* with tumor cell hybrids.

As is discussed in detail below, they interpret malignancy in terms of genetic defects that, in hybrids, sometimes are complemented by normal genes so as to give nonmalignant hybrid cells.

C. TUMORIGENICITY OF HYBRID CELLS IN VIVO

1. Testing of Malignancy

a. IMMUNOLOGICAL TRANSPLANTATION BARRIERS

As already indicated the malignancy of tumor cells or transformed cells is usually tested by injecting such cells into animals. The standard laboratory animal used in experimental tumor research is the mouse. Systematic inbreeding has produced mouse strains in which all individuals are genetically identical (*syngeneic*). With such mouse strains, normal and neoplastic tissues can usually be transplanted without eliciting immunological reactions. F_1 mice obtained by crossing a syngeneic strain with another syngeneic strain are equally receptive to tissues from other F_1 mice of the same cross and to tissues from the parental strains. If, however, a normal tissue from one mouse strain is grafted into a mouse from a genetically different strain (allogeneic), the antigens of the grafted tissue will be recognized as foreign by the immunological defence system of the recipient and the graft will be rejected. Most tumor cells also show a similar strain specific transplantation behavior. While a single tumor cell inoculated into syngeneic animals may produce a malignant tumor, large numbers of tumor cells inoculated into allogeneic animals are rejected. If a graft is to be accepted the recipient animal and the donor must be identical with respect to certain strong antigens, the *histocompatibility antigens*. If donor and host have different antigens the grafted cells will be identified as foreign and destroyed by the cytotoxic activity of host lymphocytes.

In mice the H-2 isoantigens are the strongest histocompatibility antigens. This complex of antigens is specified by a region of about 1000 cistrons located on chromosome 17 (mouse). Because of the complexity of the different types of H-2 antigens and their genes we will refer to the different antigen phenotypes as different "H-2 complexes" (H-2^a, H-2^k, etc.) and the genes as H-2 determinants. All animals belonging to the same inbred mouse strain carry the same determinants at the genetic level and the corresponding antigenic products on the cell surface. The various mouse strains, however, belong to different H-2 groups. Thus, A-strain mice carry the H-2^a complex of histocompatibility antigens,

C3H mice the H-2^k antigen complex, whereas the ACA strain carries the H-2^f antigen complex, etc. The histocompatibility antigens can be detected and measured quantitatively *in vitro* by a variety of immunological methods including absorptions and cytotoxic tests.

In addition to the H-2 histocompatibility antigens, tumor cells may exhibit tumor specific transplantation antigens which may influence the proportion of "takes" when tumor cells are inoculated into syngeneic mice. Tumors caused by oncogenic viruses may show nuclear and/or surface antigens characteristic of the virus responsible (see Chapter XV). Chemically induced tumors show other antigens not found on the cells from which the tumor cells originated. These antigens vary in different tumors.

A special case is presented by the immunoresistant Ehrlich tumor, that can grow in all mouse strains. The probable explanation of this is that for 75 years the Ehrlich tumor has been grown in a variety of genetically different mice and, thus, there has been strong selection for reduced antigen expression and rapid cell proliferation. Although there are other examples where mouse tumor cells grow in genetically unrelated mouse strains the majority of mouse tumors show expression of surface antigens and will therefore be rejected in allogeneic mice. The immunological transplantation barriers can be overcome, however, if the tumor cells are transplanted into animals where the immunological defence mechanisms have been impaired by immunosuppressive treatments (X irradiation, cortisone, antilymphocytic sera, etc.) or where these mechanisms are poorly developed (new born animals, nude mice, etc.).

b. CORRELATION BETWEEN TRANSFORMATION *IN VITRO*
 AND MALIGNANCY *IN VIVO*

Different transformed cell lines vary markedly with respect to their ability to cause tumors that kill their host animals. A correlation exists, however, between the cell line's loss of contact inhibition, or high terminal cell density *in vitro,* or ability to grow in soft agar and its capacity to induce malignant tumors *in vivo* (Vogt and Dulbecco, 1962; Aaronson and Todaro, 1968; Pollack *et al.,* 1968; Freedman and Shin, 1974). Revertants of SV40 virus-transformed 3T3 cells that have regained density-dependent inhibition of growth have been found to be less oncogenic than the original transformed cell line.

Yet, in spite of these general correlations between the transformed phenotype and oncogenic potential, many individual exceptions are found. The most common exception is that cells with a transformed phenotype fail to generate tumors *in vivo.* But conversely, cells which appear normal in regard to contact inhibition occasionally give rise to tumors. Many of these discrepancies are probably due to immunological barriers (see below) and to selection for variant cells. As previously emphasized, *in vitro* conditions select for variant cells that grow well in culture but which are not necessarily oncogenic *in vivo*. The *in vivo* environment on the other hand selects for malignant variants which may or may not grow *in vitro*. In addition to the selective pressures exerted on the cell populations it is probable that the chromosomal variability of neoplastic and transformed cells is an important factor since it generates a broad spectrum of cell types and increases the probability that some cells will be able to grow if the selection pressure suddenly changes. These considerations are important from at least two points of view. First, they explain why transformation *in vitro* does not always correlate with malignancy *in vivo*. Second, they imply that hybrid cells with their double sets of chromosomes and great intercellular variation in karyotypes may be even more prone to give rise to segregant cell populations when culture conditions are changed than are parental cells with their single chromosome sets.

2. Early Work on the Malignancy of Cell Hybrids

As already discussed in Chapter II, the first somatic cell hybrids obtained by Barski *et al.* (1960, 1961) were made by mixing two different mouse tumor cell lines. Both lines had originated in C3H mice but differed greatly in their ability to induce tumor formation in syngeneic C3H mice. One tumor line (N1) was highly malignant while the other was less so (N2). The hybrid cells called "M cells" by Barski, produced tumors at a high frequency when inoculated into C3H mice (Barski and Cornefert, 1962).

Gershon and Sachs (1963) were next to test hybrids between two malignant mouse cell lines. They noted that the hybrids had chromosome complements which contained fewer chromosomes than the sum of the two parental genomes. When malignancy was tested by inoculating the cells into appropriate F_1 mice, tumors were obtained in all mice. A few years later Scaletta and Ephrussi (1965) obtained spontaneous fusions between the N1 cells studied by Barski and normal mouse cells. The hybrid cells had a karyotype which appeared to be the sum of the two parental karyotypes and grew as a transformed cell line *in vitro*. The cells were also found to be malignant since they produced tumors in a high proportion of inoculated animals. In 1967, Silagi reported that hybrid cells formed from highly malignant (C57BL) melanoma and low-malignant A9 cells (HGPRT⁻L cells from C3H mice) produced some tumors in F1 (C3H × C57BL) mice. Defendi *et al.* (1967) fused two sublines of polyoma virus-transformed mouse embryo cells of low malignancy with normal cells derived from primary skin and lung of newborn mice. Hybrids were obtained which appeared more malignant than the transformed parental cells. Both Silagi (1967) and Defendi *et al.* (1967), however, also obtained a small number of hybrids or hybrid sublines which failed to produce tumors. In interpreting the overall results, these authors emphasized the fact that some tumors were obtained and concluded that malignancy *in vivo* behaved as a "dominant" trait in the phenotype of hybrid cells.

3. Suppression of Malignancy

Experiments which suggest that in some cases malignancy can be suppressed have been published jointly by Harris' group in Oxford and Klein's group in Stockholm (Harris *et al.*, 1969a; Harris and Klein, 1969; Klein *et al.*, 1971; Wiener *et al.*, 1971, 1974a,b). Independently Murayama and Okada (1970b; Murayama *et al.*, 1971), Jami and Ritz (1973a, 1975a,b), and Stanbridge (1976) have made similar observations.

The Harris–Klein groups fused a series of extremely malignant mouse tumors with established mouse cell lines (A9, B82) of low oncogenicity and with normal diploid cells. The A9 cells, when inoculated into X-irradiated newborn C3H mice, exhibited a low level of oncogenicity. Ehrlich tumor cells, on the other hand, were highly malignant and killed all host animals. Hybrids between these two types of cells showed a low level of malignancy similar to that of the A9 parent (Klein *et al.*, 1971). It appeared therefore that the less malignant parental cell was capable of suppressing the malignancy of the Ehrlich cell. The small number of tumors which did develop were composed of hybrid cells containing chromosomes characteristic of both parental cells, but the total number of chromosomes was reduced. At the time of injection, the Ehrlich + A9 hybrid cells had a modal chromosome number of 128 chromosomes. Most animals injected with hybrid cells did not develop tumors but in the few cases where tumors did appear, it was found that the modal chromosome numbers were 80–90. This result suggested that the vast majority of hybrid cells inoculated into the animals were nonmalignant. The tumors that did develop resulted from a selective overgrowth of a small number of hybrid cells with a reduced chromosome complement. This in turn suggested that, similar to the situation in some revertants of nonhybrid tumor cells (Codish and Paul, 1974), suppression of malignancy might depend upon repressors produced by a few specific A9 chromosomes. Alternatively, the malignant variants could have simply been a consequence of an overall re-

duction in chromosome number. Other experiments, however, made
this interpretation unlikely. When Ehrlich + A9 hybrid cells were
cultured for a prolonged period (18 months) *in vitro,* the modal chromosome number decreased to approximately the same number as that
found in Ehrlich + A9 hybrid tumors. Such hybrid cells that had
lost their chromosomes under a selection pressure quite different
from that operating *in vivo* showed a low level of malignancy. This
supported the hypothesis that malignancy in hybrid cells was due to
a loss of specific chromosomes rather than to a random reduction in
chromosome number.

A series of other hybrid cells obtained by fusing A9 cells with cells
from tumors induced by chemical carcinogens (MSWBS-tumor),
oncogenic DNA viruses (SEWA) and RNA viruses (YAC, YACIR) produced results similar to those obtained with Ehrlich A9 hybrids. The
malignancy of these tumor cells was suppressed by fusion with the low-
malignancy A9 cells. When tumors did develop, they were also shown
to have a reduced chromosome number. This result again suggested
that malignancy is related to loss of chromosomes. Neoplastic hybrid
cells with a reduced chromosome complement could have been present
as a minority in the inoculated cell population or could have been generated during the *in vivo* assay for malignancy.

Although the A9 cell line exhibits a low level of malignancy, it does
produce some tumors. One such tumor was grown *in vitro* by Wiener *et
al.* (1973) and then tested by inoculation into mice. The cells were
highly malignant, they produced tumors in 87% of the mice injected.
This subline of the original A9 line was therefore designated A9HT
(HT, High-Take incidence). The chromosome patterns of the low-
malignancy A9 line and its highly malignant A9HT subline were found
to be similar although A9HT had undergone some characteristic
changes. A9HT had a smaller number of chromosomes (4 chromosomes
and 3 chromosome arms missing) and was even more heterogeneous
than A9 (Allderdice *et al.,* 1973a). The malignant A9HT line was hybridized with a series of malignant mouse tumors: Ehrlich, SEWA
MSWBS, and YACIR; that is the same tumors which had previously
been suppressed by A9 (Klein *et al.,* 1971). The A9HT + tumor cell hybrids were all highly malignant. Unlike the A9 + tumor hybrids, the
A9HT + tumor hybrids had essentially unreduced chromosome complements. It appeared therefore that the tumors arose directly from the
A9HT hybrid cells and not from the selective overgrowth of a minority
of hybrid cells which had undergone chromosome losses.

While A9HT was highly malignant and had lost the ability to
suppress the malignancy of other tumor cells, it could itself be sup-

pressed by two other cell types. Hybrids between B82 (another L cell derivative closely related to A9) and A9HT showed reduced oncogenicity (Wiener *et al.*, 1973). Normal diploid lymphocytes also suppressed the malignant properties of A9HT while no such reduction in malignancy was noted in A9HT + malignant lymphoma hybrids (Wiener *et al.*, 1974a).

Harris and Klein interpret malignancy primarily as a genetic or stable epigenetic defect that is sometimes compensated for when the malignant cell is fused with a normal or nonmalignant cell. According to this hypothesis, suppression of malignancy would be more or less equivalent to the gene complementation observed in crosses of enzyme deficient cells (see Chapters X and XIII). To check the validity of this hypothesis and to examine the genetic defects involved in malignancy, Wiener *et al.* (1974b) hybridized a number of different malignant tumors with each other to see if nonmalignant cells could arise as a result of gene complementation. In twelve different crosses only one (MSWBS + YACIR) showed a reduced oncogenicity. In all other cases the hybrid cells were highly malignant. If these results are interpreted in terms of gene complementation, the conclusion is that with one exception the tumors crossed suffered from one and the same type of genetic defect. However, as the authors pointed out and as is discussed below, it is doubtful, not to say unlikely, that malignancy can be explained in such simple terms.

Independently of Harris *et al.* (1969a) and Klein *et al.* (1971), Murayama and Okada (1970b) observed that the tumor-forming capacity of the hyperdiploid Ehrlich line was suppressed by fusion with HGPRT⁻ L cells and that the chromosomal balance between the malignant and nonmalignant genome appears to determine the oncogenicity of hybrid cells.

Bordelon (1974) examined human + mouse hybrids to see if chromosomes of a normal human cell could suppress a mouse malignant cell. Malignant mouse (RAG) cells that showed a transformed phenotype *in vitro* were fused with the "normal" diploid human cell line WI-38. The human cells did not grow in immunosuppressed mice, they agglutinated poorly with Con A, and were rapidly overgrown by transformed cells *in vitro*. The hybrids, however, produced tumors when injected into mice and showed transformed properties *in vitro*.

4. Recessive or Dominant Expression of Malignancy

The original conclusion from the work of Barski, Ephrussi, and coworkers (Barski *et al.*, 1961; Barski and Cornefert, 1962; Scaletta and Ephrussi, 1965; Defendi *et al.*, 1967) was that malignancy in cell hybrids

behaves like a dominant trait. The strongest support for this interpretation is the recent work by Croce and Koprowski on hybrids between
SV40-transformed human cells and normal mouse macrophages (Croce
and Koprowski, 1974c, 1975; Croce *et al.*, 1975b). The Harris–Klein
group (Harris *et al.*, 1969a; Klein *et al.*, 1971, 1973; Bregula *et al.*, 1971;
Wiener *et al.*, 1971, 1973, 1974a,b), on the other hand, interpret their
studies in a different way. Malignancy is looked upon as a genetic defect which behaves like a recessive trait. In hybrids between normal
and malignant cells the normal chromosomes complement the defect.
The hybrid cells only give rise to tumors when certain critical chromosomes, most likely the normal chromosomes carrying complementing
genes, are lost. It is not yet possible to decide which of these two interpretations is correct. Possibly both may be right. It is quite conceivable,
for instance, that the mechanisms underlying viral and chemical oncogenicity are different. All tumor cells caused by a specific tumor virus
share the same virus specific antigens and appear to contain viral genes
integrated on the host cell chromosomes. Among these genes may be a
special "transformation" gene [for a discussion see the Cold Spring
Harbor Symposium on Tumor Viruses (1975], which is dominantly expressed and causes the cell to assume a transformed or malignant phenotype even if all the cellular genes are intact. Chemical and physical carcinogens, on the other hand, are not likely to add genetic information
but are known to induce a variety of chromosome breaks and gene defects, some of which result in transformation and malignancy. Tumors
induced by nonviral agents differ from each other in antigenicity and
karyotypes, each tumor showing its own individuality. Because of this
heterogeneity it is possible that in a few cases hybrid cells produced by
fusing different tumor cells with each other or with normal cells show
gene complementation and as a result assume a nonmalignant or nontransformed phenotype. But, as pointed out by Harris and Klein, their
intraspecific mouse tumor hybrids are chromosomally labile and show
chromosome segregation. Malignant segregants may, therefore, arise
when complementing chromosomes are lost.

   At present it is difficult to say whether there is a real conflict between
the results of Croce and Koprowski and those of Harris and Klein. Both
groups discuss malignancy in genetic terms, but other factors must also
be considered. The evolution of malignancy may result from a series of
potentially reversible steps which generate progressively abnormal cell
populations. Then viruses and chemicals may act at different steps in
the progression of neoplasia. Furthermore some forms of malignancy
may be errors of cell differentiation. According to this view the genetic
information content of the DNA molecules is unchanged whereas regu-

latory circuits which determine transcription and translation patterns are altered. Malignant cells as well as differentiated normal cell types would then represent a series of self-maintaining regulatory states that are quite stable and can be inherited by daughter cells. Alterations in the exterior milieu might under certain circumstances induce a transition from one state to another. Some recent work by Mintz and Illmensee (1975) indicates that malignancy can be reversed. They found normal genetically mosaic mice could be produced by injecting malignant teratocarcinoma cells into mouse blastocysts. (see Chapter III). These results must, however, be regarded as exceptional, not only because the malignant cells came from a multipotent teratoma but also because these cells had retained a diploid karyotype. Most other forms of tumors are heteroploid and show chromosome abnormalities which, most likely, can never revert back to a normal karyotype. Still, the teratoma results are most interesting and further studies of this system may help us understand cell differentiation and malignancy. Another method which may be useful in studying regulatory circuits and equilibria in the cell is to combine nuclei and cytoplasms from different types of cells by Sendai virus induced fusion of nucleate and anucleate cell fragments prepared by cytochalasin induced enucleation (see Chapter VIII).

D. PHENOTYPIC EXPRESSION IN TUMOR CELL HYBRIDS

1. Expression of H-2 Antigens and Tumor-Specific Antigens

In addition to their role in determining tissue acceptance (see Section C,1,2) the H-2 antigens of mouse cells are a useful marker system for studies of phenotypic expression in hybrid cells. The first studies of antigen expression in hybrid cells made from tumor cells were those of Gershon and Sachs (1963). These authors examined a hybrid between L cells and a polyoma-induced tumor cell line by inoculating the hybrid cells into mice of the two strains from which the parental cells had originated. The hybrid cells failed to produce tumors in these strains but killed F_1 mice obtained by crossing the two parental mouse strains. This result strongly suggests that the hybrids express H-2 antigens typical of both parental cells.

In later work Spencer et al. (1964), Klein et al. (1970), and others clearly demonstrated the coexpression of H-2 antigens in intraspecific mouse hybrids where the parental cells were derived from different mouse strains. The coexpression pattern is, however, not always found. In some types of hybrids the antigens of one of the parental cells have been extinguished (Harris et al., 1969a; Klein et al., 1970, 1972). Similar to the situation with histiotypic markers (Chapter XI) it is difficult to determine whether the extinction of antigen markers is due to losses of genetic material or to regulatory phenomena. Most hybrids which have been studied for antigen expression have been intraspecific mouse hybrids. This makes it very difficult to determine if chromosomes have been selectively lost from one of the parental genomes or if both parental chromosome sets have been reduced. In some experiments the loss of one of the parental H-2 antigens has been interpreted as a loss of the corresponding chromosome (Dalianis *et al.*, 1974). In other cases, however, segregants from the original hybrid population have shown reexpression of the extinguished antigen marker (Klein *et al.*, 1970; Grundner *et al.*, 1971). In such situations it appears that the initial loss of the marker was due to regulatory phenomena rather than to the loss of a specific structural gene. Direct selection for antigen loss variants of hybrid cells can be carried out *in vivo* by inoculating the hybrid cells into mice which react against one of the parental H-2 antigens or by selection *in vitro* with antisera and complement. This approach has been used by Klein and co-workers (1973; Klein and Harris, 1972) to study the genetic control of mouse H-2 antigens. The same technique can probably also be applied to man + mouse hybrids carrying the human histocompatibility antigens (HL-A) and mouse H-2 antigens. Chromosomal segregants of such hybrids could probably be used for mapping of human histocompatibility genes and examining regulatory mechanisms modulating antigen expression.

In addition to the normal histocompatibility antigens tumor cell hybrids have been found expressing SV40 (for references see Chapter XV), polyoma (Defendi *et al.*, 1964, 1967; Basilico and Wang, 1971; Klein and Harris, 1972; Meyer *et al.*, 1974), Epstein–Barr (Glaser *et al.*, 1973a,b; Klein *et al.*, 1974), Moloney (Fenyö *et al.*, 1971, 1973a), and Rous virus specific antigens (Machala *et al.*, 1970). It has also been possible to obtain chromosomal segregants from these hybrids. Some segregants lose tumor-specific antigens as chromosomes are lost. This, therefore, has opened up the possibility of mapping virus integration sites on human and animal chromosomes and is, therefore, discussed in Chapter XV.

2. Histopathology of Hybrid Tumors

Surprisingly few studies have been made of the histopathology of tumors induced by the injection of hybrid cells into animal hosts. So far there are only two detailed reports, one by Wiener *et al.* (1972b) and the other by Cochran *et al.* (1975). The former work describes the morphology of A9 + YACIR lymphoma hybrid tumors. The A9 parent has a fibroblastic morphology *in vitro* (Fig. XIV-4) resembling that of the L cell line (Earle, 1943) from which the A9 subline was originally established. Although A9 is an essentially nonmalignant line, tumors are sometimes obtained when A9 cells are injected into genetically compatible C3H mice. These tumors resemble sarcomas and contain fibroblast-like cells surrounded by reticulin and collagen fibers. The YACIR lymphoma, on the other hand, grows *in vitro* in the form of rounded, lymphoblastoid cells which attach poorly to glass. *In vivo* these cells give rise to tumors consisting mainly of lymphocyte-like cells (Fig. XIV-5). The A9 + lymphoma hybrids produced tumors which could be divided into three main groups: those with a sarcoma-like morphology (Fig. XIV-6), those with a lymphoma-like morphology (Fig. XIV-7), and those with intermediate characteristics. The cells of the fibroblastic tumors had a karyotype which approached the sum of the two parental cells, whereas one tumor which closely resembled the lymphoma parent had undergone drastic chromosome losses. This suggests that a fibroblastic hybrid may express a lymphoma-like morphology after losing chromosomes. Similar findings were reported in a later report by Cochran *et al.* (1975) which described the morphology of hybrids between a methylcholantrene-induced sarcoma (MSWBS) and the virus-induced YACIR lymphoma.

These two studies suggest that histopathological examination of tumors derived from hybrid cells and their segregants is a valuable technique for studying the *in vivo* differentiation of tumor cells. Even though this approach may entail problems in quantifying the complex histopathological patterns by which pathologists classify tumors and problems in obtaining representative cell samples for chromosome analysis, it may contribute new information on mechanisms of tumor progression.

E. IN VIVO FUSION OF TUMOR CELLS

Fusion of somatic mammalian cells *in vivo* is a rare phenomenon except during myogenesis (see Chapter III). Under pathological conditions caused by viruses (Chapter IV) or chemical agents (Chapter V), multinucleate giant cells may, nonetheless, form as a result of cell fusion. Most of these multinucleate cells are probably nondividing, but there is suggestive evidence that some of these fusion products may give rise to dividing mononucleate cells that may be polyploid or hybrid (Stone *et al.*, 1964; Volpe and Earley, 1970; Janzen *et al.*, 1971). For instance, Goldenberg *et al.* (1971, 1974) injected human tumor cells into hamster cheek pouches and found that these fused spontaneously with the normal hamster cells to produce malignant human + hamster hy-

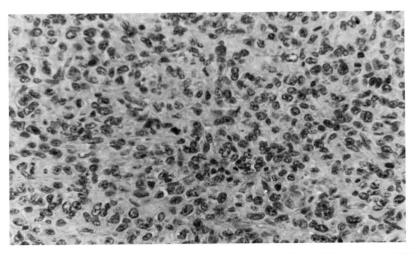

Fig. XIV-4. A9 tumor. Sarcoma with a "fibroblastic" morphology (Wiener *et al.*, 1972b).

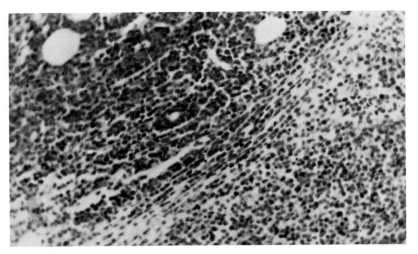

Fig. XIV-5. YACIR tumor, predominantly lymphocytic morphology (Wiener *et al.*, 1972b).

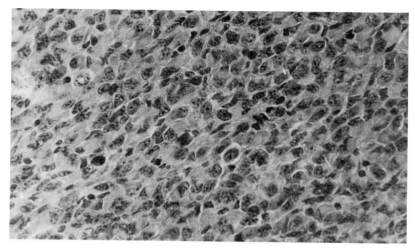

Fig. XIV-6. Sarcoma-like morphology of a YACIR + A9 hybrid cell tumor (Wiener *et al.*, 1972b).

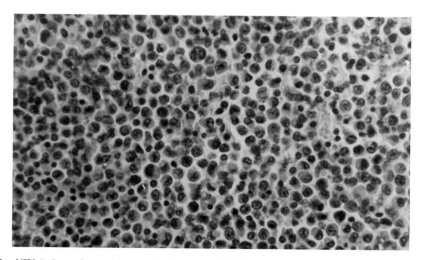

Fig. XIV-7. Lymphoma-like morphology of a YACIR + A9 hybrid cell tumor (Wiener *et al.*, 1972b).

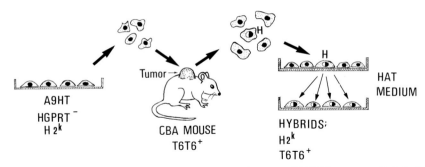

Fig. XIV-8. Schematic illustration of *in vivo* fusion between HGPRT⁻ tumor cells injected into mice (HGPRT⁺). Hybrid cells formed *in vivo* are HGPRT⁺ and have markers characteristic of the animal host.

brid cells. These contained a greatly reduced human chromosome set. It would be of great interest to identify the human chromosomes in a number of *in vivo* fused malignant human + hamster hybrid tumors. Such an analysis could provide important information about the chromosomal localization of genetic determinants for malignancy in man.

In vivo fusion between tumor cells and normal cells has also been studied by Wiener *et al.* (1972a, 1974c and e) and Fenyö *et al.* (1973b) but with a different technique. Using the malignant A9HT subline of A9 and a similar malignant derivative (B82HT) derived from B82 cells, Wiener *et al.* (1972a) obtained tumors which contained small numbers of hybrid cells (Fig. XIV-8). When the tumor cell populations were subsequently cultured *in vitro* on the selective HAT medium a majority of the cells died, as expected for unfused HGPRT⁻ (A9HT) and TK⁻ (B82HT) cells. A small minority of cells, however, gave rise to HAT-resistant colonies. On further examination these colonies were found to be composed of hybrid cells which contained chromosomes of both the tumor cells and the host animal. Absorption tests showed that the HAT-resistant cells carried H-2 antigens typical of A9 (or of B82) as well as H-2 antigens and a chromosome marker (T6T6) typical of the host animals, thus, providing additional evidence for the hybrid nature of these cells. Figure XIV-9 illustrates the appearance of a tumor cell + host cell hybrid formed *in vivo* by fusion between SEWA ascites sarcoma cells and normal host cells. Fenyö *et al.* (1973b) extended these observations by injecting two different polyoma induced ascites sarcomas and one methylcholanthrene-induced sarcoma into appropriate mice. The karyotypes and antigens of the resulting tumor cells were characteristic of both the tumor cells injected and the cells of the host. In order to determine the specific origin of the host cells which fuse with various injected tumor cells, Wiener *et al.* (1974c) inoculated mouse

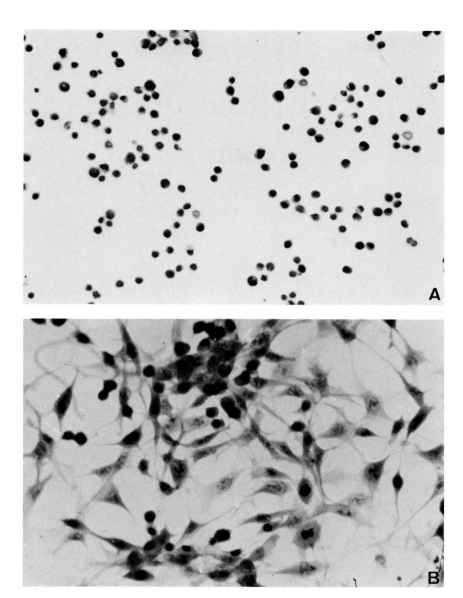

Fig. XIV-9. Morphology of (A) SEWA ascites sarcoma and (B) SEWA + host cell hybrid (Courtesy Dr. F. Wiener).

tumor cells into radiation chimeras in which the cells of the solid tissues and the repopulating hemopoietic cells carried different markers. Ascites tumor cells were found to undergo *in vivo* fusion and produce hybrids with the repopulating hemopoietic cells, whereas cells from solid tumors fused with cells of solid tissues in the irradiated host.

While the results of Wiener, Fenyö, and co-workers are highly suggestive the method used does not permit formal proof of *in vivo* fusion. Fusion could take place *in vitro* after explantation of the tumor cells since the tumor is likely to be composed of both tumor cells of the injected type and normal host cells derived from connective tissue and blood vessels in the tumor. Another question is whether or not the hybrid cells formed *in vivo* would be capable of multiplying *in vivo*. According to Harris' and Klein's hypothesis hybrids between a malignant and a normal cell should be expected to be nonmalignant and therefore be unable to grow *in vivo*. Since the hybrids would be present in the test system for malignancy when they form they would have a very small chance of multiplying and loosing chromosomes and giving rise to malignant segregants. If on the other hand the hybrids grow this would suggest that malignancy behaves as a dominant trait. For a discussion of these and other questions concerning *in vivo* fusion in mouse tumors see Wiener *et al.* (1974e).

At present it is difficult to evaluate the significance of fusion *in vivo* between experimentally introduced tumor cells and normal host cells to the usual progress of oncogenesis in animals. If this sort of process is common during tumor invasion, then it may be that it generates genetic variability and increased probability of malignant segregation because of the chromosomal instability characteristic of hybrid cells. This in turn may increase the resistance of tumors to treatment because greater variability carries with it a greater likelihood of resistance to chemotherapy and immunotherapy.

XV

Viral Infection and Somatic Cell Hybrids

Hybridization of somatic cells has been used to analyze the interaction between a number of viruses and their host cells. The most important of these applications are (a) mapping of virus integration sites; (b) virus rescue and virus detection; (c) analysis of factors determining the susceptibility of different cell types to viral infections; (d) analysis of cellular mechanisms inhibiting viral replication or modifying viral gene expression.

Most of these studies have been with tumor viruses and in order to introduce the reader to the viruses which will be mentioned in the succeeding pages, a brief summary of their host specificities and capacity to induce tumors or transform cells is given in Table XV-1. Furthermore, the following pages also review briefly the properties and molecular biology of DNA and RNA tumor viruses in order to define some of the problems which have been analyzed with cell hybrids. More specific information about these viruses and their interactions with normal, nonhybrid cells can be found in reviews and monographs by Green (1970); Temin (1971); Tooze (1973); Eckhart (1974); Fenner et al. (1974); and the proceedings of the Cold Spring Harbor Symposium on Tumor Viruses (1975).

TABLE XV-1

Tumor Viruses Studied by Cell Hybridization; Summary of Properties[a]

Virus group	Host of origin	Tumor induction in	Host cells susceptible to transformation	Mol. wt. of nucleic acid
RNA viruses				
Avian leukosis-sarcoma group (Rous sarcoma virus, RSV)	Chicken	Chicken, rodents, monkey	Chicken, rodent, monkey, human	10–13×10^6
Murine leukosis-sarcoma group (murine sarcoma virus, murine leukemia viruses)	Mouse	Mouse, rat, hamster	Mouse, rat, hamster,	10–13×10^6
DNA viruses				
Papovaviruses				
Polyoma	Mouse	Mouse, rat, hamster, rabbit, guinea pig	Mouse, rat, hamster, bovine	3×10^6
Simian virus 40 (SV40)	Monkey	Hamster	Hamster, mouse, rat, human, monkey	3×10^6
Adenoviruses (human adenovirus 12)	Human	Hamster, rat, mouse	Hamster, rat, rabbit, human	21×10^6
Herpes viruses [Epstein–Barr virus (EBV)]	Human	?	Human	54–92×10^6

[a] Based on references (Green, 1970; Tooze, 1973).

A. INTRODUCTION TO TUMOR VIRUSES

1. DNA Viruses

Four different DNA tumor viruses have been studied in hybrid cells: SV40 (monkey), polyoma (mouse), Epstein–Barr virus (human), and adenovirus (human). In what follows, emphasis will be on SV40 virus since cell hybridzation has been particularly useful in studies of this virus and because SV40 has been extensively investigated by molecular biologists. Remarkable progress has, for instance, been made in the genetic mapping of the viral genome and toward understanding the interaction between virus and host cell. It should be pointed out that although the main principles as described for SV40 also apply to other DNA tumor viruses, there are important differences between these viruses in terms of genome size, antigen properties, host specificity, mode of integration into cellular DNA, and ability to transform cells *in vitro* (for a review see Eckhart, 1974).

Infection of mammalian cells with SV40 and other DNA tumor viruses can give rise to two different types of phenomena; *transformation* and *lytic infection*. In the first case there is no virus production but a small number of viral genomes remain in the cells and induce the transformed phenotype. In the second case there is active virus multiplication which kills the cell and results in the release of new infectious particles. Which of the two phenomena, transformation or lytic infection, occurs depends on the type of host cell, the virus, and the species. Cells in which virus multiplication occurs are known as *permissive* cells while cells in which the virus fails to multiply are known as *nonpermissive*.

The infectious cycle of SV40 virus, whose natural hosts are monkey cells, is illustrated in Fig. XV-1. Following *adsorption* of the infecting virus particle to receptor sites on the cell membrane and penetration through this membrane, the virus particle is uncoated, that is, the viral protein coat is removed and the nucleic acid core exposed. Host cell RNA polymerases then transcribe the viral DNA into RNA molecules which specify the synthesis of new proteins in the infected cell. Some of these *"early proteins"* can be identified by immunological methods among them the nuclear *T antigens* (tumor antigens). The T antigen is specific for the infecting tumor virus and appears to be specified by the viral genome. Some 16 h after infection viral DNA replication can be detected in the nucleus. At the same time the host DNA synthesis is stimulated. At a somewhat later stage, there is synthesis of viral capsid proteins which may be detected immunologically (the so-called V antigens) in the infected cell. Since these antigens are specified by the viral genomes, they vary from virus to virus. New virus particles are

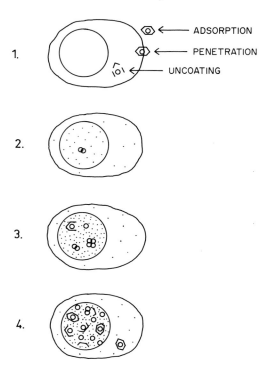

Fig. XV-1. Schematic illustration of SV40 lytic infection. (1) Virus attaches to the cell surface, penetrates, and is uncoated so as to expose the DNA. (2) The viral genome is transposed to the nucleus where it begins to replicate. The host cell dies as the nucleus (3) and the cytoplasm (4) become filled with virus.

then assembled intracellularly. By this stage the host cell shows characteristic vacuolation and other cytopathic effects. Ultimately it lyses and releases the new infectious virus particles.

If *nonpermissive* cells are exposed to SV40 they sometimes undergo transformation (Fig. XV-2). Also, isolated SV40 DNA or DNA fragments prepared by digestion with restriction enzymes can induce transformation (Graham *et al.*, 1975). As indicated in the previous chapter, this results in marked changes in growth regulation and cell surface properties. Transformed cells express early functions (e.g., synthesis of T antigens) but there is no assembly of infectious virus particles. There is now strong evidence that in transformed cells the viral genomes or a part of them become covalently linked (integrated) to the DNA of host cell chromosomes and then replicate in synchrony with the host cell nucleus without giving rise to infectious virus particles (Sambrook *et al.*, 1968). The number of viral genomes present in transformed cells varies

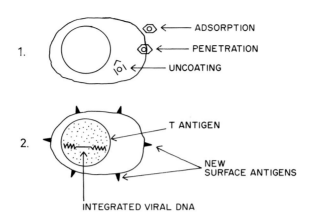

1.

2.

Fig. XV-2. Schematic summary of SV40 transformation of nonpermissive cells. Viral genomes are integrated on host cell chromosomes.

from 1–14 (Gelb *et al.*, 1971; Botchan *et al.*, 1974; Sambrook *et al.*, 1975) depending on which cell line is being analyzed. In some cases defective SV40 genomes are present (Yoshiike *et al.*, 1975; Sambrook *et al.*, 1975). That part of the viral genome which is present in transformed cells is believed to carry a "transformation gene" but the exact localization of this gene, the nature of the gene product, and the mechanism by which the transformed state is induced is not yet known.

2. RNA Viruses

A number of different RNA tumor viruses have been studied with the aid of cell hybridization. Unfortunately the classification of RNA tumor viruses is extremely complex. One commonly used classification is based on the host specificity of the viruses (avian, murine, felline) and the type of tumor induced (leukemia, sarcoma). Thus, there are avian and murine forms of leukemia and of sarcoma viruses. The individual viruses have usually been named after the person who discovered them (e.g., Gross, Friend, Moloney, Rauscher, and Graffi viruses). RNA tumor viruses have also been detected by electron microscopy and have then been referred to as "B" or "C" type particles" depending on their morphology.

Most RNA tumor viruses are spherical particles (about 100 mμ in diameter). The central nucleoid contains the viral genome and an enzyme, reverse transcriptase (Baltimore, 1970; Temin and Mizutani, 1970), which can synthesize DNA with RNA as a template. The virion is surrounded by a unit membrane of lipoprotein from which knobs or spikes of glycoprotein extend.

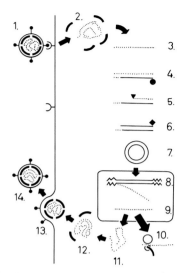

Fig. XV-3. Infectious cycle of a RNA tumor virus. (1) Attachment of the virus to receptors at the cell surface, (2) uncoating and exposure of the virus nucleic acid consisting of three 35 S RNA molecules linked together by 4 S RNA linker molecules into a 70 S RNA complex. The 35 S RNA strand, known as the plus strand (·····) is released (3) and then transcribed by reverse transcriptase (4) to form a DNA minus strand (−). The viral RNA is degraded by RNase H (5) and the DNA (−) strand is copied into a (+) strand (6). The double stranded linear DNA molecule is then circularized (7) and transposed into the nucleus where it is integrated on host cell chromosomes (8). The viral genome ("provirus") is transcribed and viral RNA (·····) moves to the cytoplasm (9) where approximately 90% specifies the synthesis of viral proteins (10) while 10% is incorporated into virions (11,12) which are released from the cell by budding (13,14).

The viral genome consists of 60–70 S RNA with a molecular weight of approximately 10^7 daltons, enough to specify 5–6 proteins. Some virus particles have defective genomes where certain genes have been deleted or otherwise become nonfunctional and, although defective virions may penetrate cells and cause transformation, they fail to give rise to new virions. In the presence of other viruses the defects can sometimes be overcome because the genome of the helper virus then specifies those proteins for which the virus is defective. As a result of this complementation complete virions are formed.

The productive infection of a susceptible cell with a nondefective RNA tumor virus can be divided into the following steps (Fig. XV-3):

    1. Binding of the virion to specific cell receptors and penetration into the cell

    2. Uncoating of the virus and transposing of the viral nucleic acid to the nucleus

3. A viral enzyme, reverse transcriptase, synthesizes a DNA strand ("minus strand") which is complementary to the viral RNA ("plus strand")

4. The single stranded complementary DNA acts as a template for the synthesis of a DNA strand with a base composition analogous to that of the viral RNA

5. The duplex viral DNA (provirus) is integrated into the DNA of the host cell; it is not known if this integration occurs at specific chromosomal sites or randomly

6. The genes of the integrated provirus are transcribed

7. The resulting RNA then migrates to the cytoplasm

8. In the cytoplasm it specifies the synthesis of virion proteins

9. Assembly of viral RNA and protein into new virions

10. Virions move to the cell surface, and bud to form free virus particles; during the budding the virions become surrounded by an outer membrane derived from the host cell.

Infection with RNA tumor viruses frequently, but not always, results in transformation of the host cell (see Chapter XIV). Chick cells transformed with RSV usually multiply while at the same time releasing virus. But if mammalian cells are transformed with RSV they usually grow without producing virus. There are also situations in which cells remain untransformed but produce virus. The outcome of the infection in each case appears to be determined by a number of factors of which only some are known. Clearly the type of RNA virus, and the species and differentiation of the infected cells are important. But even different animal strains of the same species differ in their susceptibility to a given type of virus. Thus, murine leukemia viruses (MLV) can be divided into three categories N-, B-, and NB-tropic, by their capacity to grow in different mouse strains. N-tropic viruses are those which grow in Swiss mouse cells (N) while B-tropic viruses grow in cells of Balb/c (B) strains of mice. N-cells are less sensitive to B-tropic viruses and vice versa. The pattern of sensitivity to infection in N and B cells is genetically determined essentially by a single genetic locus, the *Fv-1* locus. There are also some virus strains, NB-tropic, which grow equally well in both mouse strains.

B. MAPPING OF SV40 INTEGRATION SITES

Human + mouse hybrid cells show a strong tendency to shed human chromosomes. Weiss *et al.* (1968b; Weiss, 1970) found that this also occurs in hybrids made by crossing SV40-transformed human cells with enzyme deficient mouse cells and that there is a correlation between loss of chromosomes and loss of T antigen. These observations, thus, indicated that it might be possible to assign the gene specifying the T antigen (i.e., the SV40 genome) to a specific human chromosome. Such studies were undertaken by Croce and co-workers (1973c, 1974c,e, 1975a; Croce and Koprowski, 1974a,b,c, 1975) who fused human SV40 transformed HGPRT$^-$ cells with TK$^-$ mouse cells. Hybrid cells were isolated by selection on HAT medium. After some time of culturing, hybrids still contained practically the whole mouse complement but had lost most of the human chromosomes. Eighty-five clones were examined for karyotype and expression of SV40 specified T and V antigens. A positive correlation was found between the presence of human chromosome 7 and the expression of T antigen in different clones (Croce *et al.*, 1973c) and subclones (Croce and Koprowski, 1975).

When T$^+$ hybrid clones were fused with permissive monkey cells the hybrids developed V antigen showing that late viral functions were activated. T$^-$ hybrids, however, failed to develop V antigen, thus establishing a positive correlation between V antigen on one hand and T antigen and chromosome 7 on the other hand. This result was confirmed in a later analysis of hybrids between human SV40 transformed cells and normal mouse macrophages (Croce and Koprowski, 1975; Croce *et al.*, 1975a).

These results seem to establish chromosome 7 as an integration site for SV40 genes in at least two independent lines of human cells. Whether this is true for other human cell lines is uncertain and there are several lines of evidence which suggest that the integration sites may be multiple and vary from one cell line to another:

(a) The number of SV40 genomes per diploid amount of host cell DNA varies in different cell lines from 1 to a maximum of 20 (Hirai and Defendi, 1975). When many viral genomes are present it is possible that they have been integrated on several host cell chromosomes.

(b) Hirai *et al.* (1974) tested a number of SV40-transformed Chinese hamster cell lines for the presence of integrated viral DNA in nucleolus-associated DNA versus the rest of the nuclear DNA. In one case the SV40 genome (1 equivalent/diploid cell) was integrated in the nucleolus-associated DNA but in others the viral genomes were found in both classes of DNA with some preference for nonnucleolar DNA.

(c) Watkins (1974) found that during rescue of SV40 virus in heterokaryons, SV40 replication starts in 2–11 independent foci in the transformed nuclei suggesting that SV40 genomes are present and are being rescued in several different locations.

Thus, while it remains uncertain whether or not DNA tumor viruses integrate at unique sites in the host cell genome, the utility of hybrid cells in mapping viral integration sites has nonetheless been convincingly demonstrated by the studies of Croce and Koprowski (1974b,c, 1975) on SV40-transformed cells.

C. RESCUE OF SV40 VIRUS IN HETEROKARYONS

1. Type of Rescue Experiments

The first indications that DNA tumor viruses could be rescued from transformed cells that were free of infectious virus came from experiments in which SV40-transformed cells were cocultured with permissive cells (Gerber and Kirschstein, 1962; Sabin and Koch, 1963; Black, 1966). From such studies it also became evident that intimate contact between the transformed cells and the permissive "indicator" cells was required for virus production to occur. UV-inactivated Sendai virus was found to increase the efficiency by which virus could be detected in transformed cells (Gerber, 1966). Three different groups (Koprowski *et al.*, 1967; Knowles *et al.*, 1968, 1969, 1971; Koprowski, 1971; Watkins and Dulbecco, 1967; Dubbs and Kit, 1968) then demonstrated that reactivation of SV40 virus ("rescue") took place in heterokaryons formed by fusion of transformed cells with permissive cells. Later work showed that SV40 virus can also be rescued by fusing SV40-transformed, nonpermissive cells with enucleated cytoplasms from permissive cells (Croce and Koprowski, 1973; Poste *et al.*, 1974) or by treating SV40-transformed cells with extracts from permissive cells (Suarez *et al.*, 1972b; Lavialle *et al.*, 1974). Sometimes fusion of two different SV40-transformed cells results in heterokaryons capable of producing infectious SV40 virus (Jensen and Koprowski, 1969; Knowles *et al.*, 1968; Kit *et al.*, 1970).

2. Defective SV40 in Transformed Cells

In a number of cases, however, attempts to rescue SV40 virus by fusing transformed cells with each other or with permissive cells have failed. The reason for this appears to be that many transformed cells contain defective viral genomes which are capable of transforming cells but not of producing complete virus particles. This hypothesis is supported by the following observations:

(a) In cultures where SV40 rescue does occur after fusion of transformed and permissive cells, only a small proportion ($<10\%$) of the heterokaryons show active virus multiplication (Watkins and Dulbecco, 1967; Uchida and Watanabe, 1969). All heterokaryons, however, express at least one early viral function in the form of T antigen, thus, showing that the corresponding viral gene is both present and expressed. These findings may reflect the presence of defective viral genomes in many of the transformed cells but other explanations are also possible (see Watkins, 1975).

(b) In some fusions the heterokaryons produce a defective SV40 virus which can induce T antigen in other cells but is unable to infect permissive cells and to give rise to plaques (Botchan *et al.*, 1975). In some cases the rescued virus has a reduced capacity to infect neighboring cells and therefore gives rise to microplaques (Koprowski and Knowles, 1974; Croce *et al.*, 1974c; Huebner *et al.*, 1974, 1975).

(c) The observation that fusion of two different SV40-transformed cell lines can induce SV40 production (Knowles *et al.*, 1968; Jensen and Koprowski, 1969; Kit *et al.*, 1970) may point to complementation or recombination of two types of defective virus.

(d) DNA from transformed cells contain varying quantities of different viral DNA sequences. For example, the SVT2 line analyzed by Sambrook and co-workers (Botchan *et al.*, 1974, 1975; Sambrook *et al.*, 1975) contains one complete and five partial copies of a region which includes early viral genes. In the case of other DNA tumor viruses, transformed cells contain even less viral genetic material. Gallimore *et al.* (1974; Sharp *et al.*, 1974) analyzed rat cell lines transformed by adenovirus 2 for the presence of viral genomes. None of their lines appeared to contain the complete viral genome and some contained as little as 14%. Using sheared DNA, Graham *et al.* (1974) demonstrated that as little as 5% of the adenovirus 2 genome can transform cells and that only specific sequences of the viral genome will cause transformation.

In view of these results it appears that many cell lines, transformed with SV40 and other DNA tumor viruses, contain defective viral genomes and that this may be the explanation why in fusion experiments only some heterokaryons produce virus.

3. Cellular Changes during SV40 Rescue in Heterokaryons

The following changes have been noted in SV40-transformed cell + permissive cell heterokaryons and are illustrated schematically in Fig. XV-4:

(a) Migration of T antigen into the permissive nucleus
(b) Increase in the T antigen amount in both types of nuclei
(c) Focal synthesis of viral DNA in the transformed nucleus
(d) Nuclear enlargement, altered chromatin morphology and other

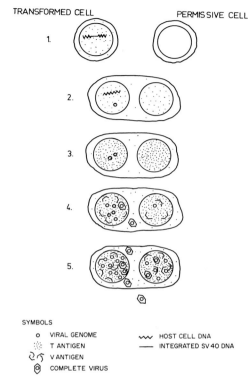

TRANSFORMED CELL   PERMISSIVE CELL

SYMBOLS

o   VIRAL GENOME   HOST CELL DNA

   T ANTIGEN   INTEGRATED SV40 DNA

   V ANTIGEN

   COMPLETE VIRUS

Fig. XV-4. Schematic summary of events during rescue of SV40 virus in heterokaryons formed by fusion of transformed + permissive cells. (1) parental cells, (2) heterokaryon, T antigen migrates into permissive nucleus, (3) viral DNA replication is initiated in transformed nucleus and T antigen expression increases, (4) virus replication and assembly of new virions in both nuclei and (5) release of complete virions, lysis and death of the heterokaryon.

nuclear "cytopathic" effects; these changes are first observed in the permissive nucleus

(e) Synthesis of viral coat antigens (V antigen) and accumulation of this antigen in both types of nuclei

(f) Formation of mature, infectious virus particles, first in the transformed nucleus, later also in the permissive nucleus

Point (a), the appearance of T antigen in the permissive as well as the transformed nucleus of the heterokaryon, seems to be characteristic of all heterokaryons, even those which later fail to produce infectious virus (Koprowski *et al.*, 1967; Steplewski *et al.*, 1968; Fogel and Defendi,

1968; Rosenqvist *et al.*, 1975). In most types of heterokaryons the non-transformed nuclei become T antigen positive within 24 h post fusion, which is before virus replication has started. Inhibitors of protein and RNA synthesis prevent the appearance of T antigen in the nontransformed nucleus, while inhibitors of DNA synthesis have no such effect (Steplewski *et al.*, 1968). Three or four days after fusion there is a marked increase in T antigen expression in a small proportion (<10%) of the heterokaryons. This increase is observed both in transformed and permissive nuclei and it is inhibited by cytosine arabinoside, an inhibitor of DNA synthesis (Uchida and Watanabe, 1969). In those heterokaryons where SV40 replication is initiated (approximately 0.5–2%) the transformed nuclei develop 2–11 foci of strong ^{3}H-thymidine incorporation (Watkins, 1974). That the DNA synthesized in these foci is viral is shown by the fact that labeled SV40 complementary RNA (cRNA) anneals at the foci.

As indicated in point (d) viral DNA synthesis is accompanied by nuclear enlargement and changes in its internal structure. The chromatin assumes a more homogeneous appearance, while nucleoli become larger and vacuolated. A "halo" can often be observed around the nuclear periphery. These changes appear to be closely coupled to viral multiplication since they are prevented by inhibitors of DNA synthesis, inhibitors which of course block viral replication (Uchida and Watanabe, 1969).

Viral coat proteins are formed soon after DNA replication begins. Immune fluorescence techniques have detected V antigen in both transformed and permissive nuclei of heterokaryons (Koprowski *et al.*, 1967; Steplewski and Koprowski, 1969; Huebner *et al.*, 1974; Watkins, 1974). If permissive cells are infected with SV40 and then fused with nonpermissive cells, preformed V antigen will also accumulate rapidly in the nontransformed nuclei.

The newly formed viral DNA and proteins are assembled into complete virus particles which can be detected by electron microscopy (Glaser and Farrugia, 1973). Although both transformed and permissive nuclei in heterokaryons show signs of virus assembly, the process appears to start in the transformed nuclei. Wever *et al.* (1970) isolated nuclei from heterokaryons formed by fusing nonpermissive transformed hamster cells with permissive monkey cells and then separated the two types of nuclei by gradient centrifugation. The virus content of the nuclei was assayed for by testing for infectious center formation after plating different nuclear fractions from the gradient on monolayers of monkey cells. Virus could be detected in transformed nuclei 40 h post fusion, and by 70 h virus also appeared in the permissive nuclei. Similar observations have been made in electron-microscope studies by Glaser and Farrugia (1973).

4. Mechanism of SV40 Rescue

Although fusion studies have contributed a considerable amount of information about virus host cell interactions in transformation, a number of important questions concerning virus rescue remain unanswered. Such questions include:

(a) What is the nature of the factor(s) contributed by the permissive cell which induce virus rescue?

(b) What is the molecular basis of permissiveness and nonpermissiveness?

(c) How are viral genomes transmitted from the transformed to the permissive nucleus?

(d) To what extent is the failure to rescue virus due to the viral genomes being defective and to what extent is it due to other factors?

The fusion experiments provide some information on the first of these questions since they show that SV40 virus is rescued in cybrids formed by fusing transformed cells with anucleate permissive cells (Croce and Koprowski, 1973; Poste *et al.*, 1974). This means that factors causing virus rescue are present in the cytoplasm of permissive cells. It is reasonable to assume that these factor(s) must migrate into the transformed nucleus in order to exert their action, but the nature of the factors and their action remains to be established. On theoretical grounds one would consider agents which effect excision of the integrated viral genome, or which promote viral replication, transcription, and assembly as likely candidates for factors inducing rescue.

D. FAILURE TO RESCUE OTHER DNA TUMOR VIRUSES IN FUSION EXPERIMENTS

Fusions of polyoma-transformed hamster cells with permissive mouse cells rarely result in virus rescue (Watkins and Dulbecco, 1967; Fogel and Sachs, 1969; Defendi *et al.*, 1968) although a few instances have been reported (Fogel and Sachs, 1969). Similarly, rescue experiments with adenovirus (Champe *et al.*, 1972), SV40-adeno hybrid virus (Knowles *et al.*, 1968), and Epstein–Barr virus (EBV) (Nyormoi *et al.*, 1973a) have met with little success. The reasons for these failures are not understood. As with SV40 there are indications that cells transformed by these viruses contain defective viral genomes. Analysis of the radiation target size for the inactivation of the transforming and reproductive abilities of polyoma virus (Benjamin, 1965; Basilico and Di-Mayorca, 1965) indicate that a considerable part of the genome in infectious virus particles is not required for the transforming activity, although it is necessary for viral multiplication. Failure to rescue EBV by cell fusion, however, is not necessarily due to defective genomes since both EBV antigen positive and negative hybrid cells can be induced to produce EBV by treating with 5-iododeoxyuridine (Glaser and O'Neill, 1972).

While the difficulties in rescuing polyoma and adenovirus could thus be due to the defectiveness of the integrated viral genomes, they could equally well be a result of a faulty excision mechanism. One difference between these virus–host systems and the SV40 system may be the presence of repressors of virus multiplication in cells nonpermissive for polyoma and adenovirus. Mouse cells are permissive for polyoma virus and if exposed will normally die as a result of lytic infection. Hamster cells, on the other hand, are nonpermissive but can be transformed. If mouse cells are infected with polyoma virus and then later fused with hamster cells, virus multiplication is inhibited in the heterokaryons. Babiuk and Hudson (1973) measured the appearance of a viral capsid antigen and found that the degree of inhibition was related to the ratio of mouse to hamster nuclei. Heterokaryons containing an excess of permissive (mouse) nuclei developed viral capsid antigens, whereas heterokaryons containing an equal number of hamster and mouse or an excess of hamster nuclei remained negative. Related studies on the permissiveness of mononucleate hybrid cells are discussed below.

E. RESCUE OF ROUS SARCOMA VIRUS (RSV) IN HETEROKARYONS

RSV can be rescued from transformed but nonproducing mammalian cells by Sendai virus-induced fusion with permissive chick cells (Svoboda et al., 1967; Svoboda and Dourmashkin, 1969; Shevlyagin et al., 1969; Vigier, 1967, 1973; Donner et al., 1974). The type of chick cells used is of some importance since fibroblasts are effective while chick lymphocytes and erythrocytes are not (Table XV-2). In contrast to the very low percentages of heterokaryons showing rescue of SV40 (see Section C,2) practically all heterokaryons derived from some fusions of RSV-transformed mammalian cells with permissive chick cells yield RSV (Machala et al., 1970). Other fusions, though, have been reported to be not nearly so productive (Donner et al., 1974).

TABLE XV-2

Rescue of RSV Virus from Transformed Mammalian Cells

RSV transformed mammalian cell	Chick cell	Production of infectious virus	References
Hamster	Fibroblast	+	Vigier, 1967
	Erythrocyte	−	Machala *et al.*, 1970; Svoboda and Dourmashkin, 1969
Rat	Fibroblast	+	Lindberg, 1974; Svoboda *et al.*, 1967; Donner *et al.*, 1974
	Lymphocyte	−	Lindberg, 1974
	Erythrocyte	−	Lindberg, 1974

F. ATTEMPTS TO RESCUE UNKNOWN VIRUSES IN HUMAN DISEASES

Encouraged by the findings that SV40 and RSV viruses can be rescued from transformed cells by fusion with permissive cells, several workers have fused cells from humans with several different diseases with monkey cells in the hope of rescuing an unknown virus. A summary of these results is given in Table XV-3. They indicate that it may be possible to use cell fusion as a diagnostic method for detecting unknown viruses in cells from patients. But a major problem in this approach is selecting a cell which will prove to be a permissive cell for unknown virus. For discussion of the medical significance of the findings made so far, the references listed in Table XV-3 should be consulted.

TABLE XV-3

Fusion of Cells from Patients with Chronic Diseases of the Nervous System[a]

Human brain cells from	Other parental cell	Type of virus isolated	References
Subacute sclerosing panencephalitis	AGMK	MDR group + papova-like[b]	Koprowski, 1971; Katz *et al.*, 1970; Koprowski *et al.*, 1970
Measles encephalitis	AGMK	Measles	terMeulen *et al.*, 1972b
Multiple sclerosis	AGMK	Parainfluenza 1	terMeulen *et al.*, 1972a
Progressive multifocal leucoencephalopathy	AGMK	SV40	Weiner *et al.*, 1972

[a] Modified from Koprowski and Knowles, 1974.
[b] SV40 isolated from human cells.

G. SUSCEPTIBILITY OF HYBRID CELLS TO VIRAL INFECTIONS

In a previous section we discussed experiments in which the immediate fusion products, the heterokaryons, were used to analyze virus–host cell interactions. Mononucleate hybrid cells formed by fusing permissive and nonpermissive cells have also been extremely useful in such studies. The strategy in these experiments has been to challenge the initial hybrid cells and their segregants with virus and then to monitor production of infectious virus. By testing hybrid cells which have preferentially lost chromosomes from the permissive or the nonpermissive parent, attempts have been made to examine genetic determinants controlling the susceptibility of mammalian cells to viral infection. These studies have shown that resistance to infection may be due to several factors, including viral dependence on host cell genes, the presence of repressors, and the absence of surface receptors. A summary of these results is given in Table XV-4. The resistance of chick, hamster, and mouse cells to poliovirus is due to a lack of cell surface receptors of a type present on permissive human cells. Enders *et al.* (1967) found that poliovirus could be introduced into resistant cells by adding UV-inactivated Sendai virus and that polio virus could multiply in resistant cells once the cell surface barrier had been broached. Presumably poliovirus was incorporated into the resistant cells during the fusion process. In more recent work cell hybrids have been used to analyze the nature of cell surface receptors (Fig. XV-5). Kusano *et al.* (1970) found that human + mouse hybrids which had lost a number of human chromosomes still carried human polio receptors and were susceptible to polio infection. When these hybrids were exposed to the virus most of the cells were killed but some survived. The survivors were resistant to poliovirus infection and had lost the polio receptors (Medrano and Green, 1973). They could, however, be infected with isolated polio RNA showing again that the resistance was due to a cell surface barrier rather than to an intracellular mechanism (Wang *et al.*, 1970). Polio-resistant survivors of man + mouse hybrids show only slight chromosomal changes when compared to the original, susceptible lines. When 3 susceptible lines containing 11–17 human chromosomes were compared with their polio-resistant sublines, a correlation was found between human chromosome 19 and sensitivity to poliovirus (Miller *et al.*, 1974). Chromosome 19 was present in susceptible lines but had been lost in all resistant sublines. Miller *et al.* (1974) therefore concluded that sensitivity to poliovirus in human cells was due to a virus receptor on the cell surface and that in man the gene coding for this receptor is located on chromosome 19. Cell hybrids have also been used to analyze

TABLE XV-4

Challenge of Hybrid Cells with Viruses

Virus	Permissive parent	Nonpermissive parent	Chromosome lost	Virus production		References
				In complete hybrid	In chromosomal segregants	
MLV	Mouse	Human	Human	−	+	Tennant and Richter, 1972
SV40	Human	Mouse	Human	−	−	Swetly et al., 1969; Knowles et al., 1971
	Monkey	Mouse	Monkey			Knowles et al., 1971
Polio	Human	Mouse	Human	+	−[a]	Wang et al., 1970; Kusano et al., 1970; Miller et al., 1974
Polyoma	Mouse	Human	Human	+	?	Pollack et al., 1971
	Mouse	Hamster	Mouse	+	−	Basilico et al., 1970; Basilico and Wang, 1971; Pollack et al., 1971
Adeno 2 + 12	Human	Mouse	Human	−	−	Pollack et al., 1971

[a] Segregants produce polio virus if chromosome 19 specifying a surface receptor is retained (Miller et al., 1974).

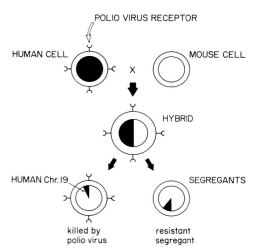

Fig. XV-5. Schematic illustration of the use of hybrid cells to study factors controlling the susceptibility of cells to viral infection.

host factors controlling the multiplication of RNA tumor viruses (Scolnick and Parks, 1974). For instance human + mouse hybrids have been used to analyze the function of a mouse gene (the *Fv-1* locus) controlling the replication of murine leukemia virus (Gazdar *et al.*, 1974). The viral restrictive functions of this gene were expressed in the hybrids. Thus, N-tropic virus only replicated in human + N-tropic mouse cell hybrids. A similar situation existed for B-tropic virus which only replicated in human + B tropic mouse cell hybrids.

H. INTERFERON SYNTHESIS IN HYBRID CELLS

Interferons are cell-coded proteins the synthesis of which can be induced by viral nucleic acid. The interferons are nontoxic for cells but are able to inhibit the multiplication of viruses (for review see Fenner *et al.*, 1974). They therefore play an important role as part of a cellular defence mechanism against viral infections. Because interferons are species specific they can be used as markers in interspecific cell fusions designed to analyze cellular defence mechanisms against viral infections.

In interferon-producing hybrids made by fusing cells of different species, it is possible to determine which of the two parental genomes is expressed. Interferon synthesis has been studied in chick + man (Guggenheim *et al.*, 1968, 1969), hamster + mouse (Carver *et al.*, 1968), monkey + mouse (Cassingena *et al.*, 1970, 1971) and mouse + human hybrids (Tan *et al.*, 1973).

Guggenheim *et al.* (1968, 1969) analyzed human fibroblast + chick erythrocyte heterokaryons and detected chick interferon synthesis. This protein is not synthesized by chick erythroid cells whose nuclei are genetically inactive. But since the erythrocyte nuclei underwent reactivation in these heterokaryons genes not normally expressed during erythropoiesis were reactivated. Hamster + mouse hybrids (Carver *et al.*, 1968) produced both hamster and mouse interferon though only the mouse parent produced interferon before fusion. Mouse + monkey hybrids studied by Cassingena *et al.* (1970, 1971; Suarez *et al.*, 1972a) produced both mouse and monkey interferon but lost the ability to produce monkey interferon as monkey chromosomes were eliminated. Similar findings were made with human + mouse hybrids by Ruddle's group (Tan *et al.*, 1973) who also used hybrid cells to assign interferon loci to human chromosomes 2 and 5 (Tan *et al.*, 1974).

XVI

Plant Cell Hybrids

If stripped of their cellulose walls, plant cells, too, can be fused. As with the animal systems already discussed, both intra- and interspecies fusions are possible, and the resulting heterokaryons may survive, regenerate walls, and reproduce themselves. The same kinds of techniques which have been applied so advantageously to animal cells—single cell culture, selective methods for isolating auxotrophs, and complementation studies—may also be applied to plant cells. Yet, plant systems offer two advantages which probably cannot be matched in the animal studies: it has become possible to culture haploid cells and to grow organisms from single somatic cells, including cell hybrids. The significance of these accomplishments may prove to be great, both for agriculture and for theoretical biology.

A. PRODUCTION OF PROTOPLASTS

The first step toward plant cell fusion is the preparation of plant protoplasts, that is, freeing plant cells from their cell walls (see Fig. XVI-1a). Botanists have been preparing protoplasts since the time of af Klercker who, in 1892, described a method for preparing them by plasmolysis and slicing. Plant tissues were treated with hypertonic media until the cell protoplasms were well withdrawn from their encapsulating cell walls then were cut with a sharp razor. The random cut traversed some of the cells without touching the shrunken protoplasm. When the tissue

290

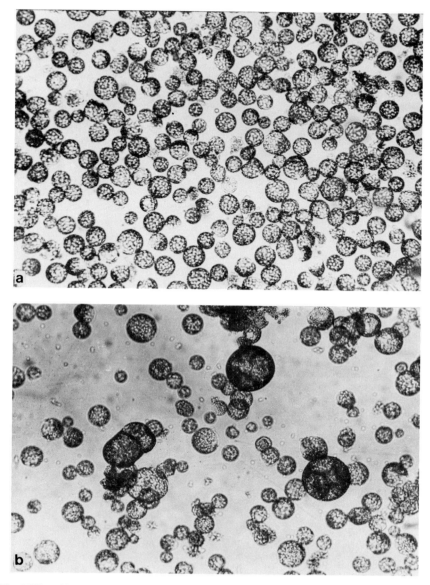

Fig. XVI-1. (a) Freshly isolated *Nicotiana tobacum* protoplasts. The protoplasts were isolated after Macerozyme digestion of leaves. (b) Aggregates and polykaryons resulting from fusion of *Nicotiana* protoplasts. Protoplasts were incubated for about 30 min at 37°C with a solution containing 0.05 M calcium chloride and 0.4 M mannitol, buffered at pH 10.5 (Melchers, 1974).

was returned to dilute medium again the protoplasm swelled, protruded from the broken end of the cell chamber, and pushed free from the wall, resulting in wall-free isolated protoplasts. Variations on this procedure have been used with success ever since (see Cocking, 1970).

Recently the more usual technique has been to digest cell walls away using mixtures of cellulase, pectinase, and hemicellulase. These enzymes* are derived commercially from flagellates or fungi (Takebe *et al.*, 1968) and are available in varying degrees of purity. As might be anticipated, digesting away the walls from plant cells is traumatic for living protoplasts. Partly this is because the enzyme preparations are contaminated by proteases and peroxidases. The extent of damage varies not only inversely with the purity and concentrations of enzyme preparations and duration of incubation, but also with the species and tissue being subjected to enzyme digestion. Cocking has found that plasmolysis of the protoplasts with 25% sucrose, which causes their withdrawal from the walls, often aids in their recovery (Cocking, 1972). But when successful, isolations produce protoplasts capable of wall regeneration, mitosis, and of course fusion.

Occasionally the enzymatic removal of cell walls will in itself produce multinucleate cells. Some, if not all, of these spontaneously formed polykaryons may not be the result of protoplast fusion, but instead the result of enlargements of preexisting intercellular connections. Adjacent plant protoplasms are often connected with one another through pores in the cell walls which are known as plasmodesmata. If freed from the constraints of the rigid carbohydrate wall, these membranous connections may simply expand to form a continuous cell encompassing more than one nucleus (Withers and Cocking, 1972; Ito and Maeda, 1973). Unless neighboring plant cells are perchance quite different in their developmental status, polykaryons generated in this way are not heterokaryons and clearly will not give rise to hybrid cells. But, as indicated in the next section, hybrid systems are generated from experimentally induced fusion of protoplasts.

B. FUSION OF PROTOPLASTS

The introduction to Michel's 1937 paper on plant protoplast fusion indicates that there had been attempts to fuse protoplasts almost from the time af Klercker first prepared them. Michel's teacher, Küster, reportedly made homokaryons as early as 1910. As noted in Chapter II, Mi-

---

* In the literature, the enzymes are often identified by their trade names (e.g., Macerozyme, Cellulase Onozuka, Onozuka P 1500, Cellulysin, Pectinol R. 10, Driselase, Rhozyme hemicellulase) reflecting preparations that different authors have found successful.

chel (1937) himself experimentally generated both homokaryons and heterokaryons from protoplasts derived from parenchymal or epidermal cells of some six genera of higher plants and a green alga. The fusions were effected by placing two protoplasts together on a microscope slide and plasmolyzing them for 10–20 min in 0.5 M sodium nitrate. Agglutination and fusion followed spontaneously.

While Michel's use of sodium nitrate was apparently motivated only by the need for a nontoxic hypertonic medium, Power *et al.* (1970) comparatively tested it and some other salts (calcium chloride, potassium nitrate, sodium nitrite, and sodium chloride) for specific effectiveness in fusing protoplasts. The test system they employed was based upon the old observation that during plasmolysis the protoplasms of some cells withdraw into two or more subprotoplasts. These frequently rejoin by fusion during a subsequent deplasmolysis, and such fusions can be facilitated or hindered by the contents of the hypotonic medium. It was found that sodium nitrate was the most effective of the tested salts. Thus, it has become standard to use 0.25 M sodium nitrate to fuse enzyme-isolated protoplasts both in this group and in others, including Carlson *et al.*, (1972), who first succeeded in generating a hybrid plant from fused protoplasts of two tobacco species.

Higher yields of fused cells have been reported to result from the use of calcium ions and high pH (Keller and Melchers, 1973). Up to 50% of the *Nicotiana* protoplast population was fused when they were centrifugally packed and incubated ($+37°C$) in a medium containing 0.05 M calcium chloride and 0.4 M mannitol buffered at pH 10.5 (Fig. XVI-1b). This result seems to put plant protoplast fusion squarely in the same category of events as animal cell fusion (see Chapter V). Melcher's group has had remarkable success in generating whole plants from the fusion products of this method (Melchers and Labib, 1974).

Binding (1974) expanded the number of tested inorganic salts and added lysozyme to the list of agents which fuse protoplasts. When tested with *Petunia* protoplasts, successful agents were 85% sea water, 0.2 M calcium nitrate at pH 8, 0.2 M calcium chloride, 0.9 M sodium nitrate, and 0.05% lysozyme in 0.6 M mannitol. Of these, the first two were the most effective.

Polyethylene glycol (PEG) has also been introduced recently as a fusing agent for plant protoplasts (Kao and Michayluk, 1974). PEG is a very water-soluble polymer with the general structural formula shown in Fig. V-2. As noted in Chapter V, its effectiveness as an agglutinating and fusing agent may be related to its ability to form ionic bonds with cell surface constituents. Kao and his co-workers suggest that its many ether links may make the molecule slightly negative in aqueous media.

It then may serve to link together positively charged surface proteins or negatively charged glycoproteins via divalent cation (Ca^{2+}) bridges (Kao and Michayluk, 1974; Constabel and Kao, 1974). In any event, if small quantities of 0.2–0.3 M PEG (molecular weight 1500 to 7500 daltons) are added to enzyme-isolated protoplasts, particularly in the presence of calcium ions, and then, after a half hour at room temperature, if it is diluted slowly away with isotonic media, large yields of polykaryons are generated [in the interspecific cross reported by Kartha *et al.* (1974), about 20% of the viable resultant protoplasts were heterokaryons]. While polykaryons made in this way have remained viable for many days, undergoing cell division and becoming synkaryons (Constabel *et al.*, 1975; Fowke *et al.*, 1975a), none has yet been reported to develop into a plant. As is the case with animal cell fusion, conditions which promote coalescence are quite different from those which promote growth. By varying culture conditions, it may prove possible to generate plants from hybrid protoplasts irrespective of the means of fusion.

C. DEVELOPMENT OF FUSED PROTOPLASTS

Isolated protoplasts, whether fused or not, may be regarded as living if they regenerate their cell walls and undergo division. However, interest in protoplasts stems from the fact that the potential for gene expression and growth is considerably greater than these two fundamental life properties. As already noted, these include the ability to regenerate whole plants.

In parallel with animal cell studies, it now seems likely that extensive growth among hybrid protoplasts is dependent upon polykaryons first becoming synkaryons. While nuclei in multinucleate protoplasts undergo simultaneous mitoses, the spindles do not regularly become conjoined and cytokinesis among daughter cells is often incomplete (Fowke *et al.*, 1975a). Perhaps in consequence of these events, multiple polykaryons do not survive long and make little contribution to populations of dividing hybrid cells. In a study on intergeneric heterokaryons, Constabel *et al.* (1975) have shown that when nuclei are distributed at some distance from one another, mitosis may result in cells of separate species again. On the other hand, if nuclei are adjacent during mitosis, the result may be synkaryons (see Fig. XVI-2). Direct nuclear fusion seems also to be a route to synkaryon formation. In both homokaryons (Fowke *et al.*, 1975a) and heterokaryons (Constabel *et al.*, 1975), internucleate bridges of about 0.2 μm diameter have been described which seem to be the forerunners to complete nuclear fusion.

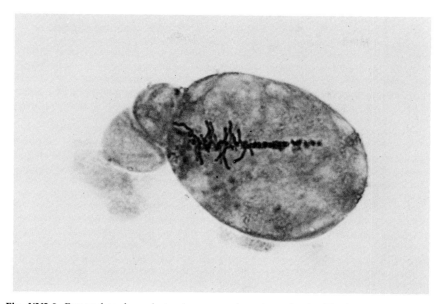

Fig. XVI-2. Pea and soybean heterokaryon undergoing mitosis. The larger pea chromosomes are seen on the left of the metaphase plate. Fusion was effected by PEG treatment of protoplasts from pea leaf and soybean cell cultures (Constabel *et al.*, 1975).

Two groups have reported success in nurturing dividing hybrid cells into whole plants (Carlson *et al.*, 1972; Melchers and Labib, 1974). Doubtless this sort of technical feat is possible with plant material because of the greater developmental plasticity of plants over animals. This plasticity not only accounts for the fact that plants have for centuries been propagated by "cuttings," but also for the fact that the technology for generating mature plants from undifferentiated parenchyma or even single cells has been available for years (see Skoog, 1944; Steward, 1970). Cocking (1972) has reviewed literature on the application of cell culture techniques specifically to plant cell protoplasts. In overview, both groups which have made somatic hybrid plants started the development by placing the dividing hybrid cells under conditions which promoted callus formation. These calli were then induced to generate shoots, leaves, and roots by treatments with various plant hormones and light regimes. In one case, the failure of *Nicotiana* interspecific somatic hybrids to develop roots was compensated for by grafting the developed shoot system to roots of one of the parental strains (Carlson *et al.*, 1972). Both groups have characterized the hybrid nature of their plants by a number of criteria (see below) and brought the specimens to flower and fruit. (Fig. XVI-3).

Fig. XVI-3. Flowering sexual and somatic hybrid *Nicotiana tobacum* shoots. Sexual hybrid is on the left, somatic on the right (Melchers, 1974).

D. SELECTION AND IDENTIFICATION OF HYBRID CELLS AND CALLI

Criteria used by botanists to establish the hybrid nature of protoplast fusion products are essentially the same as those used to identify hybrid animal cells.

The most readily accessible are morphological features, and these have been used with every stage of hybrid development, from heterokaryons to mature plants. In regard to the first, Michel (1937) fused beet cells which had red anthocyanin-filled vacuoles with colorless onion cells and identified fused cells by the dilution of the red beet pigment with the colorless onion sap. Fowke *et al.* (1975b) identified the heterokaryons of pea + *Vicia* protoplasts by the combination of two distinct sets of cytoplasmic, chloroplast, and nuclear characteristics in one cell (see Fig. XVI-4). Presence or absence of unique cytoplasmic entities like leucoplasts or chromoplasts can facilitate recognition of parental traits in somatic hybrids and would obviate the necessity of labeling cytoplasms in some special fusions, as is necessary for example in animal cell reconstitution studies (see Chapter VIII).

In characterizing their hybrid plants, Carlson's group (Carlson *et al.*, 1972) was fortunate in having available amphiploid hybrids which had

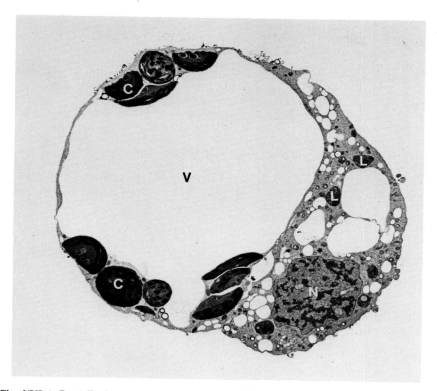

Fig. XVI-4. Partially fused protoplasts of pea and *Vicia*. The as yet unmixed cytoplasms are recognizable by the large central vacuole (V) and chloroplasts (C) of the pea and the leucoplasts (L), nucleus (N), and smaller vacuoles of *Vicia*. The upper portions of both cytoplasms are in full contact with one another, but the lower parts are still separated by agglutinated plasma membranes (L. C. Fowke and F. Constabel unpublished).

been made by sexual crosses. There was, thus, available a list of characteristics which distinguished the hybrids from the parental species. These included leaf shape, the extent of epidermal hairiness, the ease of generating tumors, the dependence or independence of callus growth upon exogenous auxins, the nature of isozymes, and chromosomal number. In regard to all these characteristics, the somatically generated hybrids were identical with the sexually generated ones. There was one distinction between the two sorts of hybrids, however: the somatic hybrid was self-fertile while the sexual was not.

Gene complementation has also provided a means for identifying hybrids, as well as for selecting them. Melchers and Labib (1974) obtained

Fig. XVI-5. Parental and hybrid *Nicotiana tobacum* plantlets regenerated from calli, grown at low light intensity. Upper row, left to right: a chlorophyll-deficient parental strain (s), two F_1 sexual hybrids of normal coloration (s × v), and a different chlorophyll-deficient parental strain (v). Middle row, left to right: strain (s), a somatic hybrid (s + v), and strain (v). Bottom row, left to right, two examples of strain(s), an (s + v), and two examples of strain (v). Somatic hybrids derived from fusions induced by calcium/high-pH treatment (Melchers, 1974).

two *Nicotiana tobacum* mutants which by themselves were deficient in chlorophyll biosynthetic pathways. Under normal conditions each failed to become typically green and grew only with difficulty (see Fig. XVI-5). F_1 sexual hybrids, however, were quite normal, green and fast-growing. Haploid plants were prepared from anther cultures of the two species, and from each of these, leaf mesophyll protoplasts were enzymically isolated. These were fused and nurtured into hybrid plants. Like the sexual hybrid, the somatic hybrids were normal in chlorophyll biosynthesis, enabling identification by their normal coloration and growth patterns (Fig. XVI-5).

An interesting case combining both nuclear and chloroplast genes in a hybrid complementation is described by Kung *et al.* (1975). In the interspecific *Nicotiana* somatic hybrid cross referred to above, these workers discovered that the large diphosphoribulose carboxylase complex of the chloroplasts (molecular weight: 550,000 daltons), the enzyme which fixes carbon dioxide photosynthetically, is a hybrid complex. The larger of two subunits of this complex is coded for by chloroplast genes, while the smaller is derived from nuclear information. It was found in the hybrid plant that the proteins of the smaller subunit were electrophoretically characteristic of both parental species, but the larger subunit was built exclusively of protein coded for by one species' chloroplast. On the surface, this appears to be a case of simultaneous coexpression and extinction. Note that a similar sort of pattern is seen in mitochondria of animal hybrid cells (see Chapter XII).

Karyotype analysis has also played a role in establishing the hybrid nature of plant materials just as they have in animal cell fusion studies. Obvious size differences of chromosomes of different species is evident in some hybrid cells (see Fig. XVI-2). As alluded to above, the number of chromosomes found in hybrid cells has been important among hybrids derived from the same species (see amphidiploidy in Carlson *et al.*, 1972; and amphihaploidy in Melchers and Labib, 1974).

E. WHY PLANT PROTOPLAST HYBRIDS?

There are several reasons why researchers, interested in problems which fall within what is broadly referred to as somatic cell genetics, might choose to use plant materials rather than animal cell cultures. While some of these have already been noted in this chapter, it seems worthwhile in this last section to underscore the advantages of plant vis-a-vis animal materials by systematically listing them. [For fuller treatments of some of these matters, the reader may wish to consult several recent reviews: Cocking (1970, 1972), Carlson (1973a,b), Chaleff and Carlson (1974), and Carlson and Polacco (1975).]

1. First, it has become apparent that many of the same kinds of questions which have been approached using hybrid animal cells may also be investigated using hybrid plant protoplasts. Paragraphs of this chapter point out the possibilities of studying gene expression and complementation and isozyme variation in plant somatic hybrids. As with zoological systems, mutant organisms are already available and can also be generated. In fact, with populations of protoplasts, it is possible to apply the same techniques of mutagenesis and selection that were developed for microorganisms and have been applied to animal cell populations.

The basic similarity of plant protoplast and animal cell systems is also shown by the fact that some of the same factors which influence fusion in one, work the same way in the other. While protoplasts are not covered with receptors for Sendai virus, agglutination, calcium ions, and high pH's, all promote fusion in both plant and animal cells. That PEG, first used for protoplast fusion, is now being applied successfully to animal cell fusion makes evident that either sort of cell may be used in studies designed to elucidate the nature of membrane and cell fusion.

2. It is easy to obtain immediately from one plant large numbers of haploid, diploid, or aneuploid protoplasts, ensuring genetic uniformity and eliminating in some instances the necessity for cloning. When compared with animal cell plating efficiency, the survivability of protoplasts is very high. Moreover, the proportion of viable polykaryons which result from PEG or calcium/high-pH fusions is also comparatively high.

3. Many normal plant protoplasts carry features which serve the function of marker properties. These include differences in kind and shapes of plastids, characteristics which are readily visible under a microscope. For some fusion experiments, this fact eliminates the necessity of dealing with mutants, isotopes, or other exogenous materials to identify a hybrid condition.

4. As noted already, the developmental plasticity of plants has made it possible to generate organisms from some fused protoplasts. In this regard, the plant system is unrivaled for studies of gene expression during development.

5. It has also been possible to study the roles of cell organelles in the biology of protoplasts in a rather unique way. Chloroplasts (Carlson, 1973a), nuclei (Potrykus and Hoffmann, 1973), and viruses (Cocking, 1970) have all been readily introduced into protoplasts. Perhaps this fact will lead to the establishment of protoplasts as ideal systems for the study of development of reconstituted cells (see Chapter VIII).

6. Aside from the foregoing issues, all of which relate to questions touching on theoretical aspects of biology, hybrid protoplast studies lend themselves very directly to what may well prove to be an important practical application. Many of the plant scientists who have contributed to the advancement in this field were motivated in the first place by the agricultural desire for crops which combine good properties of more than one species, as for example the disease resistance of one plant with the good fruiting qualities of another. This now is routinely accomplished by sexually crossing varieties or even closely related species. But the aim now is to develop technology enabling synthesis of interspecific plant hybrids. The fact that one interspecific hybrid plant has been developed from fused protoplasts suggests that this goal is attainable.

Bibliography

Aaronson, S. A., and Todaro, G. J. (1968). Basis for the acquisition of malignant potential by mouse cells cultivated *in vitro. Science* **162,** 1024–1026.

Abercrombie, M., and Ambrose, E. J. (1958). Interference microscope studies of cell contacts in tissue culture. *Exp. Cell Res.* **15,** 332–345.

Abercrombie, M., and Ambrose, E. J. (1962). The surface properties of cancer cells; a review. *Cancer Res.* **22,** 525–548.

Abercrombie, M., and Heaysman, J. E. M. (1954). Observations on the social behaviour of cells in tissue cultures. II. "Monolayering" of fibroblasts. *Exp. Cell Res.* **6,** 293–306.

Adoutte, A., and Beisson, J. (1970). Cytoplasmic inheritance of erythromycin-resistant mutations in *Paramecium aurelia. Mol. Gen. Genet.* **108,** 70–77.

af Klercker, J. (1892). Eine Methode zur Isolierung lebender Protoplasten. *Oefvers. Vet. Akad. Foerhandl. Stockholm* **9,** 463–475.

Ahkong, Q. F., Cramp, F. C., Fisher, D., Howell, J. I., and Lucy, J. A. (1972). Studies on chemically induced cell fusion. *J. Cell Sci.* **10,** 769–787.

Ahkong, Q. F., Fisher, D., Tampion, W., and Lucy, J. A. (1973). The fusion of erythrocytes by fatty acids, esters, retinol and alpha-tocopherol. *Biochem. J.* **136,** 147–155.

Ahkong, Q. F., Fisher, D., Tampion, W., and Lucy, J. A. (1975). Mechanisms of cell fusion. *Nature (London)* **253,** 194–195.

Akers, T. G., and Cunningham, C. H. (1968). Replication and cytopathogenicity of avian infections bronchitis virus in chicken embryo kidney cells. *Arch. Gesamte Virusforsch.* **25,** 30–37.

Akiba, H., Kato, T., Nakano, H., and Seiji, M. (1975). Defective DNA repair replication in *Xeroderma pigmentosum* fibroblasts and DNA repair of somatic cell hybrids after UV-irradiation. *Tohoku J. Exp. Med.* **117,** 1–13.

Albrecht, A. M., Riehm, H., and Biedler, J. (1971). Hybridization of actinomycin D and amethopterin-resistant Chinese hamster cells *in vitro. Cancer Res.* **31,** 297–307.

Allderdice, P. W., Miller, O, J., Miller, D. A., Warburton, D., Pearson, P. L., Klein, G., and Harris, H. (1973a). Chromosome analysis of two related heteroploid mouse cell lines by quinacrine fluorescence. *J. Cell Sci.* **12,** 263–274.

Allderdice, P. W., Miller, O. J., Pearson, P. L., Klein, G., and Harris, H. (1973b). Human chromosomes in 18 man–mouse somatic hybrid cell lines analysed by quinacrine fluorescence. *J. Cell Sci.* **12,** 809–830.

Alter, B. P., and Ingram, V. (1975). Globin synthesis in fibroblasts fused with erythrocytes. I. Avian fibroblasts. *Somatic Cell Genet.* **1,** 165–186.

Amano, T., Richelson, E., and Nirenberg, M. (1972). Neurotransmittor synthesis by neuroblastoma clones. *Proc. Natl. Acad. Sci. U. S. A.* **69,** 258–263.

Amano, T., Hamprecht, B., and Kemper, W. (1974). High activity of choline acetyltransferase induced in neuroblastoma × glia hybrid cells. *Exp. Cell Res.* **85,** 399–408.

Andrewes, C. H., Bang, F. B., and Burnet, F. M. (1955). A short description of the myxovirus group (Influenza and related viruses). *Virology* **1,** 176–184.

Appels, R., and Ringertz, N. R. (1975). Chemical and structural changes within chick erythrocyte nuclei introduced into mammalian cells by cell fusion. *Curr. Top. Dev. Biol.* **9,** 137–166.

Appels, R., Bolund, L., Goto, S., and Ringertz, N. R. (1974a). The kinetics of protein uptake by chick erythrocyte nuclei during reactivation in chick-mammalian heterokaryons. *Exp. Cell Res.* **85,** 182–190.

Appels, R., Bolund, L., and Ringertz, N. R. (1974b). Biochemical analysis of reactivated chick erythrocyte nuclei isolated from chick × HeLa heterokaryons. *J. Mol. Biol.* **87,** 339–355.

Appels, R., Tallroth, E., Appels, D. M., and Ringertz, N. R. (1975a). Differential uptake of protein into the chick nuclei of HeLa × chick erythrocyte heterokaryons. *Exp. Cell Res.* **92,** 70–78.

Appels, R., Bell, P. B., and Ringertz, N. R. (1975b). The first division of HeLa × chick erythrocyte heterokaryons. Transfer of chick nuclei to daughter cells. *Exp. Cell Res.* **92,** 79–86.

Appleyard, G., Westwood, J. C. N., and Zwartouw, H. T. (1962). The toxic effect of rabbitpox virus in tissue culture. *Virology* **18,** 159–169.

Arms, K. (1968). Cytonucleoproteins in cleaving eggs of *Xenopus laevis*. *J. Embryol. Exp. Morph.* **20,** 367–374.

Arrighi, F., and Hsu, T. C. (1971). Localization of heterochromatin in human chromosomes. *Cytogenetics* **10,** 81–86.

Attardi, B., and Attardi, G. (1972). Fate of mitochondrial DNA in human–mouse somatic cell hybrids. *Proc. Natl. Acad. Sci. U. S. A.* **69,** 129–133.

Auer, G., and Zetterberg, A. (1972). The role of nuclear proteins in RNA synthesis. *Exp. Cell Res.* **75,** 245–253.

Augusti-Tocco, G., and Sato, G. (1969). Establishment of functional clonal lines of neurons from mouse neuroblastoma. *Proc. Natl. Acad. Sci. U. S. A.* **64,** 311–315.

Aula, P. (1965). Virus-associated chromosome breakage. *Ann. Acad. Sci. Fenn., Ser. A4* **89,** 78.

Aula, P. (1970). Electron-microscopic observations on Sendai virus-induced chromosome pulverization in HeLa cells. *Hereditas* **65,** 163–170.

Aula, P., and Saksela, E. (1966). Early morphology of the chromosome damage induced by Sendai virus. *Hereditas* **55,** 362–366.

Azarnia, R., Larsen, W. J., and Loewenstein, W. R. (1974). The membrane junctions in communicating and noncommunicating cells, their hybrids and segregants. *Proc. Natl. Acad. Sci. U. S. A.* **71,** 880–884.

Babiuk, L. A., and Hudson, J. B. (1973). Inhibition of polyoma virus development in mouse–hamster heterokaryons. *Can. J. Microbiol.* **19,** 299–301.

Bakay, B., Croce, C. M., Koprowski, H., and Nyhan, W. L. (1973). Restoration of hypoxanthine phosphoribosyltransferase activity in mouse 1 R cells after fusion with chick embryo fibroblasts. *Proc. Natl. Acad. Sci. U. S. A.* **70,** 1998–2002.

Baker, R. M., Brunette, D. M., Mankovitz, R., Thompson, L. H., Whitmore, G. F., Siminovitch, L., and Till, J. E. (1974). Oubain-resistant mutants of mouse and hamster cells in culture. *Cell* **1,** 9–21.

Baltimore, D. (1970). RNA dependent DNA polymerase in virions of RNA tumor viruses. *Nature (London)* **226,** 1209–1211.

Baranska, W., and Koprowski, H. (1970). Fusion of unfertilized mouse eggs with somatic cells. *J. Exp. Zool.* **174,** 1–14.

Barski, G. (1970). Cell association and somatic cell hybridization. *Int. Rev. Exp. Pathol.* **9,** 151–190.

Barski, G., and Cornefert, F. (1962). Characteristics of "hybrid"-type clonal cell lines obtained from mixed cultures in vitro. *J. Natl. Cancer Inst.* **28,** 801–821.

Barski, G., Sorieul, S., and Cornefert, F. (1960). Production dans des cultures in vitro de deux souches cellulaires en association, de cellules de caractère "hybride". *C. R. Hebd. Seances Acad. Sci.* **251,** 1825–1827.

Barski, G., Sorieul, S., and Cornefert, F. (1961). "Hybrid" type cells in combined cultures of two different mammalian cell strains. *J. Natl. Cancer Inst.* **26,** 1269–1291.

Bartalos, M., and Baramki, T. A. (1967). "Medical Cytogenetics." Williams & Wilkins, Baltimore, Maryland.

Baserga, R. (1974). Minireview. Non-histone chromosomal proteins in normal and abnormal growth. *Life Sci.* **15,** 1057–1071.

Baserga, R., and Stein, G. (1971). Nuclear acidic proteins and cell proliferation. *Fed. Proc., Fed. Am. Soc. Exp. Biol.* **30,** 1752–1759.

Basilico, C. (1974). Temperature sensitive mutants of BHK-21 cells. In "Somatic Cell Hybridization" (R. L. Davidson and F. F. de la Cruz, eds.), pp. 235–238. Raven, New York.

Basilico, C., and DiMayorca, G. (1965). Radiation target size of the lytic and the transforming ability of polyoma virus. *Proc. Natl. Acad. Sci. U. S. A.* **54,** 125–127.

Basilico, C., and Wang, R. (1971). Susceptibility to superinfection of hybrids between polyoma "transformed" BHK and "normal" 3T3 cells. *Nature (London), New Biol.* **230,** 105–107.

Basilico, C., Matsuya, Y., and Green, H. (1970). The interaction of polyoma virus with mouse–hamster somatic hybrid cells. *Virology* **41,** 295–305.

Beale, G. H. (1969). A note on the inheritance of erythromycin resistance in *Paramecium aurelia. Genet. Res.* **14,** 341–342.

Beale, G. H., Knowles, J. K., and Tait, A. (1972). Mitochondrial genetics in Paramecium. *Nature (London)* **235,** 396–397.

Benda, P., and Davidson, R. L. (1971). Regulation of specific functions of glial cells in somatic hybrids. I. Control of S 100 protein. *J. Cell. Physiol.* **78,** 209–216.

Bendall, J. R. (1969). "Muscles, Molecules and Movements." Am. Elsevier, New York.

Bendich, A., Borenfreund, E., and Sternberg, S. S. (1974). Penetration of somatic mammalian cells by sperm. *Science* **183,** 857–859.

Benedetti, E. L., and Emmelot, P. (1967). Studies on plasma membranes. IV. The ultrastructural localization and content of sialic acid in plasma membranes isolated from rat liver and hepatoma. *J. Cell Sci.* **2,** 499–512.

Benedict, W. F., Nebert, D. W., and Thompson, E. B. (1972). Expression of aryl hydrocarbon hydroxylase induction and suppression of tyrosine aminotransferase induction in somatic-cell hybrids. *Proc. Natl. Acad. Sci. U. S. A.* **69,** 2179–2183.

Bengtsson, B. O., Nabholz, M., Kennett, R., Bodmer, W. F., Povey, S., and Swallow, D. (1975). Human intraspecific somatic cell hybrids: A genetic and karyotypic analysis of crosses between lymphocytes and D98/AH-2. *Somatic Cell Genet.* **1**, 41–64.

Benjamin, T. L. (1965). Relative target sizes for the inactivation of the transforming and reproductive abilities of polyoma virus. *Proc. Natl. Acad. Sci. U.S.A.* **54**, 121–124.

Benjamin, T. L. (1974). Methods of cell transformation by tumor viruses. *Methods Cell Biol.* **8**, 367–437.

Bernhard, H. P., Darlington, G. J., and Ruddle, F. H. (1973). Expression of liver phenotypes in cultured mouse hepatoma cells: Synthesis and secretion of serum albumin. *Dev. Biol.* **35**, 83–96.

Bernstein, R. M., and Mukherjee, B. B. (1972). Control of nuclear RNA synthesis in 2-cell and 4-cell mouse embryos. *Nature (London)* **238**, 457–459.

Bertolotti, R., and Weiss, M. C. (1972a). Expression of differentiated functions in hepatoma cell hybrids. II. Aldolase. *J. Cell. Physiol.* **79**, 211–224.

Bertolotti, R., and Weiss, M. C. (1972b). Expression of differentiated functions in hepatoma cell hybrids. VI. Extinction and re-expression of liver alcohol dehydrogenase. *Biochimie* **54**, 195–201.

Bertolotti, R., and Weiss, M. C. (1972c). Aldolase in hepatoma cell hybrids: Extinction of the hepatic form and its reexpression following loss of chromosomes. In "Cell Differentiation" (R. Harris, P. Allin, and D. Viza, eds.), pp. 202–205. Munksgaard, Copenhagen.

Bertolotti, R., and Weiss, M. C. (1974). Expression of differentiated functions in hepatoma cell hybrids. V. Re-expression of aldolase B *in vitro* and *in vivo*. *Differentiation* **2**, 5–17.

Billington, W. D. (1969). Immunological processes in mammalian reproduction. In "Immunology and Development" (M. Adinolfi, ed.), pp. 89–113. Spastics Int. Med. Publ., London.

Binding, H. (1974). Fusions Versuche mit isolierten Protoplasten von *Petunia hybrida* L. *Z. Pflanzenphysiol.* **72**, 422–426.

Bischoff, R., and Holtzer, H. (1969). Mitosis and the processes of differentiation of myogenic cells *in vitro*. *J. Cell Biol.* **41**, 188–200.

Black, P. H. (1966). An analysis of SV40 induced transformation of hamster kidney tissue in vitro. III. Persistence of SV40 viral genome in clones of transformed hamster cells. *J. Natl. Cancer Inst.* **37**, 487–493.

Bloom, A. D., and Nakamura, F. T. (1974). Establishment of a tetraploid, immunoglobulin-producing cell line from the hybridization of two human lymphocyte lines. *Proc. Natl. Acad. Sci. U. S. A.* **71**, 2689–2692.

Blough, H. A. (1964). Role of the surface state in the development of myxoviruses. *Cell. Biol. Myxovirus Infect., Ciba Found. Symp., 1964* pp. 120–143.

Bobrow, M., and Cross, J. (1974). Differential staining of human and mouse chromosomes in interspecific cell hybrids. *Nature (London)* **251**, 77–79.

Bobrow, M., Madan, K., and Pearson, P. L. (1972). Staining of some specific regions of human chromosomes, particularly the secondary constriction of No. 9. *Nature (London), New Biol.* **238**, 122–124.

Bolund, L., Ringertz, N. R., and Harris, H. (1969a). Changes in the cytochemical properties of erythrocyte nuclei reactivated by cell fusion. *J. Cell Sci.* **4**, 71–87.

Bolund, L., Darzynkiewicz, Z., and Ringertz, N. R. (1969b). Growth of hen erythrocyte nuclei undergoing reactivation in heterokaryons. *Exp. Cell Res.* **56**, 406–410.

Bonner, W. M. (1975a). Protein migration into nuclei. I. Frog oocyte nuclei in vivo accumulate microinjected histones, allow entry to small proteins and exclude large proteins. *J. Cell Biol.* **64,** 421–430.

Bonner, W. M. (1975b). Protein migration into nuclei. II. Frog oocyte nuclei accumulate a class of microinjected oocyte nuclear proteins and exclude a class of microinjected oocyte cytoplasmic proteins. *J. Cell Biol.* **64,** 431–437.

Bonnett, H. T., and Eriksson, T. (1974). Transfer of algal chloroplasts into protoplasts of higher plants. *Planta* **120,** 71–79.

Boone, C. M., and Ruddle, F. H. (1969a). Interspecific hybridization between human and mouse somatic cells: Enzyme and linkage studies. *Biochem. Genet.* **3,** 119–136.

Boone, C. M., Chen, T.-R., and Ruddle, F. H. (1972). Assignment of three human genes to chromosomes (*LDH-A* to 11, *TK* to 17 and *IDH* to 20) and evidence for translocation between human and mouse chromosomes in somatic cell hybrids. *Proc. Natl. Acad. Sci. U. S. A.* **69,** 510–514.

Bootsma, D. (1974). Cell fusion in the study of genetic heterogeneity of *Xeroderma pigmentosum.* In "Somatic Cell Hybridization" (R. L. Davidson and F. F. de la Cruz, eds.), pp. 265–269. Raven, New York.

Bordelon, M. R. (1974). Malignant characteristics of somatic cell hybrids of normal human and malignant mouse cells. *J. Cell Biol.* **63,** 32a.

Borek, C., and Sachs, L. (1966). *In vitro* cell transformation by X-irradiation. *Nature (London)* **210,** 276–278.

Borisy, G. G., and Olmsted, J. B. (1972). Nucleated assembly of microtubules in porcine brain extract. *Science* **177,** 1196–1197.

Bossart, W., Loeffler, H., and Bienz, K. (1975). Enucleation of cells by density gradient centrifugation. *Exp. Cell Res.* **96,** 360–366.

Botchan, M. B., Ozanne, B., Sugden, P. A., Sharp, P. A., and Sambrook, J. (1974). Viral DNA in transformed cells. III. The amounts of different regions of the SV40 genome present in a line of transformed mouse cells. *Proc. Natl. Acad. Sci. U. S. A.* **71,** 4183–4187.

Botchan, M. B., Ozanne, B., and Sambrook, J. (1975). Characterization of SV40 rescued from transformed mouse cells. *Cold Spring Harbor Symp. Quant. Biol.* **39,** 95–99.

Boyd, Y. L., and Harris, H. (1973). Correction of genetic defects in mammalian cells by the input of small amounts of foreign genetic material. *J. Cell Sci.* **13,** 841–861.

Brackett, B. G., Baranska, W., Sawicki, W., and Koprowski, H. (1971). Uptake of heterologous genome by mammalian spermatozoa and its transfer to ova through fertilization. *Proc. Natl. Acad. Sci. U. S. A.* **68,** 353–357.

Bramwell, M. E., and Handmaker, S. D. (1971). Ribosomal RNA synthesis in human–mouse hybrid cells. *Biochim. Biophys. Acta* **232,** 580–583.

Bratt, M. A., and Gallaher, W. R. (1972). Biological parameters of fusion from within and fusion from without. In "Membrane Research" (C. F. Fox, ed.), pp. 383–406. Academic Press, New York.

Bregula, U., Klein, G., and Harris, H. (1971). The analysis of malignancy by cell fusion. II. Hybrids between Ehrlich cells and normal diploid cells. *J. Cell Sci.* **8,** 673–680.

Bretscher, M. S. (1972). Assymetric lipid bilayer structure for biological membranes. *Nature (London), New Biol.* **236,** 11–12.

Bruckdorfer, K. R., Cramp, F. C., Goodall, A. H., Verrinder, M., and Lucy, J. A. (1974). Fusion of mouse fibroblasts with oleylamine. *J. Cell Sci.* **15,** 185–199.

Buckingham, M. E., Cohen, A., Gros, F., Luzzati, D., Charmot, D., and Drugeon, G. (1974). Expression of the myosin gene in a hybrid cell derived from a rat myoblast and a mouse fibroblast. *Biochimie* **56**, 1571–1573.

Buckley, P. A., and Konigsberg, I. R. (1974). Myogenic fusion and duration of the post-mitotic gap (G1). *Dev. Biol.* **37**, 193–212.

Bunn, C. L., Wallace, D. C., and Eisenstadt, J. M. (1974). Cytoplasmic inheritance of chloramphenicol resistance in mouse tissue culture cells. *Proc. Natl. Acad. Sci. U. S. A.* **71**, 1681–1685.

Burch, J. W., and McBride, O. W. (1975). Human gene expression in rodent cells after uptake of isolated metaphase chromosomes. *Proc. Natl. Acad. Sci. U. S. A.* **72**, 1797–1801.

Burger, M. M. (1969). A difference in the architecture of the surface membrane of normal and virally transformed cells. *Proc. Natl. Acad. Sci. U. S. A.* **62**, 994–1001.

Burger, M. M. (1973). Surface changes in transformed cells detected by lectins. *Fed. Proc., Fed. Am. Soc. Exp. Biol.* **32**, 91–101.

Burgerhout, W. (1975). Identification of interspecific translocation chromosomes in human–Chinese hamster hybrid cells. *Humangenetik* **29**, 229–231.

Burgerhout, W., van Someren, H., and Bootsma, D. (1973). Cytological mapping of the genes assigned to the human A1 chromosome by use of radiation-induced chromosome breakage in a human-Chinese hamster hybrid cell line. *Humangenetik* **20**, 159–162.

Burgos, M. H., and Fawcett, D. W. (1955). Studies on the fine structure of the mammalian testis. *J. Biophys. Biochem. Cytol.* **1**, 287–300.

Burns, E. R. (1971). Synchronous and asynchronous DNA synthesis in multinucleated Ehrlich ascites tumor cells compared with multinucleated cells cultured from frog lung. *Exp. Cell Res.* **66**, 152–156.

Cantell, K., Saksela, E., and Aula, P. (1966). Virological studies on chromosome damage of HeLa cells induced by myxoviruses. *Ann. Med. Exp. Biol. Fenn.* **44**, 225–229.

Capers, C. R. (1960). Multinucleation of skeletal muscle *in vitro*. *J. Biophys. Biochem. Cytol.* **7**, 559–566.

Carlson, P. S. (1973a). The use of protoplasts for genetic research. *Proc. Natl. Acad. Sci. U. S. A.* **70**, 598–602.

Carlson, P. S. (1973b). Somatic cell genetics of higher plants. *Symp. Soc. Dev. Biol.* **31**, 329–353.

Carlson, P. S., and Polacco, J. C. (1975). Plant cell cultures: Genetic aspects of crop improvement. *Science* **188**, 622–625.

Carlson, P. S., Smith, H. H., and Dearing, R. D. (1972). Parasexual interspecific plant hybridization. *Proc. Natl. Acad. Sci. U. S. A.* **69**, 2292–2294.

Carlsson, S.-A., Savage, R. E., and Ringertz, N. R. (1970). Behaviour of differentiated hen nuclei in the cytoplasm of rat myoblasts and myotubes. *Nature (London)* **228**, 869–871.

Carlsson, S.-A., Moore, G. P. M., and Ringertz, N. R. (1973). Nucleo-cytoplasmic protein migration during the activation of chick erythrocyte nuclei in heterokaryons. *Exp. Cell Res.* **76**, 234–241.

Carlsson, S.-A., Luger, O., Ringertz, N. R., and Savage, R. E. (1974a). Phenotypic expression in chick erythrocyte × rat myoblast hybrids and in chick myoblast × rat myoblast hybrids. *Exp. Cell Res.* **84**, 47–55.

Carlsson, S.-A., Ringertz, N. R., and Savage, R. E. (1974b). Intracellular antigen migration in interspecific myoblast heterokaryons. *Exp. Cell Res.* **84**, 255–266.

Carter, S. B. (1967). Effects of cytochalasins on mammalian cells. *Nature (London)* **213**, 261–266.

Carter, S. B. (1972). The cytochalasins as research tools in cytology. *Endeavour* **31**, 77–82.

Carvalho, A. P., Sanui, H., and Pace, N. (1963). Calcium and magnesium binding properties of cell membrane materials. *J. Cell. Comp. Physiol.* **62**, 311–317.

Carver, D., Seto, D., and Migeon, B. (1968). Interferon production and action in mouse hamster and somatic hybrid mouse–hamster cells. *Science* **160**, 558–559.

Cascardo, M. R., and Karzon, D. T. (1965). Measles virus giant cell inducing factor (fusion factor). *Virology* **26**, 311–325.

Caspersson, T., and Zech, L. (1973). Chromosome identification by fluorescence. *In* "Medical Genetics" (V. A. McKusick and R. Claiborne, eds.), pp. 27–38. H. P. Publ. Co., New York.

Caspersson, T., Farber, S., Foley, G. E., Kudynowski, J. Modest, E. J., Simonsson, E., Wagh, U., and Zech, L. (1968). Chemical differentiation along metaphase chromosomes. *Exp. Cell Res.* **49**, 219–222.

Caspersson, T., Zech, L., Johansson, C., and Modest, E. J. (1970). Identification of human chromosomes by DNA-binding fluorescent agent. *Chromosoma* **30**, 215–227.

Caspersson, T., Zech, L., Harris, H., Wiener, F., and Klein,G. (1971). Identification of human chromosomes in a mouse/human hybrid by fluorescence techniques. *Exp. Cell Res.* **65**, 475–478.

Cassingena, R., Chany, C., Vignal, M., Estrade, S., and Suarez, H.-G. (1970). Utilisation des cellules hybrides (Singe-Souris) pour l'étude de la régulation cellulaire de la production et de l'action de l'interférone. *C. R. Hebd. Seances Acad. Sci., Ser. D* **270**, 1189–1191.

Cassingena, R., Chany, C., Vignal, M., Suarez, H., Estrade, S., and Lazar, P. (1971). Use of monkey–mouse hybrid cells for the study of the cellular regulation of interferon production and action. *Proc. Natl. Acad. Sci. U. S. A.* **68**, 580–584.

Cassone, A., Cali'o, R., and Pesce, C. D. (1973). Interaction of Sendai virus with human erythrocytes. II. The fusion reactions. *Boll. Ist. Sieroter. Milan* **52**, 218–223.

Chalazonitis, A., Greene, L. A., and Shain, W. (1975). Excitability and chemosensitivity properties of a somatic cell hybrid between mouse neuroblastoma and sympathetic ganglion cells. *Exp. Cell Res.* **96**, 225–238.

Chaleff, R. S., and Carlson, P. S. (1974). Somatic cell genetics of higher plants. *Annu. Rev. Genet.* **8**, 267–278.

Chambers, R. (1938). The physical state of protoplasm with special reference to its surface. *Am. Natur.* **72**, 141–159.

Champe, P. C., Strohl, W. A., and Schlesinger, R. W. (1972). Demonstration of an adenovirus-inhibitory factor in adenovirus-induced hamster tumor cells. *Virology* **50**, 482–494.

Chan, T., Long, C., and Green, H. (1975). A human–mouse somatic hybrid line selected for human deoxycytidine deaminase. *Somatic Cell Genet.* **1**, 81–90.

Chan, V. L., Whitmore, G. F., and Siminovitch, L. (1972). Mammalian cells with altered forms of RNA polymerase II. *Proc. Natl. Acad. Sci. U. S. A.* **69**, 3119–3123.

Chen, T. R., and Ruddle, F. H. (1971). Karyotype analysis utilizing differentially stained constitutive heterochromatin of human and murine chromosomes. *Chromosoma* **34**, 51–72.

Cherny, A. P., Vasiliev, J. M., and Gelfand, I. M. (1975). Spreading of normal and transformed fibroblasts in dense cultures. *Exp. Cell Res.* **90**, 317–327.

Chu, E. H. Y., Brimer, P., Jacobson, K. B., and Merriam, E. V. (1969). Mammalian cell genetics. I. Selection and characterization of mutations auxotrophic for L glutamine or resistant to 8-azaguanine in Chinese hamster cells in vitro. *Genetics* **62**, 359–377.

Chu, E. H. Y., Sun, N. C., and Chang, C. C. (1972). Induction of auxotrophic mutations by treatment of Chinese hamster cells with 5-bromodeoxyuridine and black light. *Proc. Natl. Acad. Sci. U. S. A.* **69**, 3459–3463.

Chu, M. Y., and Fisher, G. A. (1965). Comparative studies of leukemic cells sensitive and resistant to cytosine arabinoside. *Biochem. Pharmacol.* **14**, 333–341.

Clarkson, B., and Baserga, R., ed. (1974). "Control of Cell Proliferation in Animal Cells." Cold Spring Harbor Lab, Cold Spring Harbor, New York.

Clayton, D. A., Teplitz, R. L., Nabholz, M., Dovey, H., and Bodmer, W. (1971). Mitochondrial DNA of human–mouse cell hybrids. *Nature (London)* **234**, 560–562.

Clements, G. B. (1972). Studies on biochemical variants of BHK 23/C13 cells. Ph.D. Thesis, University of Glasgow.

Clements, G. B. (1975). Selection of biochemically variant cells in culture. *Adv. Cancer Res.* **21**, 273–390.

Cochran, A. J., Wiener, F., Klein, G., and Harris, H. (1975). Genetic determinants of morphological differentiation in a lymphoma–sarcoma hybrid. *J. Pathol.* **115**, 1–12.

Cocking, E. C. (1970). Virus uptake, cell wall regeneration and virus multiplication in isolated plant protoplasts. *Int. Rev. Cytol.* **28**, 89–124.

Cocking, E. C. (1972). Plant cell protoplasts: Isolation and development. *Annu. Rev. Plant Physiol.* **23**, 29–50.

Codish, S. D., and Paul, B. (1974). Reversible appearance of a specific chromosome which suppresses malignancy. *Nature (London)* **252**, 610–612.

Coffino, P., Knowles, B., Nathenson, S. G., and Scharff, M. D. (1971). Suppression of immunoglobulin synthesis by cell hybridization. *Nature (London), New Biol.* **231**, 87–90.

Cold Spring Harbor Symposium on Tumor Viruses. (1975). *Cold Spring Harbor Symp. Quant. Biol.* **39**.

Collins, J. J., Destree, A. T., Marshall, C. J., and MacPherson, I. A. (1975). Selective depletion of chromosomes in a stable mouse–Chinese hamster hybrid cell line using antisera directed against species-specific cell surface antigens. *J. Cell Physiol.* **86**, 605–620.

Colten, H. R., and Parkman, R. (1972). Biosynthesis of C4 (Fourth component) of complement by hybrids of C4-deficient guinea pig cells and HeLa cells. *Science* **176**, 1029–1031.

Colwin, L. H., and Colwin, A. L. (1960). Formation of sperm entry holes in the vitelline membrane of *Hydroides hexagonus* (Annelida) and evidence of their lytic origin. *J. Biophys. Biochem. Cytol.* **7**, 315–320.

Compans, R. W., Holmes, K. V., Dales, S., and Choppin, P. W. (1964). An electron microscopic study of moderate and virulent virus–cell interactions of the parainfluenza virus SV5. *Virology* **30**, 411–426.

Constabel, F., and Kao, K. N. (1974). Agglutination and fusion of plant protoplasts by polyethylene glycol. *Can. J. Bot.* **52**, 1603–1606.

Constabel, F., Duditis, D., Gamborg, O. L., and Kao, K. N. (1975). Nuclear fusion in intergeneric heterokaryons. *Can. J. Bot.* **53**, 2092–2095.

Cook, G. M. W. (1968). Glycoproteins in membranes. *Biol. Rev. Cambridge Philos. Soc.* **43**, 363–391.

Cook, P. R. (1970). Species specificity of an enzyme determined by an erythrocyte nucleus in an interspecific hybrid cell. *J. Cell Sci.* **7**, 1–3.

Cook, P. R. (1975). Linkage of the loci for glucose-6-phosphate dehydrogenase and for inosinic acid pyrophosphorylase to the X chromosome of the field-vole *Microtus agrestis. J. Cell Sci.* **17**, 95–112.

Coon, H. G., and Weiss, M. C. (1969). A quantitative comparison of formation of spontaneous and virus-produced viable hybrids. *Proc. Natl. Acad. Sci. U. S. A.* **62,** 852–859.
Coon, H. G., Horak, I., and Dawid, I. B. (1973). Propagation of both parental mitochondrial DNAs in rat–human and mouse–human hybrid cells. *J. Mol. Biol.* **81,** 285–298.
Cotton, R. G. H., and Milstein, C. (1973). Fusion of two immunoglobulin producing myeloma cells. *Nature (London)* **244,** 42–43.
Cramp, F. C., and Lucy, J. A. (1974). Glycerol monooleate as a fusogen for the formation of heterokaryons and interspecific hybrid cells. *Exp. Cell Res.* **87,** 107–110.
Creagan, R. P., and Ruddle, F. H. (1975). The clone panel. A systematic approach to gene mapping using interspecific somatic cell hybrids, (*Rotterdam Conf., 1974*) *Birth Defects, Orig. Artic. Ser.* **11,** No. 3.
Croce, C. M., and Bakay, B. (1974). Presence of two active X chromosomes in hybrids between normal human and SV40-transformed fibroblasts from patients with Lesch–Nyhan syndrome. *Exp. Cell Res.* **87,** 422–425.
Croce, C. M., and Koprowski, H. (1973). Enucleation of cells made simple and rescue of SV40 by enucleated cells made even simpler. *Virology* **51,** 227–229.
Croce, C. M., and Koprowski, H. (1974a). Concordant segregation of the expression of SV40 T antigen and human chromosome 7 in mouse–human hybrid subclones. *J. Exp. Med.* **139,** 1350–1353.
Croce, C. M., and Koprowski, H. (1974b). Positive control of transformed phenotype in hybrids between SV40-transformed and normal human cells. *Science* **184,** 1288–1289.
Croce, C. M., and Koprowski, H. (1974c). Somatic cell hybrids between mouse peritoneal macrophages and SV40-transformed human cells. I. Positive control of the transformed phenotype by the human chromosome 7 carrying the SV40 genome. *J. Exp. Med.* **140,** 1221–1229.
Croce, C. M., and Koprowski, H. (1975). Assignment of gene(s) for transformation to human chromosome 7 carrying the Simian virus 40 genome. *Proc. Natl. Acad. Sci. U. S. A.* **72,** 1658–1660.
Croce, C. M., Sawicki, W., Kritchewsky, D., and Koprowski, H. (1971). Induction of homokaryocyte, heterokaryocyte and hybrid formation by lysolecithin. *Exp. Cell Res.* **67,** 427–435.
Croce, C. M., Koprowski, H., and Eagle, H. (1972). Effect of environmental pH on the efficiency of cellular hybridization. *Proc. Natl. Acad. Sci. U. S. A.* **69,** 1953–1956.
Croce, C. M., Litwack, G., and Koprowski, H. (1973a). Human regulatory gene for inducible tyrosine aminotransferase in rat–human hybrids. *Proc. Natl. Acad. Sci. U. S. A.* **70,** 1268–1272.
Croce, C. M., Bakay, B., Nyhan, W., and Koprowski, H. (1973b). Reexpression of the rat hypoxanthine phosphoribosyltransferase gene in rat-human hybrids. *Proc. Natl. Acad. Sci. U. S. A.* **70,** 2590–2594.
Croce, C. M., Girardi, A. J., and Koprowski, H. (1973c). Assignment of the T antigen gene of Simian virus 40 to human chromosome C-7. *Proc. Natl. Acad. Sci. U. S. A.* **70,** 3617–3620.
Croce, C. M., Kieba, I., and Koprowski, H. (1973d). Unidirectional loss of human chromosomes in rat–human hybrids. *Exp. Cell Res.* **79,** 461–463.
Croce, C. M., Knowles, B. B., and Koprowski, H. (1973e). Preferential retention of the human chromosome C-7 in human–(thymidine kinase deficient) mouse hybrid cells. *Exp. Cell Res.* **82,** 457–461.
Croce, C. M., Kieba, I., Koprowski, H., Molino, M., and Rothblat, G. H. (1974a). Restoration of the conversion of desmosterol to cholesterol in L cells after hybridization with human fibroblasts. *Proc. Natl. Acad. Sci. U. S. A.* **71,** 110–113.

Croce, C. M., Huebner, K., Girardi, A. J., and Koprowski, H. (1974c). Rescue of defective SV40 from mouse–human hybrid cells containing human chromosome 7. *Virology* **60,** 276–281.

Croce, C. M., Koprowski, H., and Litwack, G. (1974d). Regulation of the corticosteroid inducibility of tyrosine aminotransferase in interspecific hybrid cells. *Nature (London)* **249,** 839–841.

Croce, C. M., Huebner, K., and Koprowski, H. (1974e). Chromosome assignment of the T-antigen of Simian virus 40 in African green monkey cells transformed by adeno 7-SV40 hybrid. *Proc. Natl. Acad. Sci. U. S. A.* **71,** 4116–4119.

Croce, C. M., Litwack, G., and Koprowski, H. (1974f). Regulation of the inducibility of tyrosine aminotransferase in rat–human hybrids. *In* "Somatic Cell Hybrids" (R. L. Davidson and F. F. de la Cruz, eds.), pp. 173–176. Raven, New York.

Croce, C. M., Huebner, K., Girardi, A. J., and Koprowski, H. (1975a). Genetics of cell transformation by Simian virus 40. *Cold Spring Harbor Symp. Quant. Biol.* **39,** 335–343.

Croce, C. M., Aden, D., and Koprowski, H. (1975b). Somatic cell hybrids between mouse peritoneal macrophages and Simian virus 40-transformed human cells. II. Presence of human chromosome 7 carrying Simian virus 40 genome in cells of tumors induced by hybrid cells. *Proc. Natl. Acad. Sci. U. S. A.* **72,** 1397–1400.

Croce, C. M., Aden, D., and Koprowski, H. (1975c). Tumorigenicity of mouse–human diploid hybrids in nude mice. *Science* **190,** 1200–1202.

Cronemeyer, R. L., Thuillez, P. E., Shows, T. B., and Morrow, J. (1974). 6-hydroxydopamine sensitivity in mouse neuroblastoma and neuroblastoma × L cell hybrids. *Cancer Res.* **34,** 1652–1657.

Dales, S. (1973). Early events in cell–animal virus interactions. *Bacteriol. Rev.* **37,** 103–135.

Daniels, M. P., and Hamprecht, B. (1974). The ultrastructure of neuroblastoma–glioma hybrids. Expression of neuronal characteristics stimulated by dibutyryl adenosine 3' 5' cyclic monophosphate. *J. Cell Biol.* **63,** 691–699.

Darlington, G. J., Bernhard, H. P., and Ruddle, F. H. (1974a). The expression of hepatic functions in mouse hepatoma × human leucocyte hybrids. *Cytogenetics* **13,** 86–88.

Darlington, G. J. Bernhard, H. P., and Ruddle, F. H. (1974b). Human serum albumin phenotype activation in mouse hepatoma–human leukocyte cell hybrids. *Science* **185,** 859–862.

Darzynkiewicz, Z. (1971). Radiation-induced DNA synthesis in nuclei of hen erythrocytes reactivated in heterokaryons. *Exp. Cell Res.* **69,** 472–481.

Darzynkiewicz, Z., and Chelmicka-Szorc, E. (1972). Unscheduled DNA synthesis in hen erythrocyte nuclei reactivated in heterokaryons. *Exp. Cell Res.* **74,** 131–139.

Darzynkiewicz, Z., Chelmicka-Szorc, E., and Arnason, B. G. W. (1974a). Chick erythrocyte nucleus reactivation in heterokaryons: Suppression by inhibitors of proteolytic enzymes. *Proc. Natl. Acad. Sci. U. S. A.* **71,** 644–647.

Darzynkiewicz, Z., Chelmicka-Szorc, E., and Arnason, B. G. W. (1974b). Suppressive effect of protease inhibitors on heterokaryons containing chick erythrocyte nuclei. *Exp. Cell Res.* **87,** 333–345.

Das, N. K. (1962). Synthetic capacities of chromosome fragments correlated with their ability to maintain nucleolar material. *J. Cell Biol.* **15,** 121–130.

Davidson, R. L. (1969). Regulation of melanin synthesis in mammalian cells, as studied by somatic hybridization. III. A method for increasing the frequency of cell fusion. *Exp. Cell Res.* **55,** 424–426.

Davidson, R. L. (1972). Regulation of melanin synthesis in mammalian cells: Effect of gene dosage on the expression of differentiation. *Proc. Natl. Acad. Sci. U. S. A.* **69,** 951–955.

Davidson, R. L. (1974). Gene expression in somatic cell hybrids. *Annu. Rev. Genet.* **8**, 195–218.

Davidson, R. L., and Benda, P. (1970). Regulation of specific functions of glial cells in somatic hybrids. II. Control of inducibility of glycerol-3-phosphate dehydrogenase. *Proc. Natl. Acad. Sci. U. S. A.* **67**, 1870–1877.

Davidson, R. L., and Ephrussi, B. (1965). A selective system for the isolation of hybrids between L cells and normal cells. *Nature (London)* **205**, 1170–1171.

Davidson, R. L., and Ephrussi, B. (1970). Factors influencing the "effective mating rate" of mammalian cells. *Exp. Cell Res.* **61**, 222–226.

Davidson, R. L., and Yamamoto, K. (1968). Regulation of melanin synthesis in mammalian cells, as studied by somatic hybridization. II. The level of regulation of 3,4-dehydroxyphenylalanine oxidase. *Proc. Natl. Acad. Sci. U.S.A.* **60**, 894–901.

Davidson, R. L., Ephrussi, B., and Yamamoto, K. (1966). Regulation of pigment synthesis in mammalian cells as studied by somatic hybridization. *Proc. Natl. Acad. Sci. U. S. A.* **56**, 1437–1440.

Davidson, R., Ephrussi, B., and Yamamoto, K. (1968). Regulation of melanin synthesis in mammalian cells, as studied by somatic hybridization. I. Evidence for negative control. *J. Cell Physiol.* **72**, 115–127.

Davis, T. J., and Harris, H. (1975). Haemoglobin synthesis in fused cells. *J. Cell Sci.* **18**, 207–216.

Deák, I., Sidebottom, E., and Harris, H. (1972). Further experiments on the role of the nucleolus in the expression of structural genes. *J. Cell Sci.* **11**, 379–391.

DeChamps, M., de Saint-Vincent, B., Evrard, C., Sassi, M., and Buttin, G. (1974). Studies on 1β-D-arabinofuranosyl-cytosine (Ara-c) resistant mutants of Chinese hamster fibroblasts. II. High resistance to Ara-C as a genetic marker for cellular hybridization. *Exp. Cell Res.* **86**, 269–279.

Defendi, V., Ephrussi, B., and Koprowski, H., (1964). Expression of polyoma-induced cellular antigen in hybrid cells. *Nature (London)* **203**, 495–496.

Defendi, V., Ephrussi, B., Koprowski, H., and Yoshida, M. C. (1967). Properties of hybrids between polyoma-transformed and normal mouse cells. *Proc. Natl. Acad. Sci. U. S. A.* **57**, 299–305.

Defendi, V., Jensen, F., Frey, H., and Ferry, P. (1968). Attempt to rescue infectious polyoma virus from polyoma transformed cells. *Fed. Proc., Fed. Am. Soc. Exp. Biol.* **27**, 262.

Deisserroth, A., Burk, R., Picciano, D., Minna, J., Anderson, W. F., and Nienhuis, A. (1975a). Hemoglobin synthesis in somatic cell hybrids: globin gene expression in hybrids between mouse erythroleukemia and human marrow cells or fibroblasts. *Proc. Natl. Acad. Sci. U.S.A.* **72**, 1102–1106.

Deisserroth, A., Barker, J., Anderson, W. F., and Nienhuis, A. (1975b). Hemoglobin synthesis in somatic cell hybrids: coexpression of mouse with human or Chinese hamster globin genes in interspecific somatic cell hybrids of mouse erythroleukemia cells. *Proc. Natl. Acad. Sci. U.S.A.* **72**, 2682–2686

DelVillano, B. C., and Defendi, V. (1973). Characterization of the SV40 T antigen. *Virology* **51**, 34–46.

De Mars, R., and Hooper, J. L. (1960). A method selecting for auxotrophic mutants of HeLa cells. *J. Exp. Med.* **111**, 559–573.

Dendy, P. R., and Harris, H. (1973). Sensitivity to diphteria toxin as a species-specific marker in hybrid cells. *J. Cell Sci.* **12**, 831–837.

de Saint-Vincent, R., and Buttin, G. (1973). Studies on 1-β-arabino furanosyl-cytosine-resistant mutants of Chinese–hamster fibroblasts. *Eur. J. Biochem.* **37**, 481–488.

Deschatrette, J., and Weiss, M. C. (1975). Extinction of liver-specific functions in hybrids between differentiated and dedifferentiated rat hepatoma cells. *Somatic Cell Genet.* **1**, 279–292.

de Weerd-Kastelein, E. A., Keijzer, W., and Bootsma, D. (1972). Genetic heterogeneity of *Xeroderma pigmentosum* demonstrated by somatic cell hybridization. *Nature (London), New Biol.* **238**, 80–83.

de Weerd-Kastelein, E. A., Kleijer, W. J., Sluyter, M. L., Keijzer, W. (1973). Repair replication in heterokaryons derived from different repair-deficient *Xeroderma pigmentosum* strains. *Mutat. Res.* **19**, 237–243.

de Weerd-Kastelein, E. A., Keijzer, W., and Bootsma, D. (1974). A third complementation group in *Xeroderma pigmentosum. Mutat. Res.* **22**, 87–91.

Deys, B. F. (1972). Demonstration of X-linkage of *G6PD, HGPRT* and *PGK* in the horse by means of mule–mouse somatic cell hybridization. Ph.D. Thesis, Leiden University, Leiden.

Diacumakos, E. G. (1973). Microsurgically fused human somatic cell hybrids: Analysis and cloning. *Proc. Natl. Acad. Sci. U.S.A.* **70**, 3382–3386.

Diacumakos, E. G., (1975). Methods for microsurgical production of mammalian somatic cell hybrids and their analysis and cloning. *Methods Cell Biol.* **10**, 147–156.

Diacumakos, E. G., and Tatum, E. L. (1972). Fusion of mammalian somatic cells by microsurgery. *Proc. Natl. Acad. Sci. U.S.A.* **69**, 2959–2962.

Dingle, J. T. (1968). Vacuoles, vesicles and lysosomes. *Br. Med. Bull.* **24**, 141–145.

Dingle, J. T. (1969). The extracellular secretion of lysosomal enzymes. *In* "Lysosomes in Biology and Pathology" (J. T. Dingle and H. B. Fell, eds.) Vol. 2, p. 421. North-Holland Publ., Amsterdam.

Djordjevic, B., and Szybalski, W. (1960). Genetics of human cell lines. *J. Exp. Med.* **112**, 509–531.

Domnina, L. V., Ivanova, O. Y., Margolis, L. B., Olshevskaya, L. V., Rovensky, Y. A., Vasiliev, J. M., and Gelfand, I. M. (1972). Defective formation of the lamellar cytoplasm by neoplastic fibroblasts. *Proc. Natl. Acad. Sci. U.S.A.* **69**, 248–252.

Donner, L., Sainerová, H., Svoboda, J., and Scherneck, S. (1974). Potentiation of RSV rescue from RSV-transformed "poorly" virogenic cell lines by 5-bromodeoxyuridine treatment before fusion with chick embryo fibroblasts. *Int. J. Cancer* **13**, 37–42.

Douglas, G. R., Gee, P. A., and Hamerton, J. L. (1973a). Chromosome identification in Chinese hamster/human somatic cell hybrids. *In* "Chromosome Identification" (T. Caspersson and L. Zech, eds.), pp. 170–176. Academic Press, New York.

Douglas, G. R., McAlpine, P. J., and Hamerton, J. L. (1973b). Regional localization of loci for human *PGM* and *6PGD* on human chromosome one by use of hybrids of Chinese hamster–human somatic cells. *Proc. Natl. Acad. Sci. U.S.A.* **70**, 2737–2740.

Drets, M. E., and Shaw, M. W. (1971). Specific banding patterns of human chromosomes. *Proc. Natl. Acad. Sci. U.S.A.* **68**, 2073–2077.

Dubbs, D. R., and Kit, S. (1964). Effect of halogenated pyrimidines and thymidine on growth of L cells and a subline lacking thymidine kinase. *Exp. Cell Res.* **33**, 19–28.

Dubbs, D. R., and Kit, S. (1968). Isolation of defective lysogens from simian virus 40-transformed mouse kidney cultures. *J. Virol.* **2**, 1272–1282.

Dubbs, D. R., and Kit, S. (1976). Reactivation of chick erythrocyte nuclei in heterokaryons with temperature-sensitive Chinese hamster cells. *Somatic Cell Genet.* **2**, 11–19.

Dulbecco, R. (1969). Cell transformation by viruses. *Science* **166**, 962–968.

Dulbecco, R. (1970a). Topoinhibition and serum requirement of transformed and untransformed cells. *Nature (London)* **227**, 802–806.

Dulbecco, R. (1970b). Behaviour of tissue culture cells infected with polyoma virus. *Proc. Natl. Acad. Sci. U.S.A.* **67**, 1214–1220.

Dupuy-Coin, A. M., Bouteille, M., Ege, T., and Ringertz, N. R. (1976). Ultrastructure of chick erythrocyte nuclei undergoing reactivation in heterokaryons and enucleated cells. *Exp. Cell Res.* **101**, 355–370.

Earle, W. R. (1943). Production of malignancy *in vitro*. IV. The mouse fibroblast cultures and changes seen in living cells. *J. Natl. Cancer Inst.* **4**, 165–212.

Eckhart, W. (1974). Genetics of DNA tumor viruses. *Annu. Rev. Genet.* **8**, 301–317.

Edelman, G. M., Yahara, I., and Wang, J. L. (1973). Receptor mobility and receptor–cytoplasmic interaction in lymphocytes. *Proc. Natl. Acad. Sci. U.S.A.* **70**, 1442–1446.

Edidin, M., and Weiss, A. (1972). Antigen cap formation in cultured fibroblasts: A reflection of membrane fluidity and of cell motility. *Proc. Natl. Acad. Sci. U.S.A.* **69**, 2456–2459.

Ege, T., and Ringertz, N. R. (1974). Preparation of microcells by enucleation of micronucleate cells. *Exp. Cell Res.* **87**, 378–382.

Ege, T., and Ringertz, N. R. (1975). Viability of cells reconstituted by virus-induced fusion of minicells with anucleate cells. *Exp. Cell Res.* **94**, 469–473.

Ege, T., Carlsson, S-. A., and Ringertz, N. R. (1971). Immune microfluorimetric analysis of the distribution of species specific nuclear antigens in HeLa–chick erythrocyte heterokaryons. *Exp. Cell Res.* **69**, 472–477.

Ege, T., Zeuthen, J., and Ringertz, N. R. (1973). Cell fusion with enucleated cytoplasms. *In* "Chromosome Identification" (T. Caspersson and L. Zech, eds.), pp. 189–194. Academic Press, New York.

Ege, T., Hamberg, H., Krondahl, U., Ericsson, J., and Ringertz, N. R. (1974a). Characterization of minicells (nuclei) obtained by cytochalasin enucleation. *Exp. Cell Res.* **87**, 365–377.

Ege, T., Krondahl, U., and Ringertz, N. R. (1974b). Introduction of nuclei and micronuclei into cells and enucleated cytoplasms by Sendai virus induced fusion. *Exp. Cell Res.* **88**, 428–432.

Ege, T., Zeuthen, J., and Ringertz, N. R. (1975). Reactivation of chick erythrocyte nuclei after fusion with enucleated cells. *Somatic Cell Genet.* **1**, 65–80.

Ege, T., Sidebottom, E., and Ringertz, N. R. (1976). Preparation of microcells. *Methods in Cell Biol.* **15** (in press).

Elgin, S. C., and Bonner, J. (1970). Limited heterogeneity of the major nonhistone chromosomal proteins. *Biochemistry* **9**, 4440–4447.

Eliceiri, G. L. (1972). The ribosomal RNA of hamster–mouse hybrid cells. *J. Cell Biol.* **53**, 177–184.

Eliceiri, G. L. (1973a). The mitochondrial DNA of hamster–mouse hybrid cells. *FEBS Lett.* **36**, 232–234.

Eliceiri, G. L. (1973b). Cytoplasmic ribosomal RNA of hamster–mouse hybrid cells. *Biochim. Biophys. Acta* **312**, 737–741.

Eliceiri, G. L. (1973c). Synthesis of mitochondrial RNA in hamster–mouse hybrid cells. *Nature (London), New Biol.* **241**, 233–234.

Eliceiri, G. L., and Green, H. (1969). Ribosomal RNA synthesis in human–mouse hybrid cells. *J. Mol. Biol.* **41**, 253–260.

Elsevier, S. M., Kucherlapati, R. S., Nichols, E. A., Creagan, R. P., Giles, R. E., Ruddle, F. H., Willecke, K., and McDougall, J. K. (1974). Assignment of the gene for galactokinase to human chromosome 17 and its regional localization to band q 21–22. *Nature (London)* **251**, 633–636.

Enders, A. C., and Schlafke, S. (1971). Penetration of the uterine epithelium during implantation in the rabbit. *Am. J. Anat.* **132**, 219–240.

Enders, J. F., and Peebles, T. C. (1954). Propagation in tissue cultures of cytopathogenic agents from patients with measles. *Proc. Soc. Exp. Biol. Med.* **86**, 277–286.

Enders, J. F., Holloway, A., and Grogan, E. A. (1967). Replication of poliovirus I in chick embryo and hamster cells exposed to Sendai virus. *Proc. Natl. Acad. Sci. U.S.A.* **57**, 637–644.

Engel, E., McGee, B. J., and Harris, H. (1969a). Cytogenetic and nuclear studies on A9 and B82 cells fused together by Sendai virus: The early phase. *J. Cell Sci.* **5**, 93–120.

Engel, E., McGee, B. J., and Harris, H. (1969b). Recombination and segregation in somatic cell hybrids. *Nature (London)* **223**, 152–155.

Engel, E., Empson, J., and Harris, H. (1971). Isolation and karyotypic characterization of segregants of intraspecific hybrid somatic cells. *Exp. Cell Res.* **68**, 231–234.

Ephrussi, B. (1972). "Hybridization of Somatic Cells." Princeton Univ. Press, Princeton, New Jersey.

Ephrussi, B., and Weiss, M. C. (1965). Interspecific hybridization of somatic cells. *Proc. Natl. Acad. Sci. U.S.A.* **53**, 1040–1042.

Ericsson, R. J., Buthala, D. A., and Norland, J. F. (1971). Fertilization of rabbit ova *in vitro* by sperm with adsorbed Sendai virus. *Science* **173**, 54–55.

Fell, H. B., and Hughes, A. F. (1949). Mitosis in the mouse: A study of living and fixed cells in tissue culture. *Quart. J. Microsc. Sci.* [ns] **90**, 355–380.

Fenner, F., McAuslan, B. R., Mims, C. A., Sambrook, J., and White, D. O. (1974). "The Biology of Animal Viruses." Academic Press, New York.

Fenyö, E. M., Grundner, G., Klein, G., Klein, E., and Harris, H. (1971). Surface antigens and release of virus in hybrid cells produced by the fusion of A9 fibroblasts with Moloney lymphoma cells. *Exp. Cell Res.* **68**, 323–331.

Fenyö, E. M., Grundner, G., Wiener, F., Klein, E., Klein, G., and Harris, H. (1973a). The influence of the partner cell on the production of L virus and the expression of viral surface antigen in hybrid cells. *J. Exp. Med.* **137**, 1240–1255.

Fenyö, E. M., Wiener, F., Klein, G., and Harris, H. (1973b). Selection of tumor–host cell hybrids from polyoma virus- and methylcholanthrene- induced sarcomas. *J. Natl. Cancer Inst.* **51**, 1865–1875.

Finch, B. W., and Ephrussi, B. (1967). Retention of multiple developmental potentialities by cells of a mouse testicular teratocarcinoma during prolonged culture *in vitro* and their extinction upon hybridization with cells of permanent lines. *Proc. Natl. Acad. Sci. U.S.A.* **57**, 615–621.

Fogel, M., and Defendi, V. (1968). Internuclear transfer of SV40 T-antigen in absence of infectious virus. *Virology* **34**, 370–373.

Fogel, M., and Sachs, L. (1969). The activation of virus synthesis in polyoma-transformed cells. *Virology* **37**, 327–334.

Foster, D. O., and Pardee, A. B. (1969). Transport of amino acids by confluent and non-confluent 3T3 and polyoma virus-transformed 3T3 cells growing on glass cover slips. *J. Biol. Chem.* **244**, 2675–2681.

Fougère, C., Ruiz, F., and Ephrussi, B. (1972). Gene dosage dependence of pigment synthesis in melanoma × fibroblast hybrids. *Proc. Natl. Acad. Sci. U.S.A.* **69**, 330–334.

Fowke, L. C., Bech-Hansen, C. W., Gamborg, O. L., and Constabel, F. (1975a). Electron microscope observations of mitosis and cytokinesis in multinucleate protoplasts of soybean. *J. Cell Sci.* **18**, 491–507.

Fowke, L. C., Rennie, P. J., Kirkpatrick, J. W., and Constabel, F. (1975b). Ultrastructural characteristics of intergeneric protoplast fusion. *Can. J. Bot.* **53**, 272–278.

Francke, U., Hammond, D. S., and Schneider, J. A. (1973). The band patterns of twelve D 98-AH-2 marker chromosomes and their use for identification of intraspecific cell hybrids. *Chromosoma* **41**, 111–121.

Freed, J. J., and Mezger-Freed, L. (1970). Stable haploid cultured cell lines from frog embryos. *Proc. Natl. Acad. Sci. U.S.A.* **65**, 337–344.

Freedman, V. H., and Shin, S. (1974). Cellular tumorigenicity in nude mice: Correlation with cell growth in semi-solid medium. *Cell* **3**, 355–359.

Friend, C., Scher, W., Holland, J. G., and Sato, T. (1971). Hemoglobin synthesis in murine virus-induced leukemic cells *in vitro*: Stimulation of erythroid differentiation by dimethyl sulfoxide. *Proc. Natl. Acad. Sci. U.S.A.* **68**, 378–382.

Friend, K. K., Dorman, B. P., Kucherlapati, R. S., and Ruddle, F. H. (1976). Detection of interspecific translocations in mouse–human hybrids by alkaline Giemsa staining. *Exp. Cell Res.* **99**, 31–36.

Frye, L. D., and Edidin, M. (1970). The rapid intermixing of cell surface antigens after formation of mouse–human heterokaryons. *J. Cell Sci.* **7**, 319–335.

Furusawa, M., Nishimura, T., Yamaizumi, M., and Okada, Y. (1974). Injection of foreign substances into single cells by cell fusion. *Nature, (London)* **249**, 449–450.

Fujisawa, T., and Eliceiri, G. L. (1975). Ribosomal proteins of hamster, mouse and hybrid cells. *Biochim. Biophys. Acta* **402**, 238–243.

Gabara, B., Gledhill, B. L., Croce, C. M., Cesarini, J. P., and Koprowski, H. (1973). Ultrastructure of rabbit spermatozoa after treatment with lysolecithin and in the presence of hamster somatic cells. *Proc. Soc. Exp. Biol. Med.* **143**, 1120–1124.

Gahmberg, C. G., and Hakomori, S.-I. (1973). Altered growth behavior of malignant cells associated with changes in externally labeled glycoprotein and glycolipid. *Proc. Natl. Acad. Sci. U.S.A.* **70**, 3329–3333.

Galjaard, H., Hoogeveen, A., de Wit-Verbeek, H. A., Reuser, A. J. J., Keijzer, W., Westerveld, A., and Bootsma, D. (1974a). Tay-Sachs and Sandhoffs disease: Intergenic complementation after somatic cell hybridization. *Exp. Cell Res.* **87**, 444–448.

Galjaard, H., Hoogeveen, A., Keijzer, W., de Wit-Verbeek, E., and Vlek-Noot, C. (1974b). The use of quantitative cytochemical analyses in rapid prenatal detection and somatic cell genetic studies of metabolic diseases. *Histochem. J.* **6**, 491–509.

Galjaard, H., Hoogeveen, A., Keijzer, W. de Wit-Verbeek, H. A., Reuser, A. J. J., Ho, M. W., and Robinson, D. (1975). Genetic heterogeneity in GMl-gangliosidosis. *Nature (London)* **251**, 60–62.

Gallimore, P. H., Sharp, P. A., and Sambrook, J. (1974). Viral DNA in transformed cells. II. A study of the sequences of adenovirus 2 DNA in nine lines of transformed rat cells using specific fragments of the viral genome. *J. Mol. Biol.* **89**, 49–72.

Ganshow, R. (1966). Glucuronidase gene expression in somatic hybrids. *Science* **153**, 84–85.

Gazdar, A. F., Russell, E. K., and Minna, J. D. (1974). Replication of mouse–tropic and xenotropic strains of murine leukemia virus in human × mouse hybrid cells. *Proc. Natl. Acad. Sci. U.S.A.* **71**, 2642–2645.

Gehring, U., Mohit, B., and Tomkins, G. M., (1972). Glucocorticoid action on hybrid clones derived from cultured myeloma and lymphoma cell lines. *Proc. Natl. Acad. Sci. U.S.A.* **69**, 3124–3127.

Gelb, L. D., Kohne, D. E., and Martin, M. A. (1971). Quantitation of simian virus 40 sequences in African green monkey, mouse and virus-transformed cell genomes. *J. Mol. Biol.* **57**, 129–145.

Gerber, P. (1966). Studies on the transfer of subviral infectivity from SV40-induced hamster tumor cells to indicator cells. *Virology* **28**, 501–509.

Gerber, P., and Kirschstein, R. L. (1962). SV40 induced ependymomas in newborn hamsters. I. Virus–tumor relationships. *Virology* **18**, 582–588.

Gershon, D., and Sachs, L. (1963). Properties of a somatic hybrid between mouse cells with different genotypes. *Nature (London)* **198**, 912–913.

Giannelli, F., and Croll, P. (1971). Complementation *in vitro* between fibroblasts from normal subjects and patients with *Xeroderma pigmentosum*. *Clin. Sci.* **40**, 27p.

Giannelli, F., and Pawsey, S. A. (1974). DNA repair synthesis in human heterokaryons. II. A test for heterozygosity in *Xeroderma pigmentosum* and some insight into the structure of the defective enzyme. *J. Cell Sci.* **15**, 163–176.

Giannelli, F., Croll, P. M., and Lewin, S. A. (1973). DNA repair synthesis in human heterokaryons formed by normal and UV-sensitive fibroblasts. *Exp. Cell Res.* **78**, 175–185.

Gibbons, I. R. (1968). The biochemistry of motility. *Annu. Rev. Biochem.* **37**, 521–546.

Gilden, R. V., Carp, R. I., Taguch, F., and Defendi, V. (1965). The nature and localization of the SV40-induced complement-fixing antigen. *Proc. Natl. Acad. Sci. U.S.A.* **53**, 684–692.

Giles, R. E., and Ruddle, F. H. (1973a). Production of Sendai virus for cell fusion. *In Vitro* **9**, 103–107.

Giles, R. E., and Ruddle, F. H. (1973b). Production and characterization of proliferating somatic cell hybrids. *In* "Tissue Culture: Methods and Applications" (P. F. Kruse, Jr. and M. K. Patterson, Jr., eds.), pp. 475–500. Academic Press, New York.

Gillin, F. D., Roufa, D. J., Beaudet, A. L., and Caskey, C. T. (1972). 8-Azaguanin resistance in mammalian cells. *Genetics* **72**, 239–252.

Glaser, R., and Farrugia, R. (1973). Observations on the rescue of simian virus 40 induced by cell fusion from heterokaryon cultures using electron autoradiography. *Intervirology* **1**, 135–140.

Glaser, R., and O'Neill, F. J. (1972). Hybridization of Burkitt lymphoblastoid cells. *Science* **176**, 1245–1247.

Glaser, R., Nonoyama, M., Decker, B., and Rapp, F. (1973a). Synthesis of Epstein–Barr virus antigens and DNA in activated Burkitt somatic cell hybrids. *Virology* **55**, 62–69.

Glaser, R., Decker, B., Farrugia, R., Shows, T., and Rapp, F. (1973b). Growth characteristics of Burkitt somatic cell hybrids *in vitro*. *Cancer Res.* **33**, 2026–2029.

Gledhill, B. L., Sawicki, W., Croce, C. M., and Koprowski, H. (1972). DNA synthesis in rabbit spermatozoa after treatment with lysolecithin and fusion with somatic cells. *Exp. Cell Res.* **73**, 33–40.

Goldenberg, D. M., Bhan, R. D., and Pavia, R. A. (1971). *In vivo* human–hamster somatic cell fusion indicated by glucose-6-phosphate dehydrogenase and lactate dehydrogenase profiles. *Cancer Res.* **31**, 1148–1152.

Goldenberg, D. M., Pavia, R. A., and Tsao, M. C. (1974). *In vivo* hybridization of human tumor and normal hamster cells. *Nature (London)* **250**, 649–651.

Goldman, R. D., and Pollack, R. (1974). Use of enucleated cells. *Methods Cell Biol.* **8**, 123–143.

Goldman, R. D., Pollack, R., and Hopkins, N. (1973). Preservation of normal behaviour by enucleated cells in culture. *Proc. Natl. Acad. Sci. U.S.A.* **70**, 750–754.

Goldman, R. D., Pollack, R., Chang, C. M., and Bushnell, A. (1975). Properties of enucleated cells. III. Changes in cytoplasmic architecture of enucleated BHK 21 cells following trypsinization and replating. *Exp. Cell Res.* **93**, 175–183.

Goldstein, L. (1972). Nucleocytoplasmic interactions in amoebae. *In* "The Biology of Amoeba" (K. W. Jeon, ed.), pp. 479–504. Academic Press, New York.

Goldstein, L. (1974). Movement of molecules between nucleus and cytoplasm. *In* "The Cell Nucleus" (H. Busch, ed.), Vol. I, pp. 387–438. Academic Press, New York.

Goldstein, L., and Prescott, D. M. (1967). Protein interactions between nucleus and cytoplasm. *In* "The Control of Nuclear Activity" (L. Goldstein, ed.), pp. 3–17. Prentice-Hall, Englewood Cliffs, New Jersey.

Goldstein, L., and Prescott, D. M. (1968). Proteins in nucleocytoplasmic interactions. II.

Turnover and changes in nuclear protein distribution with time and growth. *J. Cell Biol.* **36,** 53–61.

Goldstein, L., and Ron, A. (1969). On the possibility of nuclear protein involvement in the control of DNA synthesis in *Amoeba proteus. Exp. Cell Res.* **55,** 144–146.

Goldstein, S., and Lin, C. C. (1972a). Rescue of senescent human fibroblasts by hybridization with hamster cells *in vitro. Exp. Cell Res.* **70,** 436–439.

Goldstein, S., and Lin, C. C. (1972b). Somatic cell hybrids between cultured fibroblasts from the Galapagos tortoise and the golden hamster. *Exp. Cell Res.* **73,** 266–269.

Good, N. E., Winget, G. D., Winter, W., Connolly, T. N., Izawa, S., and Singh, R. M. M. (1966). Hydrogen ion buffers for biological research. *Biochem.* **5,** 467–477.

Gopalakrishnan, T. V., and Thompson, E. B. (1975). A method for enucleating cultured mammalian cells. *Exp. Cell Res.* **96,** 435–439.

Gordon, S. (1973). Regulation of differentiated phenotype in heterokaryons. *In* "Genetic Mechanisms in Development" (F. H. Ruddle, ed.), pp. 269–293. Academic Press, New York.

Gordon, S., and Cohn, Z. (1970). Macrophage–melanocyte heterokaryons. I. Preparation and properties. *J. Exp. Med.* **131,** 981–1003.

Gordon, S., and Cohn, Z. (1971a). Macrophage–melanocyte heterokaryons. II. The activation of macrophage DNA synthesis. Studies with inhibitors of RNA synthesis. *J. Exp. Med.* **133,** 321–338.

Gordon, S., and Cohn, Z. (1971b). Macrophage–melanoma cell heterokaryons. III. The activation of macrophage DNA synthesis. Studies with inhibitors of protein synthesis and with synchronized melanoma cells. *J. Exp. Med.* **134,** 935–946.

Gordon, S., and Cohn, Z. (1971c). Macrophage–melanoma cell heterokaryons. IV. Unmasking the macrophage-specific membrane receptor. *J. Exp. Med.* **134,** 947–962.

Gordon, S., Ripps, C., and Cohn, Z. (1971). The preparation and properties of macrophage L cell hybrids. *J. Exp. Med.* **134,** 1187–1200.

Goss, S. J., and Harris, H. (1975). New method for mapping genes in human chromosomes. *Nature (London)* **255,** 680–684.

Goto, S., and Ringertz, N. R. (1974). Preparation and characterization of chick erythrocyte nuclei from heterokaryons. *Exp. Cell Res.* **85,** 173–181.

Graham, C. F. (1969). The fusion of cells with one- and two-cell mouse embryos. *Wistar Inst. Symp. Monogr.* **9,** 19–33.

Graham, C. F. (1971). Virus assisted fusion of embryonic cells. *Acta Endocrinol. (Copenhagen), Suppl.* **153,** 154–167.

Graham, C. F. (1974). Fusion of mouse blastomeres. *In* "Somatic Cell Hybridization" (R. L. Davidson and F. F. de la Cruz, eds.), pp. 277–281. Raven, New York.

Graham, C. F., Arms, K., and Gurdon, J. B. (1966). The induction of DNA synthesis by frog egg cytoplasm. *Dev. Biol.* **14,** 349–381.

Graham, F. L., van der Eb, A. J., and Heijneker, H. L. (1974). Size and location of the transforming region in human adenovirus type 5 DNA. *Nature (London)* **251,** 687–691.

Graham, F. L., Abrahams, P. J., Mulder, C., Heijneker, H. L., Warnaar, S. O., de Vries, F. A. J., Fiers, W., and van der Eb, A. J. (1975). Studies on *in vitro* transformation by DNA and DNA fragments of human adenoviruses and simian virus 40. *Cold Spring Harbor Symp. Quant. Biol.* **39,** 637–650.

Graham, J. M., and Wallach, D. F. H. (1971). Protein conformational transitions in the erythrocyte membrane. *Biochim. Biophys Acta* **241,** 180–194.

Gravel, R. A., Mahoney, M. J., Ruddle, F. H., and Rosenberg, L. E. (1975). Genetic complementation in heterokaryons of human fibroblasts defective in cobalamin metabolism. *Proc. Natl. Acad. Sci. U.S.A.* **72,** 3181–3185.

Graves, J. A. M. (1972a). DNA synthesis in heterokaryons formed by fusion of mammalian cells from different species. *Exp. Cell Res.* **72**, 393–403.

Graves, J. A. M. (1972b). Cell cycles and chromosome replication patterns in interspecific somatic hybrids. *Exp. Cell Res.* **73**, 81–94.

Green, E. L., ed. (1966). "Biology of the Laboratory Mouse." McGraw-Hill, New York.

Green, H., Ephrussi, B., Yoshida, M., and Hamerman, D. (1966). Synthesis of collagen and hyaluronic acid by fibroblast hybrids. *Proc. Natl. Acad. Sci. U.S.A.* **55**, 41–44.

Green, H., Wang, R., Kehinde, O., and Meuth, M. (1971). Multiple human TK chromosomes in human–mouse somatic cell hybrids. *Nature (London), New Biol.* **234**, 138–140.

Green, M. (1970). Oncogenic viruses. *Annu. Rev. Biochem.* **39**, 701–756.

Greene, L. A., Shain, W., Breakfield, X. O., and Chalazonitis, A. (1974). Sympathetic ganglion cell × neuroblastoma somatic cell hybrids. *J. Cell Biol.* **63**, 122*a*.

Greene, L. A., Shain, W., Chalazonitis, A., Breakfield, X., Minna, J., Coon, H. G., and Nirenberg, M. (1975). Neuronal properties of hybrid neuroblastoma × sympathetic ganglion cells. *Proc. Natl. Acad. Sci. U.S.A.* **72**, 4923–4927.

Grundner, G., Fenyö, E. M., Klein, G., Klein, E., Bregula, U., and Harris, H. (1971). Surface antigen expression in malignant sublines derived from hybrid cells of low malignancy. *Exp. Cell Res.* **68**, 315–322.

Grzeschik, K. H. (1973a). Utilization of somatic cell hybrids for genetic studies in man. *Hum. Genet.* **19**, 1–40.

Grzeschik, K. H. (1973b). Syrian hamster–human somatic cell hybrids: Isolation and characterization. *Hum. Genet.* **20**, 211–218.

Grzeschik, K. H., Allderdice, P. W., Grzeschik, A., Opitz, J. M., Miller, O. J., and Siniscalco, M. (1972). Cytological mapping of human X-linked genes by use of somatic cell hybrids involving an X-autosome translocation. *Proc. Natl. Acad. Sci. U.S.A.* **69**, 69–73.

Guggenheim, M. A., Friedman, R. M., and Rabson, A. S. (1968). Interferon: Production by chick erythrocytes activated by cell fusion. *Science* **159**, 542–543.

Guggenheim, M. A., Friedman, R. M., and Rabson, A. S. (1969). Interferon action in heterokaryons. *Proc. Soc. Exp. Biol. Med.* **130**, 1242–1245.

Gurdon, J. B. (1974). "The Control of Gene Expression in Animal Development." Oxford Univ. Press (Clarendon), London and New York.

Gurdon, J. B., and Woodland, H. R. (1968). The cytoplasmic control of nuclear activity in animal development. *Biol. Rev. Cambridge Philos. Soc.* **43**, 233–267.

Gurdon, J. B., and Woodland, H. R. (1970). On the long-term control of nuclear activity during cell differentiation. *Curr. Top. Dev. Biol.* **5**, 39–70.

Ham, A. W. (1974). "Histology," 7th ed. Lippincott, Philadelphia, Pennsylvania.

Hamerton, J. L., and Cook, P. J. L. (1975). Report of the committee on the genetic constitution of chromosomes 1 and 2 (Rotterdam Conference 1974), *Birth Defects, Orig. Artic. Ser.* **11**, 3–12.

Handmaker, S. D. (1971). Cytogenetic analysis of a Chinese hamster–mouse hybrid cell. *Nature (London)* **233**, 416–419.

Harris, H. (1965). Behaviour of differentiated nuclei in heterokaryons of animal cells from different species. *Nature (London)* **206**, 583–588.

Harris, H. (1967). The reactivation of the red cell nucleus. *J. Cell Sci.* **2**, 23–32.

Harris, H. (1968), "Nucleus and Cytoplasm." Oxford Univ. Press, London and New York.

Harris, H. (1970a). "Nucleus and Cytoplasm." Oxford Univ. Press, London and New York.

Harris, H. (1970b). "Cell Fusion, The Dunham Lectures." Oxford Univ. Press, London and New York.

Harris, H. (1974). "Nucleus and Cytoplasm." Oxford Univ. Press, London and New York.

Harris, H., and Cook, P. R. (1969). Synthesis of an enzyme determined by an erythrocyte nucleus in a hybrid cell. *J. Cell Sci.* **5**, 121–134.

Harris, H., and Klein, G. (1969). Malignancy of somatic cell hybrids. *Nature (London)* **224**, 1315–1316.

Harris, H., and Watkins, J. F. (1965). Hybrid cells derived from mouse and man: Artificial heterokaryons of mammalian cells from different species. *Nature (London)* **205**, 640–646.

Harris, H., Watkins, J. F., Campbell, G. L. M., Evans, E. P., and Ford, C. E. (1965). Mitosis in hybrid cells derived from mouse and man. *Nature (London)* **207**, 606–608.

Harris, H., Watkins, J. F., Ford, C. E., and Schoefl, G. I. (1966). Artificial heterokaryons of animal cells from different species. *J. Cell Sci.* **1**, 1–30.

Harris, H., Miller, O. J., Klein, G., Worst, P., and Tachibana, T. (1969a). Suppression of malignancy by cell fusion. *Nature (London)* **223**, 363–368.

Harris, H., Sidebottom, E., Grace, D. M., and Bramwell, M. E. (1969b). The expression of genetic information: A study with hybrid animal cells. *J. Cell Sci.* **4**, 499–525.

Harris, M. (1964). "Cell Culture and Somatic Variation." Holt, New York.

Harris, M. (1972). Effect of X-irradiation of one partner on hybrid frequency in fusions between Chinese hamster cells. *J. Cell. Physiol.* **80**, 119–128.

Harris, M. (1973). Phenotypic expression of drug resistance in hybrid cells. *J. Natl. Cancer Inst.* **50**, 423–429.

Harter, D. H., and Choppin, P. W. (1967). Cell-fusing activity of Visna virus particles. *Virology* **31**, 279–288.

Hartmann, J. X., Kao, K. N., Gamborg, O. L., and Miller, R. A. (1973). Immunological methods for the agglutination of protoplasts from cell suspension cultures of different genera. *Planta* **112**, 45–56.

Hatanaka, M., Huebner, R. J., and Gilden, R. V. (1969). Alterations in the characteristics of sugar uptake by mouse cells transformed by murine sarcoma viruses. *J. Natl. Cancer Res.* **43**, 1091–1096.

Hatanaka, M., Augl, C., and Gilden, R. V. (1970). Evidence for a functional change in the plasma membrane of murine sarcoma virus-infected mouse embryo cells. Transport and transport-associated phosphorylation of ^{14}C-2-deoxy-D-glucose. *J. Biol. Chem.* **245**, 714–717.

Haydon, D. A., and Taylor, J. (1963). The stability and properties of bimolecular lipid leaflets in aqueous solutions. *J. Theor. Biol.* **4**, 281–296.

Hayflick, L. (1965). Limited *in vitro* lifetime of human diploid cell strains. *Exp. Cell Res.* **37**, 614–636.

Hayflick, L., and Moorhead, P. S. (1961). The serial cultivation of human diploid cell strains. *Exp. Cell Res.* **25**, 585–621.

Haythorn, S. R. (1929). Multinucleated giant cells. *Arch. Pathol.* **7**, 651–713.

Hecht, F., Wyandt, H. E., and Magenis, R. E. H. (1974). The human cell nucleus: Quinacrine and other differential stains in the study of chromatin and chromosomes. *In* "The Cell Nucleus" (H. Busch, ed.), Vol. 2, pp. 33–121. Academic Press, New York.

Heneen, W. K. (1971). *In situ* analysis of cultured *Potorous* cells. I. DNA synthesis and mitotic dynamics in bi- and multinucleate cells. *Hereditas* **67**, 221–250.

Heneen, W. K., Nichols, W. W., Levan, A., and Norrby, E. (1970). Polykaryocytosis and mitosis in a human cell line after treatment with measles virus. *Hereditas* **64**, 53–84.

Henle, G., Deinhardt, F., and Girardi, A. (1954). Cytolytic effects of mumps virus in tissue cultures of epithelial cells. *Proc. Soc. Exp. Biol. Med.* **87**, 386–393.

Hilwig, I., and Gropp, A. (1972). Staining of constitutive heterochromatin in mammalian chromosomes with a new fluorochrome. *Exp. Cell Res.* **75**, 122–126.

Hirai, K., and Defendi, V. (1975). Factors affecting the process and extent of integration of the viral genome *Cold Spring Harbor Symp. Quant. Biol.* **39,** 325–333.

Hirai, K., Henner, D., and Defendi, V. (1974). Hybridization of Simian virus 40 complementary RNA with nucleolus-associated DNA isolated from Simian virus 40-transformed Chinese hamster cells. *Virology* **60,** 588–591.

Hirst, G. K. (1941). The agglutination of red cells by allantoic fluid of chick embryos infected with influenza virus. *Science* **94,** 22–23.

Hitchcock, G. (1971). A comparison of the susceptibility to fusion of cell monolayers of human origin by inactivated Sendai virus. *Exptl. Cell. Res.* **67,** 463–466.

Hitotsumachi, S., Rabinowitz, Z., and Sachs, L. (1971). Chromosomal control of reversion in transformed cells. *Nature (London)* **231,** 511–514.

Hittelman, W. N., and Rao, P. N. (1973). Prematurely condensed chromosomes for the study of aberration formation. *J. Cell Biol.* **59,** 144a.

Hittelman, W. N., and Rao, P. N. (1974a). Premature chromosome condensation. I. Visualization of X-ray induced chromosome damage in interphase cells. *Mutat. Res.* **23,** 251–258.

Hittelman, W. N., and Rao, P. N. (1974b). Premature chromosome condensation. II. The nature of chromosome gaps produced by alkylating agents and ultraviolet light. *Mutat. Res.* **23,** 259–266.

Hittelman, W. N., and Rao, P. N. (1975). The nature of adriamycin-induced cytotoxicity in Chinese hamster cells as revealed by premature chromosome condensation. *Cancer Res.* **35,** 3027–3035.

Hoggan, M. D., and Roizman, B. (1959). The isolation and properties of a variant of herpes simplex producing multinucleate giant cells in monolayer cultures in the presence of antibody. *Am. J. Hyg.* **70,** 208–219.

Holley, R. W. (1971). Studies of serum factors required by 3T3 and SV3T3 cells. *Growth Control Cell Cult., Ciba Found. Symp., 1970* p. 3.

Holley R. W. (1974). Serum factors and growth control. *In* "Control of Proliferation in Animal Cells" (B. Clarkson and R. Baserga, eds.), pp. 13–18. Cold Spring Harbor Lab., Cold Spring Harbor, New York.

Holley, R. W., and Kierman, J. A. (1968). "Contact inhibition" of cell division in 3T3 cells. *Proc. Natl. Acad. Sci. U.S.A.* **60,** 300–304.

Holmes, K. V., and Choppin, P. W. (1968). On the role of microtubules in movement and alignment of nuclei in virus-induced syncytia. *J. Cell Biol.* **39,** 526–543.

Holtzer, H., Abbott, J., and Lash, J. (1958). The formation of multi-nucleate myotubes. *Anat. Rec.* **131,** 567.

Homma, M. (1972). Trypsin action on the growth of Sendai virus in tissue culture cells. II. Restoration of the hemolytic activity of L cell-borne Sendai virus by trypsin. *J. Virol.* **9,** 829–835.

Homma, M., and Tamagawa, S. (1973). Restoration of the fusion activity of L cell-borne Sendai virus by trypsin. *J. Gen. Virol.* **19,** 423–426.

Honjo, T., and Reeder, R. H. (1973). Preferential transcription of *Xenopus laevis* ribosomal RNA in interspecies hybrids between *Xenopus laevis* and *Xenopus mulleri*. *J. Mol. Biol.* **80,** 217–228.

Horak, I., Coon, H. G., and Dawid, I. B. (1974). Interspecific recombination of mitochondrial DNA molecules in hybrid somatic cells. *Proc. Natl. Acad. Sci. U.S.A.* **71,** 1828–1832.

Horne, R. W., Waterson, A. P., Wildy, P., and Farnham, A. E. (1960). The structure and composition of the myxoviruses I. Electron microscope studies on the structure of myxovirus particles by negative staining techniques. *Virology* **11,** 79–98.

Hors-Cayla, M. C., Trébuchat, G., and Heuertz, S. (1972). Evolution caryotypique de cellules hybrides homme × homme. *Ann. Genet.* **15**, 153–158.

Hosaka, Y. (1970). Biological activities of sonically treated Sendai virus. *J. Gen. Virol.* **8**, 43–54.

Hosaka, Y., and Koshi, Y. (1968). Electron microscopic study of cell fusion by HVJ virions. *Virology* **34**, 419–434.

Howe, C., Morgan, C., de Vaux St. Cyr, C., Hsu, K. C., and Rose, H. M. (1967). Morphogenesis of type 2 parainfluenza virus examined by light and electron microscopy. *J. Virol.* **1**, 215–257.

Hsia, D. Y. Y. (1970). Study of hereditary metabolic diseases using *in vitro* techniques. *Metab., Clin. Exp.* **19**, 309–339.

Hsu, T. C. (1973). Longitudinal differentiation of chromosomes. *Annu. Rev. Genet.* **7**, 153–176.

Hsu, T. C., and Somers, C. E. (1962). Properties of L cells resistant to 5-bromodeoxyuridine. *Exp. Cell Res.* **26**, 404–410.

Hu, F., and Pasztor, L. M. (1975). *In vivo* hybridization of cultured melanoma cells and isogenic normal mouse cells. *Differentiation* **4**, 93–97.

Huebner, K., and Koprowski, H. (1974). Synthesis of SV40 capsid antigen(s) in heterokaryocytes formed by fusion of non-permissive with permissive cells. *Virology* **58**, 609–611.

Huebner, K., Croce, C. M., and Koprowski, H. (1974). Isolation of defective viruses from SV40-transformed human and hamster cells. *Virology* **59**, 570–573.

Huebner, K., Santoli, D., Croce, C. M., and Koprowski, H. (1975). Characterization of defective SV40 isolated from SV40-transformed cells. *Virology* **63**, 512–522.

Hynes, R. O. (1974). Role of surface alterations in cell transformation. The importance of proteases and surface proteins. *Cell* **1**, 147–156.

Ikeuchi, T., and Sandberg, A. A. (1970). Chromosome pulverization in virus-induced heterokaryons of mammalian cells from different species. *J. Natl. Cancer Inst.* **45**, 951–963.

Ikeuchi, T., Sanbe, M., Weinfeld, H., and Sandberg, A. A. (1971). Induction of nuclear envelopes around metaphase chromosomes after fusion with interphase cells. *J. Cell Biol.* **51**, 104–115.

Ito, M., and Maeda, M. (1973). Fusion of meiotic protoplasts in liliaceaus plants. *Exp. Cell Res.* **80**, 453–456.

Jacobson, C. O. (1968). Reactivation of DNA synthesis in mammalian neuron nuclei after fusion with cells of an undifferentiated fibroblast line. *Exp. Cell Res.* **53**, 316–318.

Jakob, H., and Ruiz, F. (1970). Preferential loss of kangaroo chromosomes in hybrids between Chinese hamster and kangaroo-rat somatic cells. *Exp. Cell Res.* **62**, 310–314.

Jami, J., and Grandchamp, S. (1971). Karyological properties of human × mouse somatic hybrids. *Proc. Natl. Acad. Sci. U.S.A.* **68**, 3097–3101.

Jami, J., and Ritz, E. (1973a). Non-malignancy of hybrids derived from two mouse malignant cells. I. Hybrids between L1210 leukemia cells and malignant L cells. *J. Natl. Cancer Inst.* **51**, 1647–1653.

Jami, J., and Ritz, E. (1973b). Expression of tumor-specific antigens in mouse somatic cell hybrids. *Cancer Res.* **33**, 2524–2528.

Jami, J., and Ritz, E. (1975a). Non-malignancy of hybrids derived from two mouse malignant cells. II. Analysis of malignancy of LM (TK⁻) C1ID parental cells. *J. Natl. Cancer Inst.* **54**, 117–122.

Jami, J., and Ritz, E. (1975b). Tumor-associated transplantation antigens in immune rejection of mouse malignant cell hybrids. *Proc. Natl. Acad. Sci. U.S.A.* **72**, 2130–2134.

Jami, J., Grandchamp, S., and Ephrussi, B. (1971). The karyologic behaviour of human × mouse cellular hybrids. *C. R. Hebd. Seances Acad. Sci., Ser. D* **272**, 323–326.

Jami, J., Failly, C., and Ritz, E. (1973). Lack of expression of differentiation in mouse teratoma–fibroblast somatic cell hybrids. *Exp. Cell Res.* **76**, 191–199.

Janzen, H. W., Millman, P. A., and Thurston, O. G. (1971). Hybrid cells in solid tumors. *Cancer* **27**, 455–459.

Jeffreys, A., and Craig, I., (1974). Differences in the products of mitochondrial protein synthesis *in vivo* in human and mouse cells and their potential use as markers for the mitochondrial genome in human–mouse somatic cell hybrids. *Biochem. J.* **144**, 161–164.

Jensen, F. C., and Koprowski, H. (1969). Absence of repressor in SV40 transformed cells. *Virology* **37**, 687–690.

Jeon, K. W., and Danielli, J. F. (1971). Micrurgical studies with large free-living amebas. *Int. Rev. Cytol.* **30**, 49–89.

Jha, K. K., and Ozer, H. L. (1976). Expression of transformation in cell hybrids. I. Isolation and application of density-inhibited Balb/3T3 cells deficient in hypoxanthine phosphoribosyl transferase and resistant to ouabain. *Somatic Cell Genet.*, **2**, 215–233.

Johnson, R. T., and Harris, H. (1969a). DNA synthesis and mitosis in fused cells. I. HeLa homokaryons. *J. Cell Sci.* **5**, 603–624.

Johnson, R. T., and Harris, H. (1969b). DNA synthesis and mitosis in fused cells. II. HeLa-chick erythrocyte heterokaryons. *J. Cell Sci.* **5**, 625–643.

Johnson, R. T., and Harris, H. (1969c). DNA synthesis and mitosis in fused cells. III. HeLa-Ehrlich heterokaryons *J. Cell Sci.* **5**, 645–697.

Johnson, R. T., and Rao, P. N. (1970). Mammalian cell fusion. Induction of premature chromosome condensation in interphase nuclei. *Nature (London)* **226**, 717–722.

Johnson, R. T., and Rao, P. N. (1971). Nucleo-cytoplasmic interactions in the achievement of nuclear synchrony in DNA synthesis and mitosis in multinucleate cells. *Biol. Rev. Cambridge Philos. Soc.* **46**, 97–155.

Johnson, R. T., and Mullinger, A. M. (1975). The induction of DNA synthesis in heterokaryons during the first cell cycle after fusion with HeLa cells. *J. Cell Sci.* **18**, 455–490.

Johnson, R. T., Rao, P. N., and Hughes, H. D. (1970). Mammalian cell fusion. III. A HeLa cell inducer of premature chromosome condensation active in cells from a variety of animal species. *J. Cell. Physiol.* **76**, 151–158.

Jones, C., and Puck, T. T. (1973). Genetics of somatic mammalian cells. XVII. Induction and isolation of Chinese hamster cell mutants requiring serine. *J. Cell. Physiol.* **81**, 299–304.

Jones, C., Wuthier, P., Kao, F., and Puck, T. T. (1972). Genetics of somatic mammalian cells. XV. Evidence for linkage between human genes for lactic dehydrogenase B and serine hydroxymethylase. *J. Cell. Physiol.* **80**, 291–297.

Jost, E., Lennox, R., and Harris, H. (1975). Affinity chromatography of DNA binding proteins from human, murine and man–mouse hybrid cell lines. *J. Cell Sci.* **18**, 41–65.

Kahan, B. and DeMars, R. (1975). Localized derepression on the human inactive X-chromosome in mouse–human cell hybrids. *Proc. Natl. Acad. Sci. U.S.A.* **72**, 1510–1514.

Kahan, B. W., and Ephrussi, B. (1970). Developmental potentialities of clonal *in vitro* cultures of mouse testicular teratoma. *J. Natl. Cancer Inst.* **44**, 1015–1036.

Kano, K., Baranska, W., Knowles, B. B., Koprowski, H., and Milgrom, F. (1969). Surface antigens on interspecies hybrid cells. *J. Immun.* **103**, 1050–1060.

Kao, F. T. (1973). Identification of chick chromosomes in cell hybrids formed between chick erythrocytes and adenine-requiring mutants of Chinese hamster cells. *Proc. Natl. Acad. Sci. U.S.A.* **70,** 2893–2898.

Kao, F. T., and Puck, T. T. (1967). Genetics of somatic mammalian cells. IV. Properties of Chinese hamster cell mutants with respect to the requirement for proline. *Genetics* **55,** 513–524.

Kao, F. T., and Puck, T. T. (1968). Genetics of somatic mammalian cells. VII. Induction and isolation of nutritional mutants in Chinese hamster cells. *Proc. Natl. Acad. Sci. U.S.A.* **60,** 1275–1281.

Kao, F. T., and Puck, T. T. (1970). Linkage studies with human–Chinese hamster cell hybrids. *Nature (London)* **228,** 329–332.

Kao, F. T., and Puck, T. T. (1972a). Genetics of somatic mammalian cells: Demonstration of a human esterase activator gene linked to the *Ade B* gene. *Proc. Natl. Acad. Sci. U.S.A.* **69,** 3273–3277.

Kao, F. T., and Puck, T. T. (1972b). Genetics of somatic mammalian cells. XIV. Genetic analysis *in vitro* of auxotrophic mutants. *J. Cell. Physiol.* **80,** 41–50.

Kao, F. T., and Puck, T. T. (1975). Mutagenesis and genetic analysis with Chinese hamster auxotrophic cell markers. *Genetics* **79** suppl: 343–352.

Kao, F. T., Chasin, L., and Puck, T. T. (1969a). Complementation analysis of glycine-requiring mutants. *Proc. Natl. Acad. Sci. U.S.A.* **64,** 1284–1291.

Kao, F. T., Johnson, R. T., and Puck, T. T. (1969b). Complementation analysis on virus-fused Chinese hamster cells with nutritional markers. *Science* **164,** 312–314.

Kao, K. N., and Michayluk, M. R. (1974). A method for high-frequency intergeneric fusion of plant protoplasts. *Planta* **115,** 355–367.

Karakoz, I., Grešíková, M., Hála, K., and Hašek, M. (1969). Attempts to induce recombinant and somatic cells in chicken erythrocyte chimeras by repeated injection of western equine encephalomyelitis virus. *Folia Biol. (Prague)* **15,** 81–86.

Kartha, K. K., Gamborg, O. L., Constable, F., and Kao, K. N. (1974). Fusion of rapeseed and soybean protoplasts and subsequent division of heterokaryocytes. *Can. J. Bot.* **52,** 2435–2436.

Kasper, C. B., and Kubinski, H. (1971). Fusion *in vitro* of membranes from animal cells. *Nature (London), New Biol.* **231,** 124–125.

Kato, H., and Sandberg, A. A. (1968a). Chromosome pulverization in Chinese hamster cells induced by Sendai virus. *J. Natl. Cancer Inst.* **41,** 1117–1123.

Kato, H., and Sandberg, A. A. (1968b). Cellular phase of chromosome pulverization induced by Sendai virus. *J. Natl. Cancer Inst.* **41,** 1125–1131.

Katz, M., Rorke, L. B., Masland, W. S., Barbanti-Brodano, G., and Koprowski, H. (1970). Subacute sclerosing panencephalitis: Isolation of a virus encephalitogenic for ferrets. *J. Infect. Dis.* **121,** 188–195.

Kavanau, J. L. (1965). "Structure and Function in Biological Membranes," Vols. 1 and 2. Holden-Day, San Francisco, California.

Keller, W. A., and Melchers, G. (1973). The effect of high pH and calcium on tobacco leaf protoplast fusion. *Z. Naturforsch., Teil C* **28,** 737–741.

Kennett, R. H., Hampshire, B., Bengtsson, B., and Bodmer, W. F. (1975). Expression and segregation of HL-A antigens in D98/AH-2 by lymphocyte and fibroblast hybrids. *Tissue Antigens* **6,** 80–92.

Kim, M. A., and Grzeschik, K. H. (1974). A method for discriminating murine and human chromosomes in somatic cell hybrids. *Exp. Cell Res.* **88,** 406–410.

Kislev, N., Spolsky, C. M., and Eisenstadt, J. M. (1973). Effect of chloramphenicol on the ultrastructure of mitochondria in sensitive and resistant strains of HeLa. *J. Cell Biol.* **57,** 571–579.

Kit, S., and Leung, W-.C. (1974). Genetic control of mitochondrial thymidine kinase in human–mouse and monkey–mouse somatic cell hybrids. _J. Cell Biol._ **61**, 35–44.

Kit, S., Dubbs, D. R., Piekarski, L. J., and Hsu, T. C. (1963). Deletion of thymidine kinase activity from L cells resistant to bromodeoxyuridine. _Exp. Cell Res._ **31**, 297–312.

Kit, S., Kurimura, T., deTorres, R. A., and Dubbs, D. R. (1969). Simian virus 40 deoxyribonucleic acid replication. I. Effect of cycloheximide on the replication of SV40 deoxyribonucleic acid in monkey kidney cells and in heterokaryons of SV40-transformed and susceptible cells. _J. Virol._ **3**, 25–32.

Kit, S., Kurimura, T., Brown, M., and Dubbs, D. R. (1970). Identification of the simian virus 40 which replicates when simian virus 40 transformed human cells are fused with simian virus 40-transformed mouse cells or superinfected with simian virus 40 deoxyribonucleic acid. _J. Virol._ **6**, 69–77.

Kit, S., Leung, W.-C., Jorgensen, G., Trkula, D., and Dubbs, D. R. (1974). Acquisition of chick cytosol thymidine kinase activity by thymidine kinase-deficient mouse fibroblast cells after fusion with chick erythrocytes. _J. Cell Biol._ **63**, 505–514.

Klebe, R. J., Chen, T. R., and Ruddle, F. H. (1970). Controlled production of proliferating somatic cell hybrids. _J. Cell Biol._ **45**, 74–82.

Klee, W. A., and Nirenberg, M. (1974). A neuroblastoma × glioma hybrid cell line with morphine receptors. _Proc. Natl. Acad. Sci. U. S. A._ **71**, 3474–3477.

Klein, A., and Bonhoeffer, F. (1972). DNA replication. _Annu. Rev. Biochem._ **41**, 301–332.

Klein, E., and Wiener, F. (1971). Loss of surface-bound immunoglobin in mouse A9 and human lymphoblast hybrid cells. _Exp. Cell Res._ **67**, 251 (abstr.).

Klein, G., and Harris, H. (1972). Expression of polyoma-induced transplantation antigen in hybrid cell lines. _Nature (London), New Biol._ **237**, 163–164.

Klein, G., Gars, U., and Harris, H. (1970). Isoantigen expression in hybrid mouse cells. _Exp. Cell Res._ **62**, 149–160.

Klein, G., Bregula, U., Wiener, F., and Harris, H. (1971). The analysis of malignancy by cell fusion. 1. Hybrids between tumor cells and L cell derivatives. _J. Cell Sci._ **8**, 659–672.

Klein, G., Friberg, S., Jr., and Harris, H. (1972). Two kinds of antigen suppression in tumor cells revealed by cell fusion. _J. Exp. Med._ **135**, 839–849.

Klein, G., Friberg, S., Jr., Wiener, F., and Harris, H. (1973). Hybrid cells derived from fusion of TA3-HA ascites carcinoma with normal fibroblasts. 1. Malignancy, karyotype, and formation of isoantigenic variants. _J. Natl. Cancer Inst._ **50**, 1259–1268.

Klein, G., Wiener, F., Zech, L., zur Hausen, H., and Reedman, B. (1974). Segregation of the EBV-determined nuclear antigen (EBNA) in somatic cell hybrids derived from the fusion of a mouse fibroblast and a human Burkitt lymphoma line. _Int. J. Cancer_ **14**, 54–64.

Klinger, H. P., and Shin, S.-I. (1974). Modulation of the activity of an avian gene transferred into a mammalian cell by cell fusion. _Proc. Natl. Acad. Sci. U.S.A._ **71**, 1398–1402.

Knowles, B. B., Jensen, F. C., Steplewski, Z., and Koprowski, H. (1968). Rescue of infectious SV 40 after fusion between different SV40-transformed cells. _Proc. Natl. Acad. Sci. U.S.A._ **61**, 42–45.

Knowles, B. B., Steplewski, Z., Swetly, P., Barbanti-Brodano, G., and Koprowski, H. (1969). Cell hybridization and tumor viruses. _Wistar Inst. Symp. Monogr._ **9**, 37–47.

Knowles, B. B., Barbanti-Brodano, G., and Koprowski, H. (1971). Susceptibility of primate–mouse hybrid cells to SV40. _J. Cell. Physiol._ **78**, 1–8.

Kohn, A. (1965). Polykaryocytosis induced by Newcastle disease virus in monolayers of animal cells. *Virology* **26,** 228–245.

Kohn, A., and Klibansky, C. (1967). Studies on the inactivation of cell-fusing property of Newcastle disease virus by phospholipase A. *Virology* **31,** 385–388.

Konigsberg, I. R. (1963). Clonal analysis of myogenesis. *Science* **140,** 1272–1284.

Konigsberg, I. R., McElvain, N., Tootle, M., and Herrmann, H. (1960). The dissociability of deoxyribonucleic acid synthesis from the development of multinuclearity of muscle cells in culture. *J. Biophys. Biochem. Cytol.* **8,** 333–343.

Koprowski, H. (1971). Cell fusion and virus rescue. *Fed. Proc., Fed. Am. Soc. Exp. Biol.* **30,** 914–920.

Koprowski, H., and Knowles, B. (1974). Viruses, immune functions and antigenic determinants in heterokaryons and hybrids. In "Somatic Cell Hybridization" (R. L. Davidson and F. F. de la Cruz, eds.), pp. 71–100. Raven, New York.

Koprowski, H., Jensen, F. C., and Steplewski, Z. (1967). Activation of production of infectious tumor virus SV40 in heterokaryon cultures. *Proc. Natl. Acad. Sci. U.S.A.* **58,** 127–133.

Koprowski, H., Barbanti-Brodano, G., and Katz, M. (1970). Interaction between papovalike virus and paramyxovirus in human brain cells. A hypothesis. *Nature (London)* **225,** 1045–1047.

Korn, E. D. (1966). Structure of biological membranes. *Science* **153,** 1491–1498.

Kornberg, R. D., and McConnell, H. M. (1971a). Inside-outside transitions of phospholipids in vesicle membranes. *Biochemistry* **10,** 1111–1120.

Kornberg, R. D., and McConnel, H. M. (1971b). Lateral diffusion of phospholipids in a vesicle membrane. *Proc. Natl. Acad. Sci. U.S.A.* **68,** 2564–2568.

Koyama, H., and Ono, T. (1970). Initiation of a differentiated function (hyaluronic acid synthesis) by hybrid formation in culture. *Biochim. Biophys. Acta* **217,** 477–487.

Koyama, H., Yatabe, I., and Ono, T. (1970). Isolation and characterization of hybrids between mouse and Chinese hamster cell lines. *Exp. Cell Res.* **62,** 455–463.

Kraemer, K. H., Coon, H. G., and Robbins, J. H. (1973). Cell-fusion analysis of different inherited mutations causing defective DNA repair in *Xeroderma Pigmentosum* fibroblasts. *J. Cell Biol.* **59,** 176a.

Kraemer, K. H., Coon, H. G., Petinga, P. A., Barrett, S. F., Rahe, A. E., and Robbins, J. H. (1975). Genetic heterogeneity in *Xeroderma pigmentosum:* Complementation groups and their relationship to DNA repair rates. *Proc. Natl. Acad. Sci. U.S.A.* **72,** 59–63.

Kroon, A. M., and Succone, C. (1973). "Biogenesis of Mitochondria." Academic Press, New York.

Krooth, R. S., and Sell, E. K. (1970). The action of mendelian genes in human diploid cell strains. *J. Cell. Physiol.* **76,** 311–330.

Kucherlapati, R. S., Creagan, R. P., and Ruddle, F. H. (1974). Progress in human gene mapping by somatic cell hybridization. In "The Cell Nucleus" (H. Busch, ed.), Vol. 2, pp. 209–222. Academic Press, New York.

Kung, S. D., Gray, J. C., Wildman, S. G., and Carlson, P. S. (1975). Polypeptide composition of fraction 1 protein from parasexual hybrid plants in the genus *Nicotiana. Science* **187,** 353–355.

Kusano, T., Wang, R., Pollack, R., and Green, H. (1970). Human–mouse hybrid cell lines and susceptibility to polio virus. II. Polio sensitivity and the chromosome constitution of the hybrids. *J. Virol.* **5,** 682–685.

Kusano, T., Long, C., and Green, H. (1971). A new reduced human–mouse somatic cell hybrid containing the human gene for adenine phosphoribosyltransferase. *Proc. Natl. Acad. Sci. U.S.A.* **68,** 82–86.

Kuter, D. J., and Rodgers, A. (1975). The synthesis of ribosomal protein and ribosomal RNA in rat–mouse hybrid cell line. *Exp. Cell Res.* **91,** 317–325.

Labella, T., Amati, P., and Marin, G. (1973). Relationship between the ratio of parental chromosomes and parental doubling times in Chinese hamster–mouse somatic cell hybrids. *J. Cell. Physiol.* **81,** 347–354.

Ladda, R. L., and Estensen, R. D. (1970). Introduction of a heterologous nucleus into enucleated cytoplasms of cultured mouse L cells. *Proc. Natl. Acad. Sci. U.S.A.* **67,** 1528–1533.

Langhans, T. (1868). Über Riesenzellen mit wandständigen Kernen in Tuberkeln und die fibröse Form des Tuberkels. *Arch. Pathol. Anat. Physiol. Klin. Med.* **42,** 382–404.

Lavialle, C., Suarez, H. G., Estrade, S., and Cassingena, R. (1974). Further studies on SV40 permissive cell protein factor(s) allowing rescue of infectious virus and viral DNA from non-shedder SV40-transformed cells. *Int J. Cancer* **13,** 311–318.

Le Douarin, N. M., and Rival, J. M. (1975). A biological nuclear marker in cell culture: Recognition of nuclei in single cells and heterokaryons. *Dev. Biol.* **47,** 215–221.

Lehman, J. M., Speers, W. C., Swartzendruber, D. E., and Pierce, G. B., (1974). Neoplastic differentiation: Characteristics of cell lines derived from a murine teratocarcinoma. *J. Cell. Physiol.* **84,** 13–28.

Lenard, J., and Compans, R. W. (1974). The membrane structure of lipid-containing viruses. *Biochim. Biophys. Acta* **344,** 51–94.

Leung, W. C., Chen, T. R., Dubbs, D. R., and Kit, S. (1975). Identification of chick thymidine kinase determinant in somatic cell hybrids of chick erythrocytes and thymidine kinase-deficient mouse cells. *Exp. Cell Res.* **95,** 320–326.

Levan, A. (1954). Colchicine-induced C-mitosis in two mouse ascites tumors. *Hereditas* **40,** 1–64.

Levan, A., Fredga, K., and Sandberg, A. A. (1964). Nomenclature for centromeric position of chromosomes. *Hereditas* **52,** 201–220.

Levisohn, S. R., and Thompson, E. B. (1973). Contact inhibition and gene expression in HTC/L cell hybrid lines. *J. Cell. Physiol.* **81,** 225–232.

Levy, N. L., and Ladda, R. L. (1971). Restoration of haemolytic complement activity in C5-deficient mice by gene complementation in hybrid cells. *Nature (London), New Biol.* **229,** 51–52.

Lewis, W. H. (1927). The formation of giant cells in tissue culture and their similarity to those in tuberculous lesions. *Am. Rev. Tuberc.* **15,** 616–628.

Lin, M. S., and Davidson, R. L. (1975). Replication of human chromosomes in human–mouse hybrids: Evidence that the timing of DNA synthesis is determined independently in each human chromosome. *Somatic Cell Genet.* **1,** 111–122.

Lin, T. P., Florence, J., and Oh, J. O. (1973). Cell Fusion induced by a virus within the *Zona pellucida* of mouse eggs. *Nature (London)* **242,** 47–49.

Lindberg, L. G. (1974). Ultrastructural study of heterokaryons from Rous rat sarcoma cells and normal chicken cells. *Acta Pathol. Microbiol. Scand.* **82,** 299–310.

Ling, H., and Ling, M. (1974). Genetic control of somatic cell fusion in a myxomycete. *Heredity* **32,** 95–104.

Linnane, A. W., Saunders, G. W., Gingold, E. B., and Lukins, H. B. (1968). The biogenesis of mitochondria. V. Cytoplasmic inheritance of erythromycin resistance in *Saccharomyces cerevisiae. Proc. Natl. Acad. Sci. U.S.A.* **59,** 903–910.

Little, J. B., Richardson, U. I., and Tashjian, A. H., Jr. (1972). Unexpected resistance to X-irradiation in a strain of hybrid mammalian cells. *Proc. Natl. Acad. Sci. U.S.A.* **69,** 1363–1365.

Littlefield, J. W. (1963). The inosinic acid pyrophosphorylase activity of mouse fibroblasts partially resistant to 8-azaguanine. *Proc. Natl. Acad. Sci. U.S.A.* **50**, 568–576.

Littlefield, J. W. (1964a). Selection of hybrids from matings of fibroblasts *in vitro* and their presumed recombinants. *Science* **145**, 709–710.

Littlefield, J. W. (1964b). Three degrees of guarylic acid-inosinic acid pyrophosphorylase deficiency in mouse fibroblasts. *Nature (London)* **203**, 1142–1144.

Littlefield, J. W. (1965). Studies on thymidine kinase in cultured mouse fibroblasts. *Biochim. Biophys. Acta* **95**, 14–22.

Littlefield, J. W. (1966). The use of drug-resistant markers to study the hybridization of mouse fibroblasts. *Exp. Cell Res.* **41**, 190–196.

Littlefield, J. W. (1969). Hybridization of hamster cells with high and low folate reductase activity. *Proc. Natl. Acad. Sci. U. S. A.* **62**, 88–95.

Lobban, P. E., and Siminovitch, L. (1975). Alpha-amanitin resistance: A dominant mutation in CHO cells. *Cell* **4**, 167–172.

Loyter, A., Zakai, N., and Kulka, R. G. (1975). Ultramicroinjection of macromolecules or small particles into animal cells. A new technique based on virus-induced cell fusion. *J. Cell Biol.* **66**, 292–304.

Lucas, J. J., Szekely, E., and Kates, J. R. (1976). The regeneration and division of mouse L-cell karyoplasts. *Cell* **7**, 115–122.

Lucy, J. A. (1970). The fusion of biological membranes. *Nature (London)* **227**, 815–817.

Lyons, L. B., Cox, R. P., and Dancis, J. (1973). Complementation analysis of maple syrup urine disease in heterokaryons derived from cultured human fibroblasts. *Nature (London)* **243**, 533–535.

McBride, O. W., and Ozer, H. L. (1973). Transfer of genetic information by purified metaphase chromosomes. *Proc. Natl. Acad. Sci. U.S.A.* **70**, 1258–1262.

McDougall, J. K., Kucherlapati, R. S., and Ruddle, F. H. (1973). Localization and induction of the human thymidine kinase gene by adenovirus 12. *Nature (London), New Biol.* **245**, 172–175.

Machala, O., Donner, L., and Svoboda, J. (1970). A full expression of the genome of Rous sarcoma virus in heterokaryons formed after fusion of virogenic mammalian cells and chicken fibroblasts. *J. Gen. Virol.* **8**, 219–229.

McKusick, V. A. (1971). "Mendelian Inheritance in Man." Johns Hopkins Press, Baltimore, Maryland.

McMorris, F. A., and Ruddle, F. H. (1974). Expression of neuronal phenotypes in neuroblastoma cell hybrids. *Dev. Biol.* **39**, 226–246.

McMorris, F. A., Kolber, A. R., Moore, B. W., and Perumal, A. S. (1974). Expression of the neuron-specific protein, 14-3-2, and steroid sulfatase in neuroblastoma cell hybrids. *J. Cell. Physiol.* **84**, 473–480.

MacPherson, I., and Montagnier, L. (1964). Agar suspension culture for the selective assay of cells transformed by polyoma virus. *Virology* **23**, 291–294.

Malawista, S. E., and Weiss, M. C. (1974). Expression of differentiated functions in hepatoma cell hybrids. High frequency of induction of mouse albumin production in rat hepatoma × mouse lymphoblast hybrids. *Proc. Natl. Acad. Sci. U.S.A.* **71**, 927–931.

Margulis, L. (1973). Colchicine-sensitive microtubules. *Int. Rev. Cytol.* **34**, 333–361.

Mariano, M., and Spector, W. G. (1974). The formation and properties of macrophage polykaryons (inflammatory giant cells). *J. Pathol.* **113**, 1–19.

Marin, G. (1969). Selection of chromosomal segregants in a "hybrid" line of Syrian hamster fibroblasts. *Exp. Cell Res.* **57**, 29–36.

Marin, G. (1971). Segregation of morphological revertants in polyoma transformed hybrid clones of hamster fibroblasts. *J. Cell Sci.* **9**, 61–69.

Marin, G., and Colletta, G. (1974). Patterns of late chromosomal DNA replication in unbalanced Chinese hamster–mouse somatic cell hybrids. *Exp. Cell Res.* **89,** 368–376.

Marin, G., and Littlefield, J. W. (1968). Selection of morphologically normal cell lines from polyoma-transformed BHK21/13 hamster fibroblasts. *J. Virol.* **2,** 69–77.

Marin, G., and Manduca, P. (1972). Synchronous replication of the parental chromosomes in a Chinese hamster–mouse somatic hybrid. *Exp. Cell Res.* **75,** 290–293.

Marin, G., and Pugliatti-Crippa, L. (1972). Preferential segregation of homospecific groups of chromosomes in heterospecific somatic cell hybrids. *Exp. Cell Res.* **70,** 253–256.

Marshall, C. J. (1975). A method for analysis of chromosomes in hybrid cells employing sequential G-banding and mouse specific C-banding. *Exp. Cell Res.* **91,** 464–469.

Marshall, C. J., Handmaker, S. D., and Bramwell, M. E. (1975). Synthesis of ribosomal RNA in synkaryons and heterokaryons formed between human and rodent cells. *J. Cell Sci.* **17,** 307–325.

Marshall, J. M., and Nachmias, V. T. (1965). Cell surface and pinocytosis. *J. Histochem. Cytochem.* **13,** 92–104.

Martin, G. R., and Evans, M. J. (1975). Differentiation of clonal lines of teratocarcinoma cells: Formation of embryoid bodies *in vitro. Proc. Natl. Acad. Sci. U.S.A.* **72,** 1441–1445.

Mathai, C. K., Ohno, S., and Beutler, E. (1966). Sex linkage of the *G6PD* gene in equidae. *Nature (London)* **210,** 115–116.

Matsui, S. (1973). "Prophasing" as a possible cause of chromosome translocation in virus-fused cells. *Nature (London), New Biol.* **243,** 208–209.

Matsui, S., Weinfeld, H., and Sandberg, A. A. (1971). Dependence of chromosome pulverization in virus-fused cells on events in the G2 period. *J. Natl. Cancer Inst.* **47,** 401–411.

Matsui, S., Weinfeld, H., and Sandberg, A. A. (1972a). Fate of chromatin of interphase nuclei subjected to "prophasing" in virus-fused cells. *J. Natl. Cancer Inst.* **49,** 1621–1630.

Matsui, S., Yoshida, H., Weinfeld, H., and Sandberg, A. A. (1972b). Induction of prophase in interphase nuclei by fusion with metaphase cells. *J. Cell Biol.* **54,** 120–132.

Matsuya, Y., Green, H., and Basilico, C. (1968). Properties and uses of human–mouse hybrid cell lines. *Nature (London)* **220,** 1199–1202.

Mazia, D. (1961). Mitosis and the physiology of cell division. In "The Cell" (J. Brachet and A. E. Mirsky, eds.), Vol. 3, Chapter 2, pp. 77–412. Academic Press, New York.

Mazia, D. (1963). Synthetic activities leading to mitosis. *J. Cell. Comp. Physiol.* **62,** 123–140.

Medrano, L., and Green, H. (1973). Picornavirus receptors and picornavirus multiplication in human–mouse hybrid cell lines. *Virology* **54,** 515–524.

Meiss, H. K., and Basilico, C. (1972). Temperature sensitive mutants of BHK 21 cells. *Nature (London), New Biol.* **239,** 66–68.

Melchers, G. (1974). Summation: Haploid research in higher plants. *In* "Haploids in Higher Plants: Advances and Potential" (K. J. Kasha, ed.). Crop. Sci. Dept., University of Guelph, Canada.

Melchers, G., and Labib, G. (1974). Somatic hybridization of plants by fusion of protoplasts. I. Selection of light resistant hybrids of "haploid" light sensitive varieties of tobacco. *Mol. Gen. Genet.* **135,** 277–294.

Melnick, J. L. (1973). Classification and nomenclature of viruses. *Prog. Med. Virol.* **15,** 380–384.

Merriam, R. W. (1969). Movement of cytoplasmic proteins into nuclei induced to enlarge and initiate DNA or RNA synthesis. *J. Cell Sci.* **5**, 333–349.

Meyer, G., Berebbi, M., and Klein, G. (1974). Expression of polyoma-induced antigens in low malignant hybrids derived from fusion of a polyoma-induced tumor with a fibroblast line. *Nature (London)* **249**, 47–49.

Mezger-Freed, L. (1972). Effect of ploidy and mutagens on bromodeoxyuridine resistance in haploid and diploid frog cells. *Nature (London), New Biol.* **235**, 245–246.

Michaelis, G., Petrochilo, E., and Slonimski, P. R. (1973). Mitochondrial genetics III. Recombined molecules of mitochondrial DNA obtained from crosses between cytoplasmic *petite* mutants of *Saccharomyces cerevisiae*. Physical and genetic characterization. *Mol. Gen. Genet.* **123**, 51–65.

Michel, W. (1937). Über die experimentelle Fusion pflanzlicher Protoplasten. *Arch. Exp. Zellforsch. Besonders Gewebezuecht.* **20**, 230–252.

Migeon, B. R. (1968). Hybridization of somatic cells derived from mouse and Syrian hamster: Evolution of karyotype and enzyme studies. *Biochem. Genet.* **1**, 305–322.

Migeon, B. R. (1972). Using human somatic cell/mouse cell hybrids no evidence for reactivation of the silent X chromosome in human somatic cell has been obtained. *Nature (London)* **239**, 87–89.

Migeon, B. R., and Miller, C. S. (1968). Human–mouse somatic cell hybrids with single human chromosome (group E): Link with thymidine kinase activity. *Science* **162**, 1005–1006.

Migeon, B. R., Norum, R. A., and Corsaro, C. M. (1974). Isolation and analysis of somatic hybrids derived from two human diploid cells. *Proc. Natl. Acad. Sci. U.S.A.* **71**, 937–941.

Miggiano, V., Nabholz, M., and Bodmer, W. (1969). Hybrids between human leukocytes and a mouse cell line: Production and characterization. *Wistar Inst. Symp. Monogr.* **9**, 61–76.

Miller, D. A., Miller, O. J., Dev, V. G., Hashmi, S., and Tantravahi, R. (1974). Human chromosome 19 carries a poliovirus receptor gene. *Cell* **1**, 167–170.

Miller, O. J., Allderdice, P. W., Miller, D. A., Breg, W. R., and Migeon, B. R. (1971). Human thymidine kinase gene locus: Assignment to chromosome 17 in a hybrid of man and mouse cells. *Science* **173**, 244–245.

Miller, O. J., Miller, D. A., and Warburton, D. (1973). Application of new staining techniques to the study of human chromosomes. *Prog. Med. Genet.* **9**, 1–47.

Minna, J., Nelson, P., Peacock, J., Glazer, D., and Nirenberg, M. (1971). Genes for neuronal properties expressed in neuroblastoma × L cell hybrids. *Proc. Natl. Acad. Sci. U.S.A.* **68**, 234–239.

Minna, J., Glazer, D., and Nirenberg, M. (1972). Genetic dissection of neural properties using somatic cell hybrids. *Nature (London), New Biol.* **235**, 225–231.

Minna, J. D., and Coon, H. G. (1974). Human × mouse hybrid cells segregating mouse chromosomes and isozymes. *Nature (London)* **252**, 401–404.

Minna, J. D., Lueders, K. K., and Kuff, E. L. (1974a). Expression of genes for intracisternal A-particle antigen in somatic cell hybrids. *J. Natl. Cancer Inst.* **52**, 1211–1217.

Minna, J. D., Gazdar, A. F., Iverson, G. M., Marshall, T. H., Stromberg, K., and Wilson, S. H. (1974b). Oncornavirus expression in human × mouse hybrid cells segregating mouse chromosomes. *Proc. Natl. Acad. Sci. U.S.A.* **71**, 1695–1700.

Minna, J. D., Yavelow, J., and Coon, H. G. (1975). Expression of phenotypes in hybrid somatic cells derived from the nervous system. *Genetics* **79**, suppl: 373–383.

Minor, P. D., and Roscoe, D. H. (1975). Colchicine resistance in mammalian cell lines. *J. Cell Sci.* **17**, 381–396.

Mintz, B. (1962). Formation of genotypically mosaic mouse embryos. *Amer. Zoologist* **2**, 432.

Mintz, B. (1971). Genetic mosaicism *in vivo:* Development and disease in allophenic mice. *Fed. Proc., Fed. Am. Soc. Exp. Biol.* **30**, 935–943.

Mintz, B., and Baker, W. W. (1967). Normal mammalian muscle differentiation and gene control of isocitrate dehydrogenase. *Proc. Natl. Acad. Sci. U.S.A.* **58**, 592–598.

Mintz, B., and Illmensee, K. (1975). Normal genetically mosaic mice produced from malignant teratocarcimoma cells. *Proc. Natl. Acad. Sci. U.S.A.* **72**, 3585–3589.

Mohit, B. (1971). Immunoglobulin G and free kappa chain synthesis in different clones of a hybrid cell line. *Proc. Natl. Acad. Sci. U.S.A.* **68**, 3045–3048.

Mohit, B., and Fan, K. (1971). Hybrid cell line from a cloned immunoglobulin-producing mouse myeloma and a nonproducing mouse lymphoma. *Science* **171**, 75–77.

Morgan, C., and Howe, C. (1968). Structure and development of viruses as observed in the electron microscope. IX. Entry of parainfluenza I (Sendai) virus. *J. Virol.* **2**, 1122–1132.

Morris, J. A., Blount, R. F., and Savage, R. E. (1956). Recovery of cytopathogenic agent from chimpanzees with coryza. *Proc. Soc. Exp. Biol. Med.* **92**, 544–549.

Moscona, A. (1966). The development *in vitro* of chimeric aggregates of dissociated embryonic chick and mouse cells. *In* "Molecular and Cellular Aspects of Developments" (E Bell, ed.), pp. 55–64. Harper, New York.

Moser, F. G., Dorman, B. P., and Ruddle, F. H. (1975). Mouse–human heterokaryon analysis with A 33258 Hoechst–Giemsa technique. *J. Cell Biol.* **66**, 676–680.

Moses, E., and Kohn, A. (1963). Polykaryocytosis induced by Rous sarcoma virus in chick fibroblasts. *Exp. Cell Res.* **32**, 182–186.

Mukherjee, A. B., Dev, G. V., Miller, O. J. (1971). Sendai virus-induced cell sorting leading to apparent preferential fusion of like cells. *Exptl. Cell. Res.* **68**, 130–136.

Murayama, F., and Okada, Y. (1965). Effect of calcium on the cell fusion reaction caused by HVJ. *Biken J.* **8**, 103–105.

Murayama, F., and Okada, Y. (1970a). Efficiency of hybrid cell formation from heterokaryons fused by HVJ. *Biken J.* **13**, 1–9.

Murayama, F., and Okada, Y. (1970b). Appearance, characteristics and malignancy of somatic hybrid cells between L and Ehrlich ascites tumor cells formed by artificial fusion with UV-HVJ. *Biken J.* **13**, 11–23.

Murayama-Okabayashi, F., Okada, Y., and Tachibana, T. (1971). A series of hybrid cells containing different ratios of parental chromosomes formed by two steps of artificial fusion. *Proc. Natl. Acad. Sci. U.S.A.* **68**, 38–42.

Nabholz, M., Miggiano, V., and Bodmer, W. (1969). Genetic analysis with human–mouse somatic cell hybrids. *Nature (London)* **223**, 358–363.

Nachmias, V. T., Huxley, H. E., and Kessler, D. (1970). Electron microscope observations on actomyosin and actin preparations from *Physarum polycephalum,* and on their interaction with heavy meromyosin subfragment I from muscle myosin. *J. Mol. Biol.* **50**, 83–90.

Nadler, H. L., Chacko, C. M., and Rachmeler, M. (1970). Interallelic complementation in hybrid cells derived from human diploid strains deficient in galactose-1-phosphate uridyl transferase activity. *Proc. Natl. Acad. Sci. U.S.A.* **67**, 976–982.

Naha, P. M. (1969). Temperature sensitive conditional mutants of monkey kidney cells. *Nature (London)* **223**, 1380–1381.

Neff, J. M., and Enders, S. F. (1968). Polio virus replication and cytopathogenicity in monolayer hamster cultures fused with beta propiolactone inactivated Sendai virus. *Proc. Soc. Exp. Biol. Med.* **127**, 260–267.

Neurath, A. R. (1965). Study on the adenosine diphosphatase (adenosine triphosphatase) associated with Sendai virus. *Acta Virol. (Engl. Ed.)* **9,** 313–322.

Neurath, A. R., and Sokol, F. (1963). Association of myxoviruses with an adenosine diphosphatase/adenosine triphosphatase/ as revealed by chromatography on DEAE-cellulose and by density gradient centrifugation. *Z. Naturforsch., Teil B* **18,** 1050–1052.

New Haven Conference 1973. (1974). First international workshop on human gene mapping, *Birth Defects; Orig. Artic. Ser.* **10,** No 3.

Nichols, W. W. (1963). Relationships of viruses, chromosomes and carcinogenesis. *Hereditas* **50,** 53–80.

Nichols, W. W., Levan, A., Aula, P., and Norrby, E. (1964). Extreme chromosome breakage induced by measles virus in different *in vitro* systems. *Hereditas* **51,** 380–382.

Nichols, W. W., Levan, A., Aula, P., and Norrby, E. (1965). Chromosome damage associated with the measles virus *in vitro*. *Hereditas* **54,** 101–118.

Nichols, W. W., Aula, P., Levan, A., Heneen, W., and Norrby, E. (1967). Radioautography with tritiated thymidine in measles and Sendai virus-induced chromosome pulverizations. *J. Cell Biol.* **35,** 257–262.

Nicolson, G. L. (1971). Difference in topology of normal and tumor cell membranes shown by different surface distributions of ferritin-conjugated concanavalin A. *Nature (London), New Biol.* **233,** 244–246.

Norrby, E., Levan, A., and Nichols, W. W. (1965). The correlation between the chromosome pulverization effect and other biological activities of measles virus preparations. *Exp. Cell Res.* **41,** 483–491.

Norum, R. A., and Migeon, B. R. (1974). Non-random loss of human markers from man–mouse somatic cell hybrid. *Nature (London)* **251,** 72–74.

Norwood, T. H., Pendergrass, W. R., Sprague, C. A., and Martin, G. M. (1974). Dominance of the senescent phenotype in heterokaryons between replicative and post-replicative human fibroblast-like cells. *Proc. Natl. Acad. Sci. U.S.A.* **71,** 2231–2235.

Norwood, T. H., Pendergrass, W. R., and Martin, G. M. (1975). Reinitiation of DNA synthesis in senescent human fibroblasts upon fusion with cells of unlimited growth potential. *J. Cell Biol.* **64,** 551–556.

Nuzzo, F., Dambrosio, F., Marini, A., and Decarli, L. (1971). Chromosomal chimerism in a patient transfused *in utero*. *Ann. Ostet. Ginecol.* **92,** 615–617.

Nyormoi, O., Klein, G., Adams, A., and Dombos, L. (1973a). Sensitivity to EBV superinfection and IUdR inducibility of hybrid cells formed between a sensitive and a relatively resistant Burkitt lymphoma cell line. *Int. J. Cancer* **12,** 396–408.

Nyormoi, O., Coon, H. G., and Sinclair, J. H. (1973b). Proliferating hybrid cells formed between rat spermatids and an established line of mouse fibroblast. *J. Cell Sci.* **13,** 863–878.

Obara, Y., Yoshida, H., Chai, L. S., Weinfeld, H., and Sandberg, A. A. (1973). Contrast between the environmental pH dependencies of prophasing and nuclear membrane formation in interphase–metaphase cells. *J. Cell Biol.* **58,** 608–617.

Obara, Y., Chai, L. S., Weinfeld, H., and Sandberg, A. A. (1974a). Synchronization of events in fused interphase–metaphase binucleate cells: Progression of the telophase-like nucleus. *J. Natl. Cancer Inst.* **53,** 247–259.

Obara, Y., Chai, L. S., Weinfeld, H., and Sandberg, A. A. (1974b). Prophasing of interphase nuclei and induction of nuclear envelopes around metaphase chromosomes in HeLa and Chinese hamster homo- and heterokaryons. *J. Cell Biol.* **62,** 104–113.

Obara, Y., Weinfeld, H., and Sandberg, A. A. (1975). Detection by means of cell fusion of macromolecular synthesis involved in the reconstruction of the nuclear envelope in mitosis. *J. Cell Biol.* **64**, 378–388.

Oftebro, R., and Wolf, I. (1967). Mitosis of bi- and multinucleate HeLa cells. *Exp. Cell Res.* **48**, 39–52.

Ohnishi, T. (1962). Extraction of actin- and myosin-like proteins from erythrocyte membrane. *J. Biochem. (Tokyo)* **52**, 307–308.

Ohno, S. (1967). "Sex Chromosomes and Sex-linked Genes." Springer-Verlag, Berlin and New York.

Ohno, S., Poole, J., and Gustavsson, I. (1965). Sex-linkage of erythrocyte glucose-6-phosphate dehydrogenase in two species of wild hares. *Science* **150**, 1737–1738.

Okada, Y. (1958). The fusion of Ehrlich's tumor cells caused by HVJ virus *in vitro*. *Biken's J* **1**, 103–110.

Okada, Y. (1962a). Analysis of giant polynuclear cell formation caused by HVJ virus from Ehrlich's ascites tumor cells I. Microscopic observation of giant polynuclear cell formation. *Exp. Cell Res.* **26**, 98–107.

Okada, Y. (1962b). Analysis of giant polynuclear cell formation caused by HVJ virus from Ehrlich's ascites tumor cells. III. Relationship between cell condition and fusion reaction or cell degeneration reaction. *Exp. Cell Res.* **26**, 119–128.

Okada, Y. (1969). Factors in fusion of cells by HVJ. *Curr. Top. Microbiol. Immunol.* **48**, 102–128.

Okada, Y. (1972). Fusion of cells by HVJ (Sendai virus). *In* "Membrane Research" (C. F. Fox, ed.) pp. 371–382. Academic Press, New York.

Okada, Y., and Murayama, F. (1965). Multinucleated giant cell formation by fusion between cells of two different strains. *Exp. Cell Res.* **40**, 154–158.

Okada, Y., and Murayama, F. (1966). Requirement of calcium ions for the cell fusion reaction of animal cells by HVJ. *Exp. Cell Res.* **44**, 527–551.

Okada, Y., and Murayama, F. (1968). Fusion of cells by HVJ: Requirement of concentration of virus particles at the site of contact of two cells for fusion. *Exp. Cell Res.* **52**, 34–42.

Okada, Y., and Tadokoro, J. (1962). Analysis of giant polynuclear cell formation caused by HVJ virus from Ehrlich's tumor cells. II. Quantitative analysis of giant polynuclear cell formation. *Exp. Cell Res.* **26**, 108–118.

Okada, Y., and Tadokoro, J. (1963). The distribution of cell fusion capacity among several cell strains or cells caused by HVJ. *Exp. Cell Res.* **32**, 417–430.

Okada, Y., Suzuki, T., and Hosaka, Y. (1957). Interaction between influenza virus and Ehrlich's tumor cells. III. Fusion phenomenon of Ehrlich's tumor cells by the action of HVJ, Z strain. *Med. J. Osaka Univ.* **7**, 709–717.

Okada, Y., Murayama, F., and Yamada, K. (1966). Requirement of energy for the cell fusion reaction of Ehrlich ascites tumor cells. *Virology* **28**, 115–130.

Okazaki, K., and Holtzer, H. (1966). Myogenesis: Fusion, myosin synthesis and the mitotic cycle. *Proc. Natl. Acad. Sci. U.S.A.* **56**, 1484–1490.

Olson, M. O. J., and Busch, H. (1974). Nuclear proteins. *In* "The Cell Nucleus" (H. Busch, ed.), Vol. 3, pp. 211–268. Academic Press, New York.

Orgel, L. E. (1973). Ageing of clones of mammalian cells. *Nature (London)* **243**, 441–445.

Orkin, S. H., Buchanan, P. D., Yount, W. J., Reisner, H., and Littlefield, J. W. (1973). Lambda-chain production in human lymphoblast–mouse fibroblast hybrids. *Proc. Natl. Acad. Sci. U.S.A.* **70**, 2401–2405.

Orkin, S. H., Harosi, F. E., and Leder, P. (1975). Differentiation in erythroleukemic cells and their somatic hybrids. *Proc. Natl. Acad. Sci. U.S.A.* **72**, 98–102.

Ossowski, L., Unkeless, J. C., Tobia, A., Quigley, J. P., Rifkin, D. B., and Reich, E.

(1973). An enzymatic function associated with transformation of fibroblasts by oncogenic viruses. II. Mammalian fibroblast cultures transformed by DNA and RNA tumor viruses. *J. Exp. Med.* **137**, 112–126.

Pagano, R. E., Huang, L., and Wey, C. (1974). Interaction of phospholipid vesicles with cultured mammalian cells. *Nature (London)* **252**, 166–167.

Paine, P. L., and Feldherr, C. M. (1972). Nucleocytoplasmic exchange of macromolecules. *Exp. Cell Res.* **74**, 81–98.

Papahadjopoulos, D., Poste, G., and Schaeffer, B. E. (1973). Fusion of mammalian cells by unilamellar lipid vesicles. Influence of lipid surface charge, fluidity and cholesterol. *Biochim. Biophys. Acta* **323**, 23–42.

Papahadjopoulos, D., Mayhew, E., Poste, G., Smith, S., and Vail, W. J. (1974). Incorporation of lipid vesicles by mammalian cells provides a potential method for modifying cell behaviour. *Nature (London)* **252**, 163–166.

Pardue, M. L., and Gall, J. G. (1970). Chromosomal localization of mouse satellite DNA. *Science* **168**, 1356–1358.

Paris Conference. (1972). Standardization in human cytogenetics. *Birth Defects, Orig. Artic. Ser. 8, 7.*

Parkman, R., Hagemeier, A., and Merler, E. (1971). Production of a human γ-globulin fragment by human thymocyte-mouse fibroblast hybrid cells. *Fed. Proc., Fed. Am. Soc. Exp. Biol.* **30**, p. 530.

Pasternak, C. A., and Micklem, K. J. (1973). Permeability changes during cell fusion. *J. Membr. Biol.* **14**, 293–303.

Paterson, M. C., and Lohman, P. H. (1975). Use of enzymatic assay to evaluate UV-induced DNA repair in human and embryonic chick fibroblasts and multinucleate heterokaryons derived from both. *Basic Life Sci.* **5B**, 735–745.

Paterson, M. C., Lohman, P. H., Westerveld, A., and Sluyter, M. L. (1974a). DNA repair monitored by an enzymatic assay in multinucleate *Xeroderma Pigmentosum* cells after fusion. *Nature (London)* **248**, 50–52.

Paterson, M. C., Lohman, P. H., Westerveld, A., and Sluyter, M. L. (1974b). DNA repair in human/embryonic chick heterokaryons. Ability of each species to aid the other in the removal of ultraviolet-induced damage. *Biophys. J.* **14**, 835–845.

Patterson, D. (1975). Biochemical genetics of Chinese hamster cell mutants with deviant purine metabolism: Biochemical analysis of eight mutants. *Somatic Cell Genet.* **1**, 91–110.

Patterson, D., Kao, F. T., and Puck, T. T. (1974). Genetics of somatic mammalian cells: Biochemical genetics of Chinese hamster cell mutants with deviant purine metabolism. *Proc. Natl. Acad. Sci. U.S.A.* **71**, 2057–2061.

Paul, J., and Hickey, I. (1974). Haemoglobin synthesis in inducible, uninducible and hybrid Friend cell clones. *Exp. Cell Res.* **87**, 20–30.

Peacock, J. H., McMorris, F. A., and Nelson, P. G. (1973). Electrical excitability and chemosensitivity of mouse neuroblastoma × mouse or human fibroblast hybrids. *Exp. Cell Res.* **79**, 199–212.

Pearson, P. L., Sanger, R., and Brown, J. A. (1975). Report of the committee on the genetic constitution of the X-chromosome. *Cytogenet. Cell Genet.* **14**, 190–195.

Pederson, T. (1972). Chromatin structure and the cell cycle. *Proc. Natl. Acad. Sci. U.S.A.* **69**, 2224–2228.

Pederson, T., and Robbins, E. (1972). Chromatin structure and the cell division cycle: Actinomycin binding in synchronized HeLa cells. *J. Cell Biol.* **55**, 322–327.

Periman, P. (1970). IgG synthesis in hybrid cells from an antibody-producing mouse myeloma and a L cell substrain. *Nature (London)* **228**, 1086–1087.

Peterson, J. A., and Weiss, M. C. (1972). Expression of differentiated functions in hepatoma cell hybrids: Induction of mouse albumin production in rat hepatoma–mouse fibroblast hybrids. *Proc. Natl. Acad. Sci. U.S.A.* **69,** 571–575.

Phillips, S. G., and Phillips, D. M. (1969). Sites of nucleolus production in cultured Chinese hamster cells. *J. Cell Biol.* **40,** 248–268.

Phillips, S. G., Phillips, D. M., Dev, V. G., Miller, D. A., van Diggelen, O. P., and Miller, O. J. (1976). Spontaneous cell hybridization of somatic cells present in sperm suspensions. *Exp. Cell Res.* **98,** 424–443.

Pollack, R. E., Green, H., and Todaro, G. J. (1968). Growth control in cultured cells: Selection of sublines with increased sensitivity to contact inhibition and decreased tumor producing ability. *Proc. Natl. Acad. Sci. U.S.A.* **60,** 126–133.

Pollack, R. E., Salas, J., Wang, R., Kusano, T., and Green, H. (1971). Human–mouse hybrid cell lines and susceptibility to species-specific viruses. *J. Cell. Physiol.* **77,** 117–119.

Pollack, R., E., Risser, R., Conlon, S., and Rifkin, D. (1974). Plasminogen activator production accompanies loss of anchorage regulation in transformation of primary rat embryo cells by Simian virus 40. *Proc. Natl. Acad. Sci. U.S.A.* **71,** 4792–4796.

Pollard, T. D., and Ito, S. (1970). Cytoplasmic filaments of *Amoeba proteus.* I. The role of filaments in consistency changes and movement. *J. Cell Biol.* **46,** 267–289.

Pontecorvo, G. (1956). The parasexual cycle in fungi. *Annu. Rev. Microbiol.* **10,** 393–400.

Pontecorvo, G. (1971). Induction of directional chromosome elimination in somatic cell hybrids. *Nature (London)* **230,** 367–369.

Pontecorvo, G. (1976). Production of indefinitely multiplying mammalian somatic cell hybrids by polyethylene glycol (PEG) treatment. *Somatic. Cell Genet.* **1,** 397–400.

Poole, A. R., Howell, J. I., and Lucy, J. A. (1970). Lysolecithin in cell fusion. *Nature (London)* **227,** 810–814.

Poste, G. (1970). Virus-induced polykaryocytosis and the mechanism of cell fusion. *Adv. Virus Res.* **16,** 303–356.

Poste, G. (1971). Sublethal autolysis. Modification of cell periphery by lysosomal enzymes. *Exp. Cell Res.* **67,** 11–16.

Poste, G. (1972). Enucleation of mammalian cells by cytochalasin B. I. Characterization of anucleate cells. *Exp. Cell Res.* **73,** 273–286.

Poste, G. (1973). Anucleate mammalian cells: Applications in cell biology and virology. *Methods Cell Biol.* **7,** 211–250.

Poste, G., and Allison, A. C. (1971). Membrane fusion reaction: A theory. *J. Theor. Biol.* **32,** 165–184.

Poste, G., and Reeve, P. (1971). Formation of hybrid cells and heterokaryons by fusion of enucleated and nucleated cells. *Nature (London), New Biol.* **229,** 123–125.

Poste, G., and Reeve, P. (1972). Enucleation of mammalian cells by cytochalasin B. II. Formation of hybrid cells and heterokaryons by fusion of anucleate and nucleated cells. *Exp. Cell Res.* **73,** 287–294.

Poste, G., Schaeffer, B., Reeve, P., and Alexander, D. J. (1974). Rescue of Simian virus 40 (SV40) from SV40-transformed cells by fusion with anucleate monkey cells and variation in the yield of virus rescued by fusion with replicating or nonreplicating monkey cells. *Virology* **60,** 85–95.

Potrykus, I., and Hoffmann, F. (1973). Transplantation of nuclei into protoplasts of higher plants. *Z. Pflanzenphysiol.* **69,** 287–289.

Power, J. B., Cummins, S. E., and Cocking, E. C. (1970). Fusion of isolated plant protoplasts. *Nature (London)* **225,** 1016–1018.

Prescott, D. M., and Bender, M. A. (1962). Synthesis of RNA and protein during mitosis in mammalian tissue culture cells. *Exp. Cell Res.* **26,** 260–268.

Prescott, D. M., and Kirkpatrick, J. B. (1973). Mass enucleation of cultured animal cells. *Methods Cell Biol.* **7**, 189–202.

Prescott, D. M., Myerson, D., and Wallace, J. (1972). Enucleation of mammalian cells with cytochalasin B¹. *Exp. Cell Res.* **71**, 480–485.

Puck, T. T. (1974). Hybridization and somatic cell genetics. *In* "Somatic Cell Hybridization" (R. L. Davidson and F. F. de la Cruz, eds.), pp. 27–32. Raven, New York.

Puck, T. T., and Kao, F. (1967). Genetics of somatic mammalian cells. V. Treatment with 5-bromodeoxyuridine and visible light for the isolation of nutritionally deficient mutants. *Proc. Natl. Acad. Sci. U.S.A.* **58**, 1227–1234.

Rabinowitz, Z., and Sachs, L. (1970). Control of the reversion of properties in transformed cells. *Nature (London)* **225**, 136–139.

Rao, P. N., and Johnson, R. T. (1970). Mammalian cell fusion: Studies on the regulation of DNA synthesis and mitosis. *Nature (London)* **225**, 159–164.

Rao, P. N., and Johnson, R. T. (1971). Mammalian cell fusion. IV. Regulation of chromosome formation from interphase nuclei by various chemical compounds. *J. Cell. Physiol.* **78**, 217–224.

Rao, P. N., and Johnson, R. T. (1972a). Cell fusion and its application to studies on the regulation of the cell cycle. *Methods Cell Physiol.* **5**, 75–126.

Rao, P. N., and Johnson, R. T. (1972b). Premature chromosome condensation: A mechanism for the elimination of chromosomes in virus-fused cells. *J. Cell Sci.* **10**, 495–513.

Rao, P. N., and Johnson, R. T. (1974). Regulation of cell cycle in hybrid cells. *In* "Control of Proliferation in Animal Cells" (B. Clarkson and R. Baserga, eds.), pp. 785–800. Cold Spring Harbor Lab., Cold Spring Harbor, New York.

Rao, P. N., and Johnson, R. T. (1974). Induction of chromosome condensation in interphase cells. *Advan. Cell Molec. Biol.* **3**, 135–189.

Rao, P. N., Hittelman, W. N., and Wilson, B. A. (1975). Mammalian cell fusion. VI. Regulation of mitosis in binucleate HeLa cells. *Exp. Cell Res.* **90**, 40–46.

Raper, J. R. (1966). "Genetics of Sexuality in Higher Fungi," p. 601. Ronald Press, New York.

Reeve, P., Hewlett, G., Watkins, H., Alexander, D. J., and Poste, G. (1974). Virus-induced cell fusion enhanced by phytohaemagglutinin. *Nature (London)* **249**, 355–356.

Renwick, J. H. (1971a). The Rhesus syntenic group in man. *Nature (London)* **234**, 475.

Renwick, J. H. (1971b). The mapping of human chromosomes. *Annu. Rev. Genet.* **5**, 81–120.

Ricciuti, F. C., and Ruddle, F. H. (1971). Biochemical and cytological evidence for a triple hybrid cell line formed from fusion of three different cells. *Science* **172**, 470–472.

Ricciuti, F. C., and Ruddle, F. H. (1973a). Assignment of nucleoside phosphorylase to D-14 and localization of X-linked loci in man by somatic cell genetics. *Nature (London), New Biol.* **241**, 180–182.

Ricciuti, F. C., and Ruddle, F. H. (1973b). Assignment of three gene loci (*PGK, HGPRT, G6PD*) to the long arm of the human X chromosome by somatic cell genetics. *Genetics* **74**, 661–678.

Richardson, B. J., Czuppon, A. B., and Sharman, G. B. (1971). Inheritance of G6PD variation in kangaroos. *Nature (London), New Biol.* **230**, 154–155.

Ringertz, N. R. (1969). Cytochemical properties of nuclear proteins and deoxyribonucleoprotein complexes in relation to nuclear function. *In* "Handbook of Molecular Cytology" (A. Lima-de-Faria, ed.), Chapter 27, pp. 656–684. North-Holland Publ., Amsterdam.

Ringertz, N. R. (1974). Regulation and gene expression in heterokaryons. *In* "Somatic Cell Hybridization (R. L. Davidson and F. de la Cruz, eds.), pp. 239–264. Raven, New York.

Ringertz, N. R., and Bolund, L. (1974a). Reactivation of chick erythrocyte nuclei by somatic cell hybridization. *Int. Rev. Exp. Pathol.* **8,** 83–116.

Ringertz, N. R., and Bolund, L. (1974b). The nucleus during avian erythroid differentiation. *In* "The Cell Nuclus" (H. Busch, ed.), Vol. 3, pp. 417–446. Academic Press, New York.

Ringertz, N. R., Carlsson, S.-A., Ege, T., and Bolund, L. (1971). Detection of human and chick nuclear antigens in nuclei of chick erythrocytes during reactivation in heterokaryons with HeLa cells. *Proc. Natl. Acad. Sci. U.S.A.* **68,** 3228–3232.

Rintoul, D., Colofiore, J., and Morrow, J. (1973). Expression of differentiated properties in foetal liver cells and their somatic cell hybrids. *Exp. Cell Res.* **78,** 414–422.

Rizki, R. M., Rizki, T. M., and Andrews, C. A. (1975). Drosophila cell fusion induced by wheat germ agglutinin. *J. Cell Sci.* **18,** 113–121.

Robbins, J. H., Kraemer, K. H., Lutzner, M. A., Festoff, B. W., and Coon, H. G. (1974). *Xeroderma Pigmentosum.* An inherited disease with sun sensitivity, multiple cutaneous neoplasms and abnormal DNA repair. *Ann. Intern. Med.* **80,** 221–248.

Röhme, D. (1974). Prematurely condensed chromosomes of the Indian muntjac: A model system for the analysis of chromosome condensation and banding. *Hereditas* **76,** 251–258.

Röhme, D. (1975). Evidence suggesting chromosome continuity during the S-phase of Indian muntjac cells. *Hereditas* **80,** 145–149.

Roizman, B. (1962). Polykaryocytosis induced by viruses. *Proc. Natl. Acad. Sci. U.S.A.* **48,** 228–234.

Rosenberg, M. (1972). Fusion of frog and tadpole erythrocytes. *Nature (London)* **239,** 520–522.

Rosenblith, J. Z., Ukena, T. E., Yin, H. H., Berlin, R. D., and Karnovsky, M. J. (1973). A comparative evaluation of the distribution of concanavalin A-binding sites on the surfaces of normal, virally-transformed, and protease-treated fibroblasts. *Proc. Natl. Acad. Sci. U.S.A.* **70,** 1625–1629.

Rosenqvist, M., Stenman, S., and Ringertz, N. R. (1975). Uptake of SV40 T-antigen into chick erythrocyte nuclei in heterokaryons. *Exp. Cell Res.* **92,** 515–518.

Rosenthal, M. D., Wishnow, R. M., and Sato, G. H. (1970). *In vitro* growth and differentiation of clonal populations of multipotential mouse cells derived from a transplantable testicular teratocarcinoma. *J. Natl. Cancer Inst.* **44,** 1001–1014.

Rosenthal, A. S., Kregenow, F. M., and Moses, H. L. (1970). Some characteristics of a Ca^{2+}-dependent ATPase activity associated with a group of erythrocyte membrane proteins which form fibrils. *Biochim. Biophys. Acta* **196,** 254–262.

Rott, R., Drzeniek, R., Saber, M. S., and Reichert, E. (1966). Blood group substances, Forssman and mononucleosis antigens in lipid-containing RNA viruses. *Arch. Gesamte Virusforsch.* **19,** 271–288.

Rotterdam Conference 1974 (1975). Second international workshop on human gene mapping. *Birth Defects; Orig. Artic. Ser.* **11** No. 3.

Roy, K. L., and Ruddle, F. H. (1973). Microscale isoelectric focusing studies of mouse and human hypoxanthine-guanine phosphoribosyl transferases. *Biochem. Genet.* **9,** 175–185.

Rubin, H. (1966). A substance in conditioned medium which enhances the growth of small numbers of chick embryo cells. *Exp. Cell Res.* **41,** 138–148.

Ruddle, F. H., and Creagan, R. P. (1975). Parasexual approaches to the genetics of man. *Annu. Rev. Genet.* **9,** 407–486.

Ruddle, F. H., and Kucherlapati, R. S. (1974). Hybrid cells and human genes. *Sci. Am.* **231,** 36–44.

Ruddle, F. H., and Nichols, E. A. (1971). Starch gel electrophoretic phenotypes of mouse × human somatic cell hybrids and mouse isoenzyme polymorphisms. *In Vitro* **7,** 120–131.

Ruddle, F. H., Chen, T., Shows, T. B., and Silagi, S. (1970a). Interstrain somatic cell hybrids in the mouse. Chromosome and enzyme analyses. *Exp. Cell Res.* **60,** 139–147.

Ruddle, F. H., Chapman, V. M., Chen, T. R., and Klebe, R. J. (1970b). Linkage between human lactate dehydrogenase A and B and peptidase B. *Nature (London)* **227,** 251–257.

Ruddle, F. H., Chapman, V. M., Ricciuti, F., Murnane, M., Klebe, R., and Meera Khan, P. (1971). Linkage relationships of seventeen human gene loci as determined by man–mouse somatic cell hybrids. *Nature (London), New Biol.* **232,** 69–73.

Sabban, E., and Loyter, A. (1974). Fusion of chicken erythrocytes by phospholipase C (*Clostridium perfringens*). The requirement for hemolytic and hydrolytic factors for fusion. *Biochim. Biophys. Acta* **362,** 100–109.

Sabin, A. B., and Koch, M. A. (1963). Behaviour of noninfectious SV40 viral genome in hamster tumor cells: Induction of synthesis of infectious virus. *Proc. Natl. Acad. Sci. U.S.A.* **50,** 407–417.

Sachs, L. (1974). Regulation of membrane changes, differentiation and malignancy in carcinogenesis. *Harvey Lect.* **68,** 1–35.

Sachs, L., Inbar, M., and Shinitzky, M. (1974). Mobility of lectin sites on the surface membrane and the control of cell growth and differentiation. *In* "Control of Proliferation in Animal Cells" (B. Clarkson and R. Baserga, eds.), pp. 283–296. Cold Spring Harbor Lab., Cold Spring Harbor, New York.

Sager, R. (1972). "Cytoplasmic Genes and Organelles." Academic Press, New York.

Saksela, E., Aula, P., and Cantell, K. (1965). Chromosomal damage of human cells induced by Sendai virus. *Ann. Med. Exp. Biol. Fenn.* **43,** 132–136.

Sambrook, J., Westphal, H., Srinivasan, P. R., and Dulbecco, R. (1968). The integrated state of viral DNA in SV40-transformed cells. *Proc. Natl. Acad. Sci. U.S.A.* **60,** 1288–1295.

Sambrook, J., Botchan, M., Gallimore, P., Ozanne, B., Pettersson, U., Williams, J., and Sharp, P. A. (1975). Viral DNA sequences in cells transformed by Simian virus 40, adenovirus type 2 and adenovirus type 5. *Cold Spring Harbor Symp. on Tumor Viruses Quant. Biol.* **39,** 615–632.

Sanbe, M., Aya, T., Ikeuchi, T., and Sandberg, A. A. (1970). Electron microscopic study of fused cells, with special reference to chromosome pulverization. *J. Natl. Cancer Inst.* **44,** 1079–1089.

Sandberg, A. A., Sofuni, T., Takagi, N., and Moore, G. E. (1966). Chronology and pattern of human chromosome replication. IV. Autoradiographic studies of binucleate cells. *Proc. Natl. Acad. Sci. U.S.A.* **56,** 105–110.

Sandberg, A. A., Aya, T., Ikeuchi, T., and Weinfeld, H. (1970). Definition and morphologic features of chromosome pulverization: A hypothesis to explain the phenomenon. *J. Natl. Cancer Inst.* **45,** 615–623.

Sanford, K. K., Likely, G. D., and Earle, W. R. (1954). The development of variations in transplantability and morphology within a clone of mouse fibroblasts transformed to sarcoma-produced cells *in vitro. J. Natl. Cancer Inst.* **15,** 215–237.

Sanger, J. W., Holtzer, S., and Holtzer, H. (1971). Effects of cytochalasin B on muscle cells in tissue culture. *Nature (London), New Biol.* **229,** 121–123.

Santachiara, A. S., Nabholz, M., Miggiano, V., Darlington, A. J., and Bodmer, W. (1970). Genetic analysis with man–mouse somatic cell hybrids. *Nature (London)* **227,** 248–251.

Sanui, H., and Pace, N. (1967a). The Mg, Ca, EDTA, and ATP dependence of Na binding by rat liver microsomes. *J. Cell. Physiol.* **69**, 3–10.

Sanui, H., and Pace, N. (1967b). Effect of ATP, EDTA and EGTA on the simultaneous binding of Na, K, Mg and Ca by rat liver microsomes. *J. Cell. Physiol.* **69**, 11–20.

Sawicki, W., and Koprowski, H. (1971). Fusion of rabbit spermatozoa with somatic cells cultivated *in vitro. Exp. Cell Res.* **66**, 145–151.

Sayavedra, M. S., and Eliceiri, G. L. (1975). Regulation of species-specific ribosomal RNA in a somatic hybrid cell. *Biochim. Biophys. Acta* **378**, 216–220.

Scaletta, L. J., and Ephrussi, B. (1965). Hybridization of normal and neoplastic cells *in vitro. Nature (London)* **205**, 1169–1171.

Scaletta, L. J., Rushforth, N. B., and Ephrussi, B. (1967). Isolation and properties of hybrids between somatic mouse and Chinese hamster cells. *Genetics* **57**, 107–124.

Schatz, G., and Mason, T. L. (1974). The biosynthesis of mitochondrial proteins. *Annu. Rev. Biochem.* **43**, 51–87.

Scheid, A., and Choppin, P. W. (1974). Identification of biological activities of paramyxovirus glycoproteins. Activation of cell fusion, hemolysis and infectivity of proteolytic cleavage of an inactive precursor protein of Sendai virus. *Virology* **57**, 475–490.

Scheintaub, H. M., and Fiel, R. J. (1973). RNA polymerase in normal avian erythrocytes. *Exp. Cell Res.* **80**, 442–445.

Schlegel, R. A., and Rechsteiner, M. C. (1975). Microinjection of thymidine kinase and bovine serum albumin into mammalian cells by fusion with red blood cells. *Cell* **5**, 371–379.

Schneeberger, E. E., and Harris, H. (1966). An ultrastructural study of interspecific cell fusion induced by inactivated Sendai virus. *J. Cell Sci.* **1**, 401–406.

Schneider, J. A., and Weiss, M. C. (1971). Expression of differentiated functions in hepatoma cell hybrids. I. Tyrosine aminotransferase in hepatoma–fibroblast hybrids. *Proc. Natl. Acad. Sci. U.S.A.* **68**, 127–131.

Schor, S. L., Johnson, R. T., and Waldren, C. A. (1975a). Changes in the organization of chromosomes during the cell cycle: response to ultraviolet light. *J. Cell Sci.* **17**, 539–566.

Schor, S. L., Johnson, R. T., and Mullinger, A. M. (1975b). Perturbation of mammalian cell division. II. Studies on the isolation and characterization of human mini segregant cells. *J. Cell Sci.* **19**, 281–303.

Schwaber, J. (1975). Immunoglobulin production by a human–mouse somatic cell hybrid. *Exp. Cell Res.* **93**, 343–354.

Schwaber, J., and Cohen, E. P. (1973). Human × mouse somatic cell hybrid clone secreting immunoglobulins of both parental types. *Nature (London)* **244**, 444–447.

Schwaber, J., and Cohen, E. P. (1974). Pattern of immunoglobulin synthesis and assembly in a human–mouse somatic cell hybrid clone. *Proc. Natl. Acad. Sci. U.S.A.* **71**, 2203–2207.

Schwartz, A. G., Cook, P. R., and Harris, H. (1971). Correction of a genetic defect in a mammalian cell. *Nature (London), New Biol.* **230**, 5–8.

Scolnick, E. M., and Parks, W. P. (1974). Host range studies on xenotropic type C viruses in somatic cell hybrids. *Virology* **59**, 168–178.

Seabright, M. (1971). A rapid banding technique for human chromosomes. *Lancet* **2**, 971–972.

Sekiguchi, T., and Sekiguchi, F. (1973). Interallelic complementation in hybrid cells derived from Chinese hamster diploid clones deficient in hypoxanthine-guanine phosphoribosyl transferase activity. *Exp. Cell Res.* **77**, 391–403.

Sekiguchi, T., Sekiguchi, F., and Yamada, M. (1973). Incorporation and replication of foreign metaphase chromosomes in cultured mammalian cells. *Exp. Cell Res.* **80,** 223–236.

Sekiguchi, T., Sekiguchi, F., and Tomii, S. (1974). Complementation in hybrid cells derived from mutagen-induced mouse clones deficient in hypoxanthine-guanine phosphoribosyl-transferase activity. *Exp. Cell Res.* **88,** 410–414.

Sekiguchi, T., Sekiguchi, F., and Tomii, S. (1975a). Genetic complementation in hybrid cells derived from mutagen-induced mouse clones deficient in HGPRT activity. *Exp. Cell Res.* **93,** 207–218.

Sekiguchi, T., Sekiguchi, F., Tachibana, T., Yamada, T., and Yoshida, M. (1975b). Gene expression of foreign metaphase chromosomes integrated into cultured mammalian cells. *Exp. Cell Res.* **94,** 327–338.

Sell, E. K., and Krooth, R. S. (1972). Tabulation of somatic cell hybrids formed between lines of cultured cells. *J. Cell. Physiol.* **80,** 453–461.

Shainberg, A., Yagil, G., and Yaffe, D. (1969). Control of myogenesis *in vitro* by Ca^{2+} concentration in nutritional medium. *Exp. Cell Res.* **58,** 163–167.

Shani, M., Huberman, E., Aloni, Y., and Sachs, L. (1974). Activation of Simian virus 40 by transfer of isolated chromosomes from transformed cells. *Virology* **61,** 303–305.

Sharp, P. A., Pettersson, U., and Sambrook, J. (1974). Viral DNA in transformed cells. I. A study of the sequences of adenovirus 2 DNA in a line of transformed rat cells using specific fragments of the viral genome. *J. Mol. Biol.* **86,** 709–726.

Shay, J. W., Porter, K. R., and Prescott, D. M. (1974). The surface morphology and fine structure of CHO (Chinese hamster ovary) cells following enucleation. *Proc. Natl. Acad. Sci. U.S.A.* **71,** 3059–3063.

Shay, J. W., Gershenbaum, M. R., and Porter, K. R. (1975). Enucleation of CHO cells by means of cytochalasin B and centrifugation: The topography of enucleation. *Exp. Cell Res.* **94,** 47–55.

Sheehy, P. F., Wakonig-Vaartaja, T., Winn, R., and Clarkson, B. D. (1974). Asynchronous DNA synthesis and asynchronous mitosis in multinuclear ovarian cancer cells. *Cancer Res.* **34,** 991–996.

Shein, H. M., and Enders, J. F. (1962). Transformation induced by simian virus 40 in human renal cell cultures. I. Morphology and growth characteristics. *Proc. Natl. Acad. Sci. U.S.A.* **48,** 1164–1172.

Shevliaghyn, V. J., Biryulina, T. I., Tikhoneva, Z. N., and Karazas, N. V. (1969). Activation of Rous virus in the transplanted golden hamster tumor with the aid of artificial heterokaryon formation. *Int. J. Cancer* **4,** 42–46.

Shin, S. I. (1974). Nature of mutations conferring resistance to 8-azaguanine in mouse cell lines. *J. Cell Sci.* **14,** 235–251.

Shin, S.-I., Caneva, R., Schildkraut, C. L., Klinger, N. P., and Siniscalco, M. (1973). Cells with phosphoribosyl transferase activity recovered from mouse cells resistant to 8-azaguanine. *Nature (London), New Biol.* **241,** 194–196.

Shin, S.-I., Meera Khan, P., and Cook, P. R. (1971). Characterization of hypoxanthine-guanine phosphoribosyl transferase in man–mouse somatic cell hybrids by an improved electrophoretic method. *Biochem. Genet.* **5,** 91–99.

Shows, T. B., and Brown, J. A. (1974). An(Xq −; 9p+) translocation suggests the assignment of *G6PD HPRT* and *PGK* to the long arm of the X chromosome in somatic cell hybrids. *Cytogenet. Cell Genet.* **13,** 146–149.

Sidebottom, E. (1974). Heterokaryons and their uses in studies of nuclear function. *In* "The Cell Nucleus" (H. Busch, ed.), Vol. 1, pp. 441–469. Academic Press, New York.

Sidebottom, E., and Harris, H. (1969). The role of the nucleolus in the transfer of RNA from nucleus to cytoplasm. *J. Cell Sci.* **5,** 351–364.

Silagi, S. G. (1967). Hybridization of a malignant melanoma cell line with L cells *in vitro*. *Cancer Res.* **27,** 1953–1960.

Silagi, S. G., and Bruce, S. A. (1970). Suppression of malignancy and differentiation in melanotic melanoma cells. *Proc. Natl. Acad. Sci. U.S.A.* **66,** 72–78.

Silagi, S. G., Darlington, G., and Bruce, S. A. (1969). Hybridization of two biochemically marked human cell lines. *Proc. Natl. Acad. Sci. U.S.A.* **62,** 1085–1092.

Siminovitch, L. (1974). Isolation and characterization of mutants of somatic cells. *In* "Somatic Cell Hybridization" (R. L. Davidson and F. F. de la Cruz, eds.), pp. 229–231. Raven, New York.

Siminovitch, L. (1976). On the nature of hereditable variation in cultured somatic cells. *Cell* **7,** 1–11.

Singer, D., Cooper, M., Maniatis, G. M., Marks, P. A., and Rifkind, R. A. (1974). Erythropoetic differentiation in colonies transformed by Friend virus. *Proc. Natl. Acad. Sci. U.S.A.* **71,** 2668–2670.

Singer, S. J., and Nicolson, G. L. (1972). The fluid mosaic model of the structure of cell membrances. *Science* **175,** 720–731.

Siniscalco, M. (1974). Strategies for X-chromosome mapping with somatic cell hybrids. *In* "Somatic Cell Hybridization" (R. L. Davidson and F. F. de la Cruz, eds.), pp. 35–48. Raven, New York.

Siniscalco, M., Klinger, H. P., Eagle, H., Koprowski, H., Fujimoti, W. F., and Seegmiller, J. E. (1969a). Evidence for intergenic complementation in hybrid cells derived from two human diploid strains each carrying an X-linked mutation. *Proc. Natl. Acad. Sci. U.S.A.* **62,** 793–799.

Siniscalco, M., Knowles, B. B., and Steplewski, Z. (1969b). Hybridization of human diploid strains carrying X-linked mutants and its potential in studies of somatic cell genetics. *Wistar Inst. Symp. Monogr.* **9,** 117–136.

Skoog, F. (1944). Growth and organ formation in tobacco tissue culture. *Am. J. Bot.* **31,** 19–24.

Skou, J. C. (1971). Sequence of steps in the (Na+K)-activated enzyme in relation to sodium and potassium transport. *Curr. Top. Bioenerg.* **4,** 357–398.

Smith, C. W., and Goldman, A. S. (1971). Macrophages from human colostrum. Multinucleated giant cell formation by phytohemagglutinin and concanavalin A. *Exp. Cell Res.* **66,** 317–320.

Sobel, J. S., Albrecht, A. M., Riehm, H., and Biedler, J. L. (1971). Hybridization of actinomycin D and amethopterin-resistant Chinese hamster cells *in vitro*. *Cancer Res.* **31,** 297–307.

Sonnenschein, C. (1970). Somatic cell hybridization: Pattern of chromosome replication in viable Chinese hamster + Armenian hamster hybrids. *Exp. Cell Res.* **63,** 195–199.

Sonnenschein, C., Tashjian, A. H., and Richardson, U. I. (1968). Somatic cell hybridization: Mouse–rat hybrid cell line involving a growth hormone-producing parent. *Genetics* **60,** 227–228.

Sonnenschein, C., Richardson, U. I., and Tashjian, A. H. (1971). Loss of growth hormone production following hybridization of a functional rat pituitary cell strain with a mouse fibroblast line. *Exp. Cell Res.* **69,** 336–344.

Sorieul, S., and Ephrussi, B. (1961). Karyological demonstration of hybridization of mammalian cells *in vitro*. *Nature (London)* **190,** 653–654.

Soupart, P., and Clewe, T. W. (1965). Sperm penetration of rabbit zona pellucida inhibited by treatment of ova with neuraminidase. *Fertil. Steril.* **16,** 677–689.

Sparkes, R. S., and Weiss, M. C. (1973). Expression of differentiated functions in hepatoma cell hybrids: Alanine aminotransferase. *Proc. Natl. Acad. Sci. U.S.A.* **70,** 377–381.

Spencer, R. A., Hauschka, T. S., Amos, D. B., and Ephrussi, B. (1964). Co-dominance of isoantigens in somatic hybrids of murine cells grown *in vitro*. *J. Natl. Cancer Inst.* **33,** 893–903.

Sperling, K., and Rao, P. N. (1974a). Mammalian cell fusion. V. Replication behaviour of heterochromatin as observed by premature chromosome condensation. *Chromosoma* **45,** 121–131.

Sperling, K., and Rao, P. N. (1974b). The phenomenon of premature chromosome condensation: Its relevance to basic and applied research. *Humangenetik* **23,** 235–258.

Spolsky, C. M., and Eisenstadt, J. M. (1972). Chloramphenicol-resistant mutants of human HeLa cells. *FEBS Lett.* **25,** 319–324.

Stadler, J. K., and Adelberg, E. A. (1972). Cell cycle changes and the ability of cells to undergo virus-induced fusion. *Proc. Natl. Acad. Sci. U.S.A.* **69,** 1929–1933.

Stadler, J. K., Ward, A., and Adelberg, E. A. (1975). Optimal conditions for the fusion of lymphoid cell lines. *In vitro* **11,** 224–229.

Stanbridge, E. J. (1976). Suppression of malignancy in human cells. *Nature (London)* **260,** 17–20.

Stanners, C. P., Elicieri, G. L., and Green, H. (1971). Two types of ribosome in mouse–hamster hybrid cells. *Nature (London), New Biol.* **230,** 52–54.

Stenman, S. (1971). Depression of RNA synthesis in the prematurely condensed chromatin of pulverized HeLa cells. *Exp. Cell Res.* **69,** 372–376.

Stenman, S., and Saksela, E. (1969). Susceptibility of human chromosomes to pulverization induced by myxoviruses. *Hereditas* **62,** 323–338.

Stenman, S., and Saksela, E. (1971). The relationship of Sendai virus-induced chromosome pulverization to cell cyclus in HeLa cells. *Hereditas* **69,** 1–14.

Steplewski, Z., and Koprowski, H. (1969). Development of SV40 coat protein antigen in nonpermissive nuclei in heterokaryocytes. *Exp. Cell Res.* **57,** 433–440.

Steplewski, Z., Knowles, B. B., and Koprowski, H. (1968). The mechanism of internuclear transmission of SV40 induced complement fixation antigen in heterokaryocytes. *Proc. Natl. Acad. Sci. U.S.A.* **59,** 769–776.

Steward, F. L. (1970). From cultured cells to whole plants. *Proc. Roy. Soc. London, Ser. B* **175,** 1–30.

Stich, H. F., Hsu, T. C., and Rapp, F. (1964). Virus and mammalian chromosomes. I. Localization of chromosome aberrations after infection with Herpes Simplex virus. *Virology* **22,** 439–445.

Stoker, M. (1963). Effect of X-irradiation on susceptibility of cells to transformation by polyoma virus. *Nature (London)* **200,** 756–758.

Stoker, M., O'Neill, C., Berryman, C., and Waxman, V. (1968). Anchorage and growth regulation in normal and virus-transformed cells. *Int. J. Cancer* **3,** 683–693.

Stoker, M. G. P., and Rubin, H. (1967). Density dependent inhibition of cell growth in culture. *Nature (London)* **215,** 171–172.

Stone, W. H., Friedman, J., and Fregin, A. (1964). Possible somatic cell mating in twin cattle with erythrocyte mosaicism. *Proc. Natl. Acad. Sci. U.S.A.* **51,** 1036–1044.

Stubblefield, E. (1964). DNA synthesis and chromosomal morphology of Chinese hamster cells cultured in media containing N-deacetyl-N-methylcolchicine (Colcemid). *In* "Cytogenetics of Cells in Culture" (R. J. C. Harris, ed.), Vol. 3, pp. 223–298. Academic Press, New York.

Suarez, H. G., Chany, C., Cassingena, R., Vignal, M., and Lemaitre, J. (1972a). Localisation du chromosome responsable de la synthèse de l'interféron simïen. *C. R. Hebd. Seances Acad. Sci.* **274,** 3632–3634.

Suarez, H. G., Sonenshein, G. E., Estrade, S., Bourali, M. F., Cassingena, R., and Tournier, P. (1972b). Activation of the viral genome in simian virus 40-transformed nonpermissive cells by permissive cell extracts. *Proc. Natl. Acad. Sci. U.S.A.* **69,** 1290–1293.

Summers, K. E., and Gibbons, I. R. (1971). Adenosine triphosphate-induced sliding of tubules in trypsin-treated flagella of sea-urchin sperm. *Proc. Natl. Acad. Sci. U.S.A.* **68,** 3092–3096.

Sumner, A. T. (1972). A simple technique for demonstrating centromeric heterochromatin. *Exp. Cell Res.* **75,** 304–306.

Sumner, A. T., Evans, H. J., and Buckland, R. A. (1971). New technique for distinguishing between human chromosomes. *Nature (London), New Biol.* **232,** 31–32.

Svoboda, J., and Dourmashkin, R. (1969). Rescue of Rous sarcoma virus from virogenic mammalian cells associated with chicken cells and treated with Sendai virus. *J. Gen. Virol.* **4,** 523–529.

Svoboda, J., Machala, O., and Hložanek, I. (1967). Influence of Sendai virus in RSV formation in mixed culture of virogenic mammalian cells and chicken fibroblasts. *Folia Biol. (Prague)* **13,** 155–157.

Swetly, P., Barbanti-Brodano, G., Knowles, B. B., and Koprowski, H. (1969). Response of simian virus 40-transformed cell lines and cell hybrids to superinfection with simian virus 40 and its deoxyribonucleic acid. *J. Virol.* **4,** 348–355.

Szpirer, C. (1974). Reactivation of chick erythrocyte nuclei in heterokaryons with rat hepatoma cells. *Exp. Cell. Res.* **83,** 47–54.

Szpirer, J., and Szpirer, C. (1975). The control of serum protein synthesis in hepatoma–fibroblast hybrids. *Cell* **6,** 53–60.

Szybalski, W., Szybalska, E. H., and Ragni, G. (1962). Genetic studies with human cell lines. *Natl. Cancer Inst., Monogr.* **7,** 75–89.

Takagi, N., Aya, T., Kato, H., and Sandberg, A. A. (1969). Relation of virus-induced cell fusion and chromosome pulverization to mitotic events. *J. Natl. Cancer Inst.* **43,** 335–347.

Takebe, I., Otsuki, Y., and Aoki, S. (1968). Isolation of tobacco mesophyll cells in intact and active state. *Plant Cell Physiol.* **9,** 115–124.

Tan, Y. H., Tischfield, J., and Ruddle, F. H. (1973). The linkage of genes for the human interferon-induced antiviral protein and indophenol oxidase-B traits to chromosome G-21. *J. Exp. Med.* **137,** 317–330.

Tan, Y. H., Creagan, R. P., and Ruddle, F. H. (1974). The somatic cell genetics of human interferon: Assignment of human interferon loci to chromosomes 2 and 5. *Proc. Natl. Acad. Sci. U.S.A.* **71,** 2251–2255.

Tappel, A. L. (1969). Lysosomal enzymes and other components. *In* "Lysosomes in Biology and Pathology" (J. T. Dingle and H. B. Fell, eds.), pp. 207–244. North-Holland Publ., Amsterdam.

Tarkowski, A. K. (1961). Mouse chimaeras developed from fused eggs. *Nature (London)* **190,** 857–860.

Taylor, J. H. (1960). Nucleic acid synthesis in relation to the cell division cycle. *Ann. N. Y. Acad. Sci.* **90,** 409–421.

Taylor, R. B., Duffus, W. P. H., Raff, M. C., and dePetris, S. (1971). Redistribution and pinocytosis of lymphocyte surface immunoglobulin molecules induced by anti-immunoglobin antibody. *Nature (London), New Biol.* **233,** 225–229.

Temin, H. M. (1971). Mechanism of cell transformation by RNA tumor viruses. *Annu. Rev. Microbiol.* **25,** 609–648.

Temin, H. M., and Mizutani, S. (1970). DNA polymerases in virions of Rous sarcoma virus. *Nature (London)* **226,** 1211–1213.

Tennant, R. W., and Richter, C. B. (1972). Murine leukemia virus: Restriction in fused permissive and nonpermissive cells. *Science* **178,** 516–518.

terMeulen, V., Iwasaki, Y., Koprowski, H., Kackell, Y. M., and Müller, D. (1972a). Fusion of cultured multiple-sclerosis brain cells with indicator cells: Presence of nucleocapsids and virions and isolation of parainfluenza-type virus. *Lancet* **2**, 1–5.

terMeulen, V., Kackell, Y., Müller, D., Katz, M., and Meyermann, R. (1972b). Isolation of infectious measles virus in measles encephalitis. *Lancet* **2**, 1172–1175.

Thomas, G. H., Taylor, H. A., Miller, C. S., Axelman, J., and Migeon, B. R. (1974). Genetic complementation after fusion of Tay–Sachs and Sandhoff cells. *Nature (London)* **250**, 580–582.

Thompson, E. B., and Gelehrter, T. D. (1971). Expression of tyrosine aminotransferase activity in somatic cell heterokaryons: Evidence for negative control of enzyme expression. *Proc. Natl. Acad. Sci. U.S.A.* **68**, 2589–2593.

Thompson, L. H., and Baker, R. M. (1973). Isolation of mutants of cultured mammalian cells. *Methods Cell Biol.* **6**, 209–281.

Thompson, L. H., Mankovitz, R., Baker, R. M., Till, J. E., Siminovitch, L., and Whitmore, G. F. (1970). Isolation of temperature-sensitive mutants of L cells. *Proc. Natl. Acad. Sci. U.S.A.* **66**, 377–384.

Thompson, L. H., Mankovitz, R., Baker, R. M., Wright, J. A., Till, J. E., Siminovitch, L., and Whitmore, G. F. (1971). Selective and nonselective isolation of temperature-sensitive mutants of mouse L cells and their characterization. *J. Cell. Physiol.* **78**, 431–440.

Thompson, L. H., Stanners, C. P., and Siminovitch, L. (1975). Selection by [³H]amino acids of CHO-cell mutants with altered leucyl- and asparagyl-transfer RNA synthetases. *Somatic Cell Genet.* **1**, 187–208.

Tischfield, J. A., and Ruddle, F. H. (1974). Assignment of the gene for adenine phosphoribosyltransferase to human chromosome 16 by mouse–human somatic cell hybridization. *Proc. Natl. Acad. Sci. U.S.A.* **71**, 45–49.

Todaro, G. J., and Green, H. (1963). Quantitative studies of the growth of mouse embryo cells in culture and their development into established lines. *J. Cell Biol.* **17**, 299–313.

Todaro, G. J., and Green, H. (1964). An assay for cellular transformation by SV40. *Virology* **23**, 117–119.

Todaro, G. J., Lazar, G. K., and Green, H. (1966). The initiation of cell division in a contact-inhibited mammalian cell line. *J. Cell. Comp. Physiol.* **66**, 325–333.

Toister, Z., and Loyter, A. (1971). Ca²⁺-induced fusion of avian erythrocytes. *Biochim. Biophys. Acta* **241**, 719–724.

Toister, Z., and Loyter, A. (1973). The mechanism of cell fusion. II. Formation of chicken erythrocyte polykaryons. *J. Biol. Chem.* **248**, 422–432.

Tomkins, G. M., Gelehrter, T. D., Granner, D., Martin, D., Samuels, H., and Thompson E. B. (1969). Control of specific gene expression in higher organisms. *Science* **166**, 1474–1480.

Toniolo, D., and Basilico, C. (1974). Complementation of a defect in the production of ribosomal RNA in somatic cell hybrids. *Nature (London)* **248**, 411–413.

Tooze, J., ed. (1973). "The Molecular Biology of Tumor Viruses." Cold Spring Harbor Lab., Cold Spring Harbor, New York.

Trisler, G. D., and Coon, H. G. (1973). Somatic cell hybrids between two vertebrate classes. *J. Cell Biol.* **59**, 347 (abstr.).

Trujillo, J. M., Walden, B., O'Neil, P., and Anstall, H. B. (1965). Sex-linkage of *G6PD* in the horse and donkey. *Science* **148**, 1603–1604.

Turleau, C., de Grouchy, J., and Klein, M. (1972). Phylogenie chromosomique de l'homme et des primates hominiens (*Pan troglodytes, Gorilla gorilla* et *Pongo pygmaeus*). Essai de reconstitution. de L'ancêtre commun. *Ann. Genet.* **15**, 225–240.

Turner, J. H., Hutchinson, D. L., and Petricciani, J. C. (1973). Chimerism following fetal transfusion: Report of leucocyte hybridization and infant with acute lymphocytic leukaemia. *Scand. J. Haematol.* **10**, 358–366.

Uchida, S., and Watanabe, S. (1969). Transformation of mouse 3T3 cells by T antigen-forming defective SV40 virions (T particles). *Virology* **39**, 721–728.

Unakul, W., Johnson, R. T., Rao, P. N., and Hsu, T. C. (1973). Giemsa banding in prematurely condensed chromosomes obtained by cell fusion. *Nature (London), New Biol.* **242**, 106–107.

van der Noordaa, J., van Haagen, A., Waloomers, J. M., and van Someren, H. (1972). Properties of somatic cell hybrids between mouse cells and Simian virus 40-transformed rat cells. *J. Virol.* **10**, 67–72.

van Heyningen, V., Craig, I., and Bodmer, W. (1973). Genetic control of mitochondrial enzymes in human–mouse somatic cell hybrids. *Nature (London)* **242**, 509–512.

Velazques, A., Payne, F. E., and Krooth, R. S. (1971). Viral-induced fusion of human cells. I. Quantitative studies on the fusion of human diploid fibroblasts induced by Sendai virus. *J. Cell. Physiol.* **78**, 93–110.

Venuta, S., and Rubin, H. (1973). Sugar transport in normal and Rous sarcoma virus-transformed chick-embryo fibroblasts. *Proc. Natl. Acad. Sci. U.S.A.* **70**, 653–657.

Veomett, G., Prescott, D. M., Shay, J., and Porter, K. R. (1974). Reconstruction of mammalian cells from nuclear and cytoplasmic components separated by treatment with cytochalasin B. *Proc. Natl. Acad. Sci. U.S.A.* **71**, 1999–2002.

Vigier, P. (1967). Persistance du génome du virus de Rous ans des cellules du hamster converties *in vitro*, et action du virus Sendai inactivé sur sa transmission aux cellules de poule. *C. R. Hebd. Seances Acad. Sci., Ser. D* **264**, 422–425.

Vigier, P. (1973). Persistence of Rous sarcoma virus in transformed nonpermissive cells: Characteristics of virus induction following Sendai virus-mediated fusion with permissive cells. *Int. J. Cancer* **11**, 478–483.

Vogel, A., and Pollack, R. (1974). Methods for obtaining revertants of transformed cells. *Methods Cell Biol.* **8**, 75–92.

Vogt, M., and Dulbecco, R. (1962). Properties of cells transformed by polyoma virus. *Cold Spring Harbor Symp. Quant. Biol.* **27**, 367–374.

Vollet, J. J., Roth, L. E., and Davidson, M. (1971). Calcium-induced fusion of a heliozoan protozoan. *Abstr., 11th Annu. Meet. Am. Soc. Cell Biol.* p. 314.

Volpe, E. P., and Earley, E. M. (1970). Somatic cell mating and segregation in chimeric frogs. *Science* **168**, 850–852.

Waldren, C. A., and Johnson, R. T. (1974). Analysis of interphase chromosome damage by means of premature chromosome condensation after X- and ultraviolet-irradiation. *Proc. Natl. Acad. Sci. U.S.A.* 1137–1141.

Wallace, D. C., Bunn, C. L., Eisenstadt, J. M. (1975). Cytoplasmic transfer of chloramphenicol resistance in human tissue culture cells. *J. Cell Biol.* **67**, 174–188.

Wallace, R. B., and Freeman, K. B. (1975). Selection of mammalian cells resistant to a chloramphenicol analog. *J. Cell Biol.* **65**, 492–498.

Wang, R. J., Pollack, R., Kusano, T., and Green, H. (1970). Human–mouse hybrid cell lines and susceptibility to poliovirus. I. Conversion from polio sensitivity to polio resistance accompanying loss of human gene dependent polioreceptors. *J. Virol.* **5**, 677–681.

Watkins, J. F. (1971). Cell fusion in the study of tumor cells. *Int. Rev. Exp. Pathol.* **10**, 115–141.

Watkins, J. F. (1974). SV40 rescue from a virogenic cell line. *In* "Somatic Cell Hybridization" (R. L. Davidson and F. F. de la Cruz, eds.), pp. 101–104. Raven, New York.

Watkins, J. F. (1975). The SV40 rescue problem. *Cold Spring Harbor Symp. Quant. Biol.* **39**, 355–362.

Watkins, J. F., and Dulbecco, R. (1967). Production of SV40 virus in heterokaryons of transformed and susceptible cells. *Proc. Natl. Acad. Sci. U.S.A.* **58**, 1396–1403.

Watkins, J. F., and Grace, D. M. (1967). Studies on the surface antigens of interspecific mammalian cell heterokaryons. *J. Cell Sci.* **2**, 193–204.

Watson, B., Gormley, I. P., Gardiner, S. E., Evans, H. J., and Harris, H. (1972). Reappearance of murine hypoxanthine guanine phosphoribosyl transferase activity in mouse A9 cells after attempted hybridization with human cell lines. *Exp. Cell Res.* **75**, 401–409.

Weber, K., Lazarides, E., Goldman, R. D., Vogel, A., and Pollack, R. (1975). Localization and distribution of actin fibers in normal transformed and revertant cells. *Cold Spring Harbor Symp. Quant. Biol.* **39**, 363–369.

Weiner, L. P., Herndon, R. M., Narayan, O., Johnson, R. T., Shah, K., Rubinstein, L. J., Preziosi, T. J., and Conley, F. K. (1972). Isolation of virus related to SV40 from patients with progressive multifocal leukoencephalopathy. *N. Engl. J. Med.* **286**, 385–390.

Weisenberg, R. C. (1972). Microtubule formation *in vitro* in solutions containing low calcium concentrations. *Science* **177**, 1104–1105.

Weiss, L. (1967). "The Cell Periphery, Metastasis and Other Contact Phenomena." North-Holland Publ., Amsterdam.

Weiss, M. C. (1970). Further studies on loss of T-antigen from somatic hybrids between mouse cells and SV40 transformed human cells. *Proc. Natl. Acad. Sci. U.S.A.* **66**, 79–86.

Weiss, M. C. (1974). Reexpression of liver specific enzymes in hepatoma cell hybrids. *In* "Somatic Cell Hybridization" (R. L. Davidson and F. F. de la Cruz, eds.), pp. 151–155. Raven, New York.

Weiss, M. C., and Chaplain, M. (1971). Expression of differentiation functions in hepatoma cell hybrids: Reappearance of tyrosine aminotransferase inducibility after the loss of chromosomes. *Proc. Natl. Acad. Sci. U.S.A.* **68**, 3026–3030.

Weiss, M. C., and Ephrussi, B. (1966a). Studies of interspecific (rat × mouse) somatic hybrids. I. Isolation, growth and evolution of the karyotype. *Genetics* **54**, 1095–1109.

Weiss, M. C., and Ephrussi, B. (1966b). Studies of interspecific (rat × mouse) somatic hybrids. II. Lactate dehydrogenase and β-glucuronidase. *Genetics* **54**, 1111–1122.

Weiss, M. C., and Green, H. (1967). Human–mouse hybrid cell lines containing partial complements of human chromosomes and functioning human genes. *Proc. Natl. Acad. Sci. U.S.A.* **58**, 1104–1111.

Weiss, M. C., Todaro, G. J., and Green, H. (1968a). Properties of a hybrid between lines sensitive and insensitive to contact inhibition of cell division. *J. Cell. Physiol.* **71**, 105–107.

Weiss, M. C., Ephrussi, B., and Scaletta, L. J. (1968b). Loss of T-antigen from somatic hybrids between mouse cells and SV40-transformed human cells. *Proc. Natl. Acad. Sci. U.S.A.* **59**, 1132–1135.

Weiss, M. C., Sparkes, R. S., and Bertolotti, R. (1975). Expression of differentiated functions in hepatoma cell hybrids: IX. Extinction and reexpression of liver specific enzymes in rat-hepatoma–Chinese hamster fibroblast hybrids. *Somatic Cell Genet.* **1**, 27–40.

Weissmann, G., Bloomgarden, D., Kaplan, R., Cohen, C., Hoffstein, S., Collins, T., Gottlieb, A., and Nagle, D. (1975). A general method for the introduction of enzymes by means of immunoglobulin-coated liposomes, into lysosomes of deficient cells. *Proc. Natl. Acad. Sci. U.S.A.* **72,** 88–92.

Weller, T. H., Witton, H. M., and Bell, E. J. (1958). The etiologic agents of varicella and herpes zoster. Isolation, propagation and cultured characteristics *in vitro. J. Exp. Med.* **108,** 843–868.

Wells, M. A. (1972). A kinetic study of the phospholipase A_2 (*Crotalus adamanteus*) catalyzed hydrolysis of 1,2-dibuturyl-*sn*-glycero-3-phosphorylcholine. *Biochemistry* **11,** 1030–1041.

Wessels, N. K., Spooner, B. S., Ash, J. F., Bradley, M. O., Luduena, M. A., Taylor, E. L., Wrenn, J. T., and Yamada, K. M. (1971). Microfilaments in cellular and developmental processes. Contractile microfilament machinery of many cell types is reversibly inhibited by cytochalasin B. *Science* **171,** 135–143.

Westerveld, A. (1971). Gene linkage in man and Chinese hamster studied in somatic cell hybrids. Ph.D. Thesis. University of Rotterdam, Rotterdam.

Westerveld, A., and Freeke, M. A. (1971). Cell cycle of multinucleate cells after cell fusion. *Exp. Cell Res.* **65,** 140–144.

Westerveld, A., Visser, R. P. L. S., Meera Khan, P. M., and Bootsma, D. (1971). Loss of human genetic markers in man-Chinese hamster somatic cell hybrids. *Nature (London), New Biol.* **234,** 20–24.

Wever, G. H., Kit, S., and Dubbs, D. R. (1970). Initial site of synthesis of virus during rescue of Simian Virus 40 from heterokaryons of Simian Virus 40-transformed and susceptible cells. *J. Virol.* **5,** 578–585.

Whaley, W. G., Dauwalder, M., and Kephart, J. E. (1972). Golgi apparatus: Influence on cell surfaces. *Science* **175,** 596–599.

Wiblin, C. N., and MacPherson, I. (1973). Reversion in hybrids between SV40-transformed hamster and mouse cells. *Int. J. Cancer* **12,** 148–161.

Wickus, G., Gruenstein, E., Robbins, P. W., and Rich, A. (1975). Decrease in membrane-associated actin of fibroblasts after transformation by Rous sarcoma virus. *Proc. Natl. Acad. Sci. U.S.A.* **72,** 746–749.

Wiener, F., Klein, G., and Harris, H. (1971). The analysis of malignancy by cell fusion. III. Hybrids between diploid fibroblasts and other tumor cells. *J. Cell Sci.* **8,** 681–692.

Wiener, F., Fenyö, E. M., Klein, G., and Harris, H. (1972a). Fusion of tumor cells with host cells. *Nature (London), New Biol.* **238,** 155–159.

Wiener, F., Cochran, A., Klein, G., and Harris, H. (1972b). Genetic determinants of morphological differentiation in hybrid tumors. *J. Natl. Cancer Inst.* **48,** 465–486.

Wiener, F., Klein, G., and Harris, H. (1973). The analysis of malignancy by cell fusion. IV. Hybrids between tumor cells and a malignant L cell derivative. *J. Cell Sci.* **12,** 253–261.

Wiener, F., Klein, G., and Harris, H. (1974a). The analysis of malignancy by cell fusion. V. Further evidence of the ability of normal diploid cells to suppress malignancy. *J. Cell Sci.* **15,** 177–183.

Wiener, F., Klein, G., and Harris, H. (1974b). The analysis of malignancy by cell fusion. VI. Hybrids between different tumor cells. *J. Cell Sci.* **16,** 189–198.

Wiener, F., Fenyö, E. M., and Klein, G. (1974c). Tumor-host cell hybrids in radiochimeras. *Proc. Natl. Acad. Sci. U.S.A.* **71,** 148–152.

Wiener, F., Dalianis, T., Klein, G., and Harris, H. (1974d). Cytogenetic studies on the mechanism of formation of isoantigenic variants in somatic cell hybrids I. Banding analyses of isoantigenic variant sublines derived from the fusion of Ta3HA carcinoma with MSWBS sarcoma cells. *J. Natl. Cancer Inst.* **52,** 1779–1796.

Wiener, F., Fenyö, E. M., Klein, G., and Harris, H. (1974e). Fusion of tumor cells with host cells. *In* "Somatic Cell Hybridization" (R. L. Davidson and F. F. de la Cruz, eds.), pp. 105–118. Raven, New York.

Wigler, M. H., and Weinstein, I. B. (1975). A preparative method for obtaining enucleated mammalian cells. *Biochem. Biophys. Res. Commun.* **63,** 669–674.

Willecke, K., and Ruddle, F. H. (1975). Transfer of human gene for hypoxanthine-guanine phosphoribosyltransferase via isolated human metaphase chromosomes into mouse L cells. *Proc. Natl. Acad. Sci. U.S.A.* **72,** 1792–1796.

Winzler, W. J. (1970). Carbohydrates in cell surfaces. *Int. Rev. Cytol.* **29,** 77–125.

Wise, G. E., and Prescott, D. M. (1973). Ultrastructure of enucleated mammalian cells in culture. *Exp. Cell Res.* **81,** 63–72.

Withers, L. A., and Cocking, E. C. (1972). Fine-structural studies on spontaneous and induced fusion of higher plant protoplasts. *J. Cell Sci.* **11,** 59–75.

Woodin, A. M., and Wieneke, M. (1964). The participation of calcium, adenosine triphosphate and adenosine triphosphatase in the extrusion of the granule proteins from the polymorphonuclear leucocyte. *Biochem. J.* **90,** 498–509.

Wright, W. E. (1973). The production of mass populations of anucleate cytoplasms. *Methods in Cell Biol.* **7,** 203–210.

Wright, W. E., and Hayflick, L. (1972). Formation of anucleate and multinucleate cells in normal and SV40 transformed WI-38 by cytochalasin B. *Exp. Cell Res.* **74,** 187–194.

Wright, W. E., and Hayflick, L. (1975a). Use of biochemical lesions for selection of human cells with hybrid cytoplasms. *Proc. Natl. Acad. Sci. U.S.A.* **72,** 1812–1816.

Wright, W. E., and Hayflick, L. (1975b). Nuclear control of cellular aging demonstrated by hybridization of anucleate and whole cultured normal human fibroblasts. *Exp. Cell Res.* **96,** 113–121.

Wullems, G. J., van der Horst, J., and Bootsma, D. (1975). Incorporation of isolated chromosomes and induction of hypoxanthine phosphoribosyltransferase in Chinese hamster cells. *Somatic Cell Genet.* **1,** 137–152.

Yaffe, D. (1969). Cellular aspects of muscle differentiation *in vitro. Curr. Top. Dev. Biol.* **4,** 37–77.

Yaffe, D., and Feldman, M. (1965). The formation of hybrid multinucleated muscle fibers from myoblasts of different genetic origin. *Dev. Biol.* **11,** 300–317.

Yamanaka, T., and Okada, Y. (1966). Cultivation of fused cells resulting from treatment of cells with HVJ. I. Synchronization of the stages of DNA synthesis of nuclei involved in fused multinucleated cells. *Biken J.* **9,** 159–175.

Yamanaka, T., and Okada, Y. (1968). Cultivation of fused cells resulting from treatment of cells with HVJ. II. Division of binucleated cells resulting from fusion of two KB cells by HVJ. *Exp. Cell Res.* **49,** 461–469.

Yanishevsky, R., and Carrano, A. V. (1975). Prematurely condensed chromosomes of dividing and nondividing cells in aging human cell cultures. *Exp. Cell Res.* **90,** 169–174.

Yanovsky, A., and Loyter, A. (1972). The mechanism of cell fusion. I. Energy requirements for virus-induced fusion of Ehrlich ascites tumor cells. *J. Biol. Chem.* **247,** 4021–4028.

Yerganian, G., and Nell, M. B. (1966). Hybridization of dwarf hamster cells by UV-inactivated Sendai virus. *Proc. Natl. Acad. Sci. U.S.A.* **55,** 1066–1073.

Yoshida, M. C., and Ephrussi, B. (1967). Isolation and karyological characteristics of seven hybrids between somatic mouse cells *in vitro. J. Cell. Physiol.* **69,** 33–44.

Yoshiike, K., Furano, A., Watanabe, S., Uchida, S., Matsubara, K., and Takagi, Y. (1975). Characterization of defective simian virus 40 DNA comparison between large plaque and small-plaque types. *Cold Spring Harbor Symp. Quant. Biol.* **39,** 85–93.

Zakai, N., Loyter, A., and Kulka, R. G. (1974a). Fusion of erythrocytes and other cells with retention of erythrocyte cytoplasm. Nuclear activation in chicken erythrocyte—melanoma heterokaryons. *J. Cell Biol.* **61,** 241–248.

Zakai, N., Loyter, A., and Kulka, R. G. (1974b). Prevention of hemolysis by bivalent metal ions during virus-induced fusion of erythrocytes with Ehrlich ascites tumor cells. FEBS *Lett.* **40,** 331–334.

Zelenin, A. V., Shapiro, I. M., Kolesnikov, V. A., and Senin, V. M. (1974). Physicochemical properties of chromatin of mouse sperm nuclei in heterokaryons with Chinese hamster cells. *Cell Differ.* **3,** 95–101.

Zepp, H. D., Conover, J. H., Hirschhorn, K., and Hodes, H. L. (1971). Human–mosquito somatic cell hybrids induced by ultraviolet-inactivated Sendai virus. *Nature (London)*, *New Biol.* **229,** 119–121.

Zurier, R. B., Hoffstein, S., and Weissmann, G. (1973). Cytochalasin B: Effect on lysosomal enzyme release from human leukocytes. *Proc. Natl. Acad. Sci. U.S.A.* **70,** 844–848.

Glossary

Acrocentric a chromosome whose centromere is close to one end.

Acrosome a vesicle at the tip of a sperm head containing lysosomal enzymes.

Actinomycin a drug which inhibits DNA-dependent RNA synthesis.

Activation in hybrid cells, the expression of genes not previously expressed in the parental cells.

Agglutination aggregation of suspended cells which occurs in the presence of some exogenous agent.

Allele one of several alternate forms of a gene.

Allogeneic of different genetic constitution.

Allophenic consisting of cells of two different genotypes (synonym: chimeric).

Aminopterin a drug (a folic acid analog) which blocks the main pathways of purine and pyrimidine synthesis.

Amphidiploid a cell or organism which is diploid for two genomes, each from a different species.

Amphihaploid a cell or organism which is haploid for two genomes, each from a different species.

Amphiploid a cell or organism which has two genomes, each from a different species, used when the ploidy of the contributing species is uncertain.

Anchorage dependence the need for a solid substratum in order for cells to divide, a characteristic of normal diploid cells which is lacking in transformed cells.

Aneuploid referring to cells or organisms whose nuclei do not contain an exact multiple of the haploid number of chromosomes.

Annealing the association of single-stranded nucleic acid (RNA or DNA) molecules with other single-stranded nucleic acid molecules with a complementary base composition.

Anther the pollen-bearing part of the stamen of a flower.

Auxin a plant growth-regulating hormone. Indoleacetic acid is a well-known example of an auxin.

Asynteny being located on different chromosomes (antonym: synteny).

Autoradiography the technique of localizing incorporated radioactive substances by covering cells or tissue sections with a photographic emulsion which is sensitive to radiation.

349

Auxotrophic dependent on exogenous growth factors not required by normal (proto-trophic) cells.

B-Tropic ability of RNA tumor viruses to grow in mice of the Balb/c strain, but not in other mouse strains.

Back selection an experimental procedure, usually based on drug selection, by which one selects for hybrid segregants that have eliminated a marker that had previously been used to isolate the hybrid cells.

Baseline activity level of enzyme activity found in noninduced or nonstimulated cells.

Blastocyst an early embryonic stage, usually of mammals, in which the embryo consists of an inner cell mass, a central cavity (the blastocoele), and trophoblast; the stage at which the embryo is implanted into the uterine wall.

Blastomere one of the cells produced by the cleavage of a fertilized egg.

C-Type virus RNA-containing viruses that in the electron microscope can be seen to bud from the cell surface with an electron-dense central nucleoid.

Callus, -i an undifferentiated tissue of thin-walled plant cells which develops naturally on wound surfaces or experimentally from culturing single cells or bits of tissue.

Capsid a protein shell of a virus constructed of units (capsomeres).

Carcinogen an agent causing cancer.

Cell line cells which can be propagated indefinitely *in vitro* (established cell line).

Cell strain normal diploid cells which may be kept in culture for some time, but which ultimately undergo senescence (i.e., ageing) and die.

Centromere a specialized region of the chromosome which associates with microtubules of the mitotic spindle during mitosis (synonym: kinetochore).

Chimera an organism composed of cells of two or more different genotypes.

Chloramphenicol an antibiotic which inhibits bacterial and mitochondrial protein synthesis.

Chromosome segregation elimination of chromosomes in hybrid cell populations.

Clone a population of cells derived from a single cell.

Codominance *See* **Coexpression.**

Coexpression simultaneous expression of homologous genes from both parental cells in hybrid cells.

Complement (noun) a complex of at least 11 serum proteins that, in conjunction with antigen–antibody interaction, exerts a lytic effect on cell membranes.

Complementation interaction which restores the wild-type phenotype in hybrids between two mutant cells.

Concordant segregation the simultaneous appearance or disappearance of gene markers in hybrid cells undergoing chromosome segregation.

Constitutive marker property which is expressed by all cells of the same genotype (e.g., ubiquitous enzymes, household enzymes).

Contact inhibition inhibition of cell motility as a normal cell population reaches confluency *in vitro*. Soon after, cell multiplication stops (synonyms: topoinhibition, density dependent inhibition of cell growth).

Cross-activation interaction in hybrid cells in which a parental cell expressing a specific marker causes the homologous marker of a nonexpressing parental cell to become active.

Cybrid cytoplasmic hybrid obtained by fusing an intact cell with an anucleate cell.

Cytokinesis division of the cytoplasm during normal cell division.

Cytoplasmic donor cell used for the preparation of anucleate cells (cytoplasms) to be used in fusion experiments.

Cytoplast anucleate cell.

Cytosine arabinoside inhibitor of DNA synthesis.

Cytotoxicity ability of immunized lymphoid cells to kill target cells *in vitro*.

Dedifferentiation loss of characteristics typical of differentiated cells.

Deplasmolysis regaining protoplasmic volume after plasmolysis (q.v.).

Deletion mutation which results in the loss of gene material, individual genes, or chromosome segments containing many genes.

Determination the state of cells' or tissues' being predestined (programmed) for specific forms of differentiation.

Dikaryon cell containing two nuclei of the same or different types.

Dimeric consisting of two subunits.

Discordant clone a clone which in gene assignment experiments with hybrid cells disagrees with the majority of hybrid clones. The disagreement may be due to chromosome rearrangments or other factors discussed in Chapter XIII.

Dominance in hybrid cells the expression of one parental gene at the same time as the homologous gene of the other parental genome is present and not expressed.

Dopa oxidase enzyme which catalyzes the conversion of tyrosine to melanin (synonym: tyrosinase).

Ectoplasm the peripheral part of the cytoplasm which is relatively free from cytoplasmic organelles (*see also* **Lamellar cytoplasm**).

Ehrlich ascites tumor malignant mouse tumor which grows in practically all mouse strains and is characterized by suspended cells in the peritoneal cavity.

Endocytosis uptake of material from the outside by cell surface invagination; may be subdivided into *pinocytosis* when the material is liquid and *phagocytosis* when the material is solid.

Endoreduplication replication of DNA and splitting of chromosomes into daughter chromosomes without breakdown of the nuclear envelope.

Enucleation removal of the nucleus from a cell.

Epigenotype that part of the total genome which, under appropriate conditions, can be expressed in a given cell type to the exclusion of those limited to other cell types (for a discussion see Ephrussi, 1972).

Exocytosis extrusion of material from the cell by secretion or excretion (antonym: endocytosis).

Extinction absence in a hybrid cell of a property expressed by one or both of the parental cells; may be due to chromosome loss or regulatory events.

Facultative marker markers expressed only by certain differentiated cell types, not by all cell types. (synonym: luxury molecule).

Feeder layer layer of X-irradiated cells embedded in or covered by agar used to stimulate overlaid cells which otherwise grow poorly *in vitro*.

Galactosemia an inherited disease caused by a defect in the galactose metabolizing pathways. Cataract formation, enlargement of the liver and the spleen, mental retardation are the main symptoms. The disease is transmitted by a single autosomal gene. Cells from galactosemia patients are deficient in galactose-1-phosphate uridyltransferase.

Gangliosidosis a group of inherited diseases (for classification see Table IX-3) also known as lysosomal storage diseases. The cells are deficient in lysosomal enzymes and therefore accumulate large quantities of glycolipids. The age of onset and the clinical manifestations varies from patient to patient. Early neurological disturbances, deafness, blindness, difficulties to breathe, distended abdomen, and bone deformities are important symptoms.

Gene dosage the number of copies of a given gene in a cell. The dosage depends on the number of chromosomes with that gene and the reiteration frequency for the gene on the individual chromosomes. In hybrid cells gene dosage is often used to describe the input of chromosomes from the two parental cells.

Genotype the genetic constitution of a cell or an organism.

Ghost vesicle consisting mainly of the cell membrane which is formed when red blood cells are lysed, thereby losing their hemoglobin and other cytoplasmic contents.

Giant cell a large multinucleate cell formed from the spontaneous fusion of macrophages or other cells in tuberculous lesions and virus infected tissues. Giant cells are also found in bone tumors and in sites of bone resorption.

Haploid the chromosome number in gametes.

Hemagglutinin an agent causing the aggregation of red blood cells.

Hepatoma a differentiated tumor of hepatic origin.

Heterophasic having cytoplasms or nuclei of more than one phase of the cell cycle.

Heterokaryon multinucleate cell containing at least two different types of nuclei (synonym: heterokaryocyte).

Heteroplasmon a cytoplasmic hybrid (term used by Wright and Hayflick, 1975a).

Heteropolymeric enzyme enzyme molecule consisting of two or more subunits wherein the subunits are specified by two different genes or genomes.

Histocompatible referring to a state in which transplanted cells or tissues are not rejected by the recipient host.

Homokaryon a multinucleate cell in which all nuclei come from cells of the same genotype and phenotype.

Homophasic having more than one cytoplasm or nucleus, all of which are of the same cell cycle phase.

Household proteins a term used by Ephrussi (1972) to describe proteins which occur in all cell types within a given organism (i.e., they are constitutive markers).

Hybrid cell or cell hybrid a cell formed by the fusion of two different cells. Although heterokaryons are also hybrids, the term has become synonymous with mononucleate hybrid cells (synkaryons).

Infectious center assay a method for quantifying virus.

Infectious unit smallest number of virus particles which will cause a lytic infection in a susceptible cell where the virus can multiply.

Integration incorporation of exogenous DNA (e.g., viral DNA) into host cell chromosomes in a covalently linked form.

Interferon cell coded proteins which inhibit the multiplication of viruses.

Intergenic complementation interaction between two different genes from two mutant genomes which restores a normal (wild-type) phenotype.

Interspecific hybrids hybrids obtained by fusing cells from two different species (e.g., mouse + man; pea + soybean).

Intragenic complementation an interaction in a hybrid cell by which the function of a defective gene may be restored, at least partially, by the formation of heteropolymeric enzymes. While the defects are part of the same gene, they are mutations at slightly different sites. The nondefective parts of the two members in the hybrid complement each other in order to restore function (see Fig. XIII-7).

Intraspecific cell fusion the parental cells come from the same species.

Inversion rotation of a gene sequence or a chromosome segment by 180°. A sequence of genes *ABCDEFG* may change to *ABFEDCG*.

Isozymes (isoenzymes) are enzymes performing the same function, but present in more than one molecular form within the same tissue or species. Isozymes differ slightly from each other in amino acid composition and may be separated from each other by electrophoresis.

Isogenic cells or organisms with the same genotype.

Karyomere subdiploid nucleus (synonym: micronucleus).

Karyoplast see **Minicell.**

Karyotype the set of chromosomes characteristic of a species, individual or cell.

Lamellar cytoplasm a thin peripheral layer of cytoplasm free of cell organelles which is prominent in normal cells adhering to a solid substratum *in vitro*. Lamellar cytoplasm is decreased in transformed cells which have a more rounded shape.

Lectin diverse group of plant proteins and glycoproteins which can bind to carbohydrate receptors on the surface of animal cells. In some cases nondividing cells are stimulated to grow by lectin binding.

Lesch–Nyhan syndrome X-linked disease in man associated with a cellular deficiency in hypoxanthine guanine phosphoribosyltransferase (HGPRT⁻). Main symptoms: mental retardation, spasticity, and increased blood level of uric acid in the urine.

Luxury molecules molecules synthesized by differentiating cells and not found in all cell types (synonym: facultative markers).

Lymphoma group of neoplastic disorders arising in cell types native to lymphoid tissues (e.g., lymphocytes, histiocytes, and their precursors).

Lymphoblastoid cells lymphoid cells which can be cultured *in vitro*. Their morphology resembles that of stimulated lymphocytes. Many lines produce immunoglobulins. Some human cell lines are probably equivalent to transformed cells and are believed to carry the genome of the Epstein–Barr virus.

Malignancy the property of tumor cells to grow *in vivo* so as to kill their hosts.

Marker any property, molecule, or antigen of a cell which can be clearly distinguished and identified.

Melanin a brown–black pigment synthesized from tyrosine by melanocytes and melanoma cells.

Melanoma pigmented tumor cells, usually from skin.

Mesophyll chlorophyll-containing leaf tissues located between the epidermal layers.

Metacentric describes a chromosome with its centromere located at the middle, dividing the chromosome into two arms of equal length.

Metastasis the transfer of cells away from a primary tumor by lymph or blood to establish secondary tumors. Such secondary tumors are known as metastases.

Micelle spherical arrangement of lipid molecules in which the hydrophobic ends of the molecules are directed toward the center and the charged groups are oriented toward the outside.

Microcell a subdiploid cell fragment formed by the enucleation or disruption of micronucleated cells.

Minicell a nucleus surrounded by a narrow rim of cytoplasm and a plasma membrane (synonym: karyoplast).

Minimal medium the simplest medium on which most normal cells will grow. Mutants which require extra nutrients in the form of amino acids, sugars, purines, or pyrimidines do not grow on this medium.

Mononucleosis human infectious disease caused by the Epstein–Barr virus. Clinical symptoms: irregular fever, pharyngitis, enlargement of lymph nodes and spleen. Greatly increased number of lymphocytes in peripheral blood.

Multiple sclerosis remitting and relapsing disease of the nervous system affecting mainly the white matter of the brain.

Mutant cell or organism with a changed or new gene.

Mycelium threadlike filaments which constitute the vegetative structure of a fungus.

Myeloma neoplasm of plasma cells which causes widespread skeletal destruction (multiple myeloma), anemia and increased susceptibility to infection. Myeloma cells are usually monoclonal (originating from one cell) and produce immunoglobulins.

Myoblast mononucleate muscle cell.

Myotube multinucleate muscle fiber formed by the spontaneous fusion of myoblasts. Synthesizes contractile proteins (e.g., actin, myosin).

Mutagen substance or physical agent which induces cells to undergo mutation.

N-Tropic RNA tumor viruses which grow in Swiss mouse cells (see also B-tropic).

Neoplasia the formation of an abnormal mass of tissue (a tumor) which grows in excess of normal tissues.

Neuraminidase an enzyme which hydrolytically removes terminal sialic acid residues from cell surface oligosaccharides.

Neuroblastoma highly malignant tumor in man and in laboratory animals. The tumor originates in the adrenal and (rarely) in the central nervous system, most likely from primitive undifferentiated neuroblasts. Many neuroblastoma cells growing *in vitro* form axon-like extensions and express characteristics of mature nerve cells.

Nude mice mice that lack a thymus and therefore are defective with respect to all T-cell functions (response to phytohemagglutinin, graft rejection, and T-cell dependent antibody production). Obtained by crossing mice heterozygous for a gene designated *nu*.

Oncogenic causing tumors.

Oncogenicity ability of cells to give rise to tumors when injected into laboratory animals.

Osteoclast a large multinucleate cell formed in association with resorption of bone.

Osteoclastoma bone tumor containing large numbers of multinucleate giant cells (also called "giant cell tumors").

Panencephalitis infection of the brain caused by measles virus (?).

Paramyxovirus group of RNA viruses which includes Newcastle disease virus (NDV), simian virus 5, Sendai virus, and other viruses.

Plating efficiency percent of cells which form clones when a tissue culture is started.

PC Chromosome prematurely condensed chromosome formed from an interphase cell after fusion with a mitotic cell.

Phagosome a particle-containing vesicle formed during phagocytosis.

Plasmocytoma see myeloma.

Plasmodium multinucleate giant cell formed by nuclear divisions in the absence of cytoplasmic divisions. Plasmodia are formed by slime molds (e.g., *Physarum polycephalum*).

Plasmodesma (-ata) a protoplasmic connection between plant cells.

Plasmogamy fusion of mononucleate haploid cells (fungi, protozoa, etc.) to form binucleate cells.

Plasmolysis a shrinkage of protoplasmic volume due to water loss (plant cells).

Plastid a cytoplasmic organelle bounded by a double membraned envelope. Depending on the color of their pigmentation, they are classified as chloroplasts, chromoplasts, or leucoplasts.

Polykaryon multinucleate cell.

Polyoma virus DNA containing mouse tumor virus.

Primary culture cell culture started from tissue taken directly from an animal or a plant.

Pronase a mixture of proteolytic enzymes from *Streptomyces griseus*.

Protoplast the wall-free protoplasm of a single plant cell. It is bounded by a plasma membrane.

Prototrophic not requiring extra nutrients above a minimal medium.

Recessiveness in hybrid cells, the nonexpression of a gene from one parental strain even though it is present, at the same time as a homologous gene from the other parental strain is expressed.

Recombination the appearance of new combinations of genes, often resulting from an exchange of genetic material between two homologous chromosomes.

Reconstituted cell cell reconstructed by the fusion of a nonviable nuclear cell fragment (minicell) with a nonviable cytoplasmic fragment (anucleate cell or cytoplast).

Reexpression the reappearance of properties characteristic of one or both of the parental cells of a hybrid after a period of extinction.

Reverse transcriptase RNA-dependent DNA polymerase present in RNA tumor viruses.

Reversion a mutation which causes a change from the mutant phenotype back to a wild-type phenotype. Reversion is also used to describe the change in a cell population which takes place when contact-inhibited cells are selected from a transformed cell population. Many of the revertant cells may be mutants, but it is also likely that changes in the chromosomal balance may have taken place. The possibility of stable epigenetic changes also cannot be excluded.

Rous sarcoma virus (RSV) an RNA tumor virus which causes tumors in chickens and which can transform mammalian cells *in vitro*.

Salvage pathway a biochemical pathway which can substitute for a blocked pathway by using added precursors which are not normally available.

Sarcoma a malignant tumor arising in connective tissue. The cells may resemble fibroblast (fibrosarcoma), cartilage cells (chondrosarcoma) or bone cells (osteosarcoma).

Saturation density cell density at which a cell population stops multiplying.

Segregation in hybrid cells, the appearance of new genotypes (or karyotypes) as a result of chromosome losses.

Selective medium culture medium which allows selected cells (e.g., hybrid cells) to survive and to multiply but which kills others (e.g., the parental cells).

Selective marker a property which makes it possible to select for cells (variants) which either express (positive selection) or have lost the marker (negative selection).

Senescence in cell culture, the phenomenon of ageing, characteristic of normal, contact-inhibited cells.

Sporophore a structure of lower plants specialized to bear spores.

SV40 Virus DNA tumor virus isolated from monkeys. Can be used to transform human and rodent cells *in vitro*. Causes tumors in hamsters.

Syncytiotrophoblast multinucleate cells in the placenta formed from fetal trophoblasts by spontaneous fusion.

Syncytium large multinucleate cell arising from cell fusion.

Syngeneic of the same genetic constitution.

Synkaryon mononucleate hybrid cell, derived from somatic cell fusion.

Synteny being on the same chromosome.

T Antigen nuclear antigen which develops in cells infected or transformed by tumor viruses. The antigen is probably specified, directly or indirectly, by the viral genome.

Telocentric a chromosome with a terminal centromere.

Teratoma tumors arising from multipotent cells, usually in the gonads. The tumors may contain a variety of differentiated tissues: muscle, bone marrow cells, cartilage, yolk sac epithelium, bone, etc.

Terminal cell density *see* **Saturation density.**

Transformation term used to describe an *in vitro* change in growth regulation which is believed to be analogous to the formation of cancer cells *in vivo*.

Trophoblasts mononucleate cells of fetal origin which can fuse to form syncytiotrophoblasts.

Wild type term used to describe the normal phenotype of cells or organisms.

V Antigen nuclear antigen induced by tumor viruses. The V antigen corresponds to virus coat proteins (a late viral function).

Zona pellucida polysaccharide coat surrounding the unfertilized egg and early embryonic cleavage stages.

Index

A
B
C
D
E
F
G
H
I
J